Electrical Transport and Optical Properties of Inhomogeneous Media
(Ohio State University, 1977)

AIP Conference Proceedings
Series Editor: Hugh C. Wolfe
No. 40

Electrical Transport and Optical Properties of Inhomogeneous Media
(Ohio State University, 1977)

Editors
J.C. Garland and D.B. Tanner
Ohio State University

American Institute of Physics
New York   1978

Copying fees: The code at the bottom of the first page of each article in this volume gives the fee for each copy of the article made beyond the free copying permitted under the 1978 US Copyright Law. (See also the statement following "Copyright" below). This fee can be paid to the American Institute of Physics through the Copyright Clearance Center, Inc., Box 765, Schenectady, N.Y. 12301.

Copyright © 1978 American Institute of Physics

Individual readers of this volume and non-profit libraries, acting for them, are permitted to make fair use of the material in it, such as copying an article for use in teaching or research. Permission is granted to quote from this volume in scientific work with the customary acknowledgment of the source. To reprint a figure, table or other excerpt requires the consent of one of the original authors and notification to AIP. Republication or systematic or multiple reproduction of any material in this volume is permitted only under license from AIP. Address inquiries to Series Editor, AIP Conference Proceedings, AIP.

L.C. Catalog Card No. 78-54319
ISBN 0-88318-139-8
DOE CONF-770925

PREFACE

The physical properties of randomly inhomogeneous materials have been of interest since Clausius and Mossotti developed their well known expression for the average dielectric constant of a two-component mixture. In recent years, however, this interest has greatly intensified, in part because of basic scientific reasons but also because of the relevance of these materials to technology. Today, the physics of random media plays a role in oil exploration, earthquake prediction, solar energy collection, the development of high field superconductors, and the manufacture of ultradense integrated circuits. Although the variety of inhomogeneous systems is very large, there is an underlying unity to the subject which has not always been recognized. Indeed, the great diversity of materials has discouraged communication, even among persons working on conceptually similar problems.

In response to the growing importance of this subject and to the need to facilitate the exchange of ideas, the first conference on the electrical transport and optical properties of inhomogeneous media was held at Ohio State University on 7-9 September, 1977. The conference, attended by over a hundred physicists from thirteen countries, was sponsored by Ohio State University, the Energy Research and Development Administration (now the Department of Energy), and the U.S. National Science Foundation. In all, 52 papers were presented, including several comprehensive review papers. We believe some of these review papers are destined to become standard reference works for persons wishing to gain an overview of the subject.

Many important questions were raised at the conference. What is the effect on galvanomagnetic coefficients of inhomogeneities smaller than a mean free path? How can geometry effects be built into the Effective Medium Approximation? What are the critical exponents of percolation conduction in real materials? What causes the huge far infrared absorption in small metal particle composites? A single theme common to all conference topics, moreover, was the need for better characterized materials. To our knowledge, there is no randomly inhomogeneous three dimensional substance which has really "clean" experimental properties. In our opinion, the development of such model materials is essential to our understanding of transport effects, the metal-insulator transition, and the optical properties of inhomogeneous media.

The conference involved the efforts of many persons. The Ohio State Organizing Committee, consisting of J. C. Garland, J. Korringa, R. L. Mills, D. Stroud, and D. B. Tanner, wishes to thank Drs. B. Abeles, M. H. Cohen, J. A. Krumhansl, J. A. Marcus, and A. J. Sievers for their assistance in organizing the conference program. We would also like to thank Dr. R. Landauer for his many helpful suggestions and constant, indeed relentless, encouragement. The Organizing Committee is grateful for the help given by Sarah Freeman, M. Boysel, G. L. Carr, D. B. Heidel, J. B. Sampsell and D. J. Van Harlingen. And we acknowledge with thanks the contributions to the social activities made by Martha Garland, Betty Rosbottom, Ruth Stroud, Marcia Tanner, and Tom Johnson. Finally, we are grateful to Lauren Werling for executing the conference logo and for typing all of the conference material and correspondence. All of the aforementioned rightfully share the credit for the successes of the conference. For any errors in this volume, for misplaced manuscripts, and for the selection of cruel and small minded referees, the editors blame each other.

                                            J. C. Garland
                                            D. B. Tanner
                                            November, 1977

CONTENTS

CHAPTER I: Invited Papers: General Theory

Electrical Conductivity in Inhomogeneous Media
Rolf Landauer.. 2

Analytic Properties of the Complex Effective
Dielectric Constant of a Composite Medium with
Applications to the Derivation of Rigorous Bounds
and to Percolation Problems
David J. Bergman.................................... 46

The Electronic Properties of Inhomogeneous
Materials: Metal-Nonmetal Transitions
Morrel H. Cohen, Joshua Jortner and Itzhak Webman.. 63

Scattering Theory and Effective Medium
Approximations to Heterogeneous Materials
J. E. Gubernatis.................................... 84

The Geometry of the Percolation Threshold
Scott Kirkpatrick................................... 99

Cooperative Phenomena in Resistor Networks and
Inhomogeneous Conductors
Joseph P. Straley................................... 118

An Exact Theory of the Electrical Transport and
Optical Properties of Inhomogeneous Media
P. DQ. Landau....................................... 128

CHAPTER II: Invited Papers: Electrical Transport Properties

Inhomogeneous Superconductors
M. Tinkham.. 130

Low-Field and High-Field Hopping Conduction in
Granular Metal Films
Ping Sheng.. 143

Magnetoresistance in Inhomogeneous Metals
A. P. Pippard....................................... 153

Calculation of the Magnetoresistance of
Polycrystalline Metals
H. Stachowiak....................................... 156

Potassium: Are the Magnetoresistance Anomalies
due to Inhomogeneities?
R. S. Newrock and P. J. Tausch...................... 169

Magnetoresistance of Potassium
J. S. Lass.. 183

CHAPTER III. Invited Papers: Optical Properties

Physical and Optical Properties of Small Metal
Particle Composites
H. G. Craighead and R. A. Buhrman.................. 193

Optical Properties of Ultrafine Gold Particles
C. G. Granqvist...................................... 196

Optical Properties of Composite Materials
B. Abeles.. 222

Optical Properties of a Microscopically
Textured Surface
G. D. Cody and R. B. Stephens...................... 225

Optical Properties of Small Particle Composites:
Theories and Applications
W. Lamb, D. M. Wood and N. W. Ashcroft............. 240

CHAPTER IV. Contributed Papers: Optical Properties

On the Anomalous Absorption of Ultrafine
Particles in the Far Infrared
D. Pramanik, R. A. Buhrman and A. J. Sievers....... 257

Optical and Infrared Reflectance of Metal-
Insulator Composites
N. E. Russell, E. M. Yam and D. B. Tanner.......... 258

Far Infrared Absorption in Metal-Insulator
Composites
N. E. Russell, G. L. Carr and D. B. Tanner......... 263

Microwave Propagation in Powdered Semiconductors
J. E. Sansonetti and J. K. Furdyna................. 269

Optical Properties of Small-Particle Composites
Ronald Fuchs.. 276

Self Consistent Theory of Electromagnetic Wave
Propagation in Composite Media
D. Stroud and F. P. Pan............................ 282

Optical Constants of Cermet Materials Including
Proximity Effects
D. R. McKenzie and R. C. McPhedran................. 283

The Structural Composition and Its Influence on
the Optical Properties of Gold Black
P. O'Neill, A. Ignatiev and C. Doland............. 288

Exact Solutions for Transport Properties of Arrays
of Spheres
R. C. McPhedran and D. R. McKenzie................ 294

The Permittivity of Cubic Arrays of Spheres
W. T. Doyle....................................... 300

On the Correlation Between Optical Properties and
Chemical/Metallurgical Constitution of Cr_2O_3/Cr
Thin Films
R. Chang and W. F. Hall........................... 305

Near Infrared Reflectivity at High-Temperature of
Layers of Silicon Containing Oxygen
S. O. Sari, K. D. Materson and H. S. Gurev........ 311

The Effect of Surface Roughness on the Optical
Functions of Real Metals
J. P. Marton...................................... 317

Far-Infrared Absorption by Electron-Hole Drops
in Germanium
T. Timusk and H. G. Zarate....................... 324

Inference of Inhomogeneous Optical Absorption on the
Near Field Dynamics of LEDs
A. Zehe... 330

Characterization of Phosphorus Doped Silicon by
Infrared Reflectivity
R. Bennaceur...................................... 336

Thermoreflectance of Palladium Hydride
Gary A. Frazier and R. Glosser.................... 342

CHAPTER V.   Contributed Papers: Electrical Transport Properties

Mean Potential and Field in a Random Dielectric
Kenneth S. Mendelson and Debra Schwacher.......... 349

Microscopic Fields and Currents in d.c. Electrical
Conductivity
R. S. Sorbello.................................... 355

Minimum Metallic Conductivity in Granular Metal
Films
B. Abeles and Ping Sheng.......................... 360

The Dielectric Constants and Losses of Explosive
Mixtures Below 100 kHz
W. D. Gregory, R. Dunn, L. Capots and L. Morelli .. 365

Conductivity of Thick Film (Cermet) Resistors as a
Function of Metallic Particle Volume Fraction
G. E. Pike 366

Percolation Processes in Mixtures of Conducting
and Insulating Particles
*H. Ottavi, J. Clerc, G. Giraud, J. Rossenq,
E. Guyon and C. D. Mitescu* 372

Diffusion on Percolation Lattices: The Labyrinthine
Ant.
C. D. Mitescu, H. Ottavi and J. Roussenq 377

An Approximate Calculation of the Dimensionality
Dependence of the Resistor Lattice Conductivity
Exponents
Paul M. Kogut and Joseph P. Straley 382

Renormalization Group Approach for the Conductivity
of Random Conductance Lattices
J. Bernasconi 388

The Hysteresis in an Inhomogeneous System
J. Bernasconi, S. Straessler and H. J. Wiesmann ... 389

Electrode Effects on Geometrical Magnetoresistance
J. B. Sampsell and J. C. Garland 395

Effect of Inhomogeneities on the Galvanomagnetic
Properties of Pure Aluminum
J. B. Sampsell and J. C. Garland 401

Application of the 'Effective Medium Theory' to
Normal-Superconducting Mixtures
L. Morelli, R. Janik and W. D. Gregory 402

Hopping Conduction and Superconductivity in Granular
Aluminum
W. L. McLean, P. Lindenfeld and T. K. Worthington .. 403

Characterization of Aged ω-Phase Precipitation in
Ti-V (19 at .%) from Analysis of Rounded Super-
conducting Transition Calorimetry Experiments
J. J. White and E. W. Collings 408

Preparation and Characterization of Low Optical
Density, Superparamagnetic, Small-Particle Iron
Oxide Composites
R. F. Ziolo, W. H. H. Günther and M. P. O'Horo.... 409

Anomalous Electrical Resistivity and Magnetic
Susceptibility Temperature Dependences in Ti-V
Alloys Exhibiting Reversible Soft-Phonon-Induced
Structural Inhomogeneities
E. W. Collings..................................... 410

Detection of Residual Stress-Induced Domains from
the Transition Rounding of Antiferromagnets
J. J. White and S. N. Bhatia...................... 416

CHAPTER I: INVITED PAPERS:

GENERAL THEORY

ELECTRICAL CONDUCTIVITY IN INHOMOGENEOUS MEDIA

Rolf Landauer
IBM Thomas J. Watson Research Center, Yorktown Heights, N.Y. 10598

ABSTRACT

The history of this field is reviewed, with emphasis on the relationship to the development of molecular field concepts in dielectric theory, in the last century, and with emphasis on the relationship to the study of disordered structures, in recent decades. A few of the many methods for calculating effective conductivities will be presented and discussed. One of these is based on the direct macroscopic application of the Clausius-Mossotti relationship. In that connection we emphasize the shortcomings of the commonly accepted Lorentz derivation for the internal field and restate a less well known existing alternative derivation. The symmetrical and unsymmetrical effective medium theories of Bruggeman are presented. Connection is made to transport in randomly chosen resistor networks, to percolation threshold problems, and to transport in magnetic fields in the presence of inhomogeneities. Two more specialized topics are also discussed. One of these is the variability in field effect transistor thresholds arising from the limited size of the samples in which threshold is determined by the onset of percolation. The other specialized topic: The occurrence of strong spatial inhomogeneities in fields and currents in metals, in the presence of lattice defects, even though the mean free path is large compared to the extent of the defect.

INTRODUCTION

Many systems are heterogeneous at a relatively coarse level. Concrete, for example, consists of cement enveloping gravel. Many modern structural materials also depend on the use of composite materials.[1] Another example of a composite: ion exchange resins fluidized by aqueous solutions of sodium chloride.[2] Polycrystalline materials have grains with differing orientations. But even simpler systems, which are supposedly uniform, are beset by fluctuations in density, in composition, and in temperature. We can often view these systems as heterogeneous media in which some parameter takes on different values in different portions of the material. Thus for example Voss and Clarke[3] have invoked resistivity changes resulting from thermal equilibrium temperature fluctuations as the source of $1/f$ noise. When we are very far from equilibrium, i.e. have high current densities, such temperature induced fluctuations can in turn affect the resistive heating enough to lead to instability,[4,5,6] but that is too far from the subject of this conference, for further discussion.

The inhomogeneities can be on a sufficiently large scale so that in each part of space the behavior of the material is controlled by macroscopic constitutive equations and we only need to find a reasonable way to average over the statistical variations of the material. In that case the same basic problems can arise in a number of different fields, such as heat flow, diffusion[7,8] and elastic properties.[9,10,11]

In this paper we will focus on the electrical conduction process, and largely on D.C. or low frequency conduction. The conductive averaging problem and the dielectric one are, of course, particularly closely related. In both cases there is a potential which must

ISSN: 0094-243X/78/002/$1.50 Copyright 1978 American Institute of Physics

be continuous across the interface between adjacent regions with differing parameters. The other required boundary condition arises from the continuity of a flux across the interface. That will be electrical current in the conductive case and the normal component of the displacement, **D**, in the dielectric case. Since **E** plays the same role in both problems, and **D** and **J** have corresponding roles, the averaging equations for ε and σ become identical. Electrical conductivity is a particularly interesting property since it is easy to find systems with many powers of ten in the spatial variation of the conductivity. Furthermore, it is also easy to get strongly anisotropic and tensorial conductive behavior, e.g. as in the case of metals in strong magnetic fields, at low temperatures.[12] We will return to that subject later.

In many cases, however, we do not just deal with the question of an effective average for the linear constitutive parameters, but the variation occurs on a sufficiently small scale to require attention to more microscopic phenomena. Then, of course, the universality of the problem disappears. In the electrical case, for example, we find tunneling phenomena,[13,14,15] or even superconducting tunneling.[16] If the tunneling occurs between very small particles, electronic charge quantization manifests itself, and the changes in electrostatic energy arising from electron transfer must be explicitly considered. In our subsequent discussion we shall emphasize the macroscopic averaging problem and only allude to one of the many more microscopic problems.

In connection with this rather general introduction and our subsequent historical discussion it is appropriate to refer the reader to a most readable and related summary given by Krumhansl at an earlier conference.[17] We have intentionally structured our discussion to be somewhat complementary to that, and have stressed other aspects of the subject.

1. HISTORY

Quantitative discussion of heterogeneous media in the 19th century was closely connected with the development of the molecular field concept, often connected with the names Clausius, Mossotti, Lorenz, and Lorentz. In the time available for the preparation of this manuscript I was able to reach and read only a few of these early papers, and am dependent on secondary sources[18-22] for much of this early history.

Faraday, in 1837, proposed a model of a dielectric which consisted of a series of metallic globules separated from each other by insulating material. This is clearly a heterogeneous medium, and in the absence of real knowledge of atomic structure, about as good as could have been done. (Avogadro had made somewhat related proposals in 1806 and 1807.[23,24]) Mossotti analyzed the interaction between these polarizable entities, invoking the first of many cavity considerations in the derivation of the effective field for a dielectric. Mossotti's analysis was based on earlier closely related work by Poisson on magnetic media, and also on Mossotti's own earlier 1836 paper describing the polarization of a single molecule. Mossotti's title for his internal field paper[24] (translated into English): *Analytical discussion of the influence which the action of a dielectric medium exerts on the distribution of the electricity on the surfaces of several electric bodies dispersed in it.* The relationship to heterogeneous media is, therefore, very explicit. Mossotti's paper has been summarized in English.[25] Unfortunately the work is based on a picture of interactions between conducting bodies and ethereal atmospheres that is hard for a modern reader to follow.

Mossotti's paper was actually submitted in 1846 and not published until 1850. Professor Mirone of the University of Modena, in a letter, explains the delay: "In the meantime there was the first war for Italian independence (1848-49), and Ottavanio Fabrizio Mossotti (1791-1863) fought at the head of a battalion formed by students of the University of Pisa (where he was professor), and was made prisoner by the Austrians at the battle of Curtatone (May 29, 1848)."

A subsequent step was taken by Lorenz, whose 1869 and 1875 papers in Danish were succeeded by one in German, in 1880.[26] This 1880 paper is intimately based on Lorenz's own approach to electromagnetic theory, involving a wave equation for, "...the components of the optical vibrations." He assigns a refractive index to the interior of the molecules which differs from that of the surroundings and then proceeds in a spirit akin to the Coherent Potential Approximation and other modern wave propagation theories in stochastic media.[17] In the presence of the irregularly distributed molecules Lorenz asks for the value of the propagation constant of light such that the deviations from this well behaved sine wave average out to zero, and do not build up.

Clausius, in his 1879 book,[27] gives a derivation which a modern reader can read quickly and easily. He assumes, as did Mossotti and Faraday, that the molecules are conducting spheres. He then takes each molecule as contained in a small sphere, cut out of the continuous surrounding medium, which is characterized by the final dielectric constant of the material. In contrast to some of the other derivations, Clausius' sphere fits tightly around the molecule under consideration. It is an argument very much akin to that of Bruggeman's effective medium theories, to be discussed later.

The discussion of the effective field contained in most modern textbooks is due to Lorentz.[28] The original year of Lorentz's work has been given as 1868 by Larmor[18] and elsewhere[29] as 1870. (Lorentz matriculated at Leiden in 1870[30].) Lorentz's derivation invokes the well known *large* sphere cut out of the dielectric, and centered about the molecule under consideration. The large sphere separates the far away molecules, which can be treated as a continuum, from the close molecules which have to be taken into account more explicitly. This derivation, at best, seems somewhat contrived and unmotivated, but in fact can be criticized more specifically. The derivation assumes that the molecules acting on the one in question are point dipoles, and assumes this at two points that we will specify shortly. If, however, we assume that the molecules occupy a small enough fraction of space so that this point dipole assumption becomes reasonable, then the very effect Lorentz is trying to assess, namely the effect of the molecules on each other, becomes unimportant. The point dipole assumption matters in two places, only one of which is generally recognized. First of all the molecules within the sphere, near the one under consideration, are treated as point dipoles. If the molecules are uniformly polarized spheres, even though they occupy an appreciable fraction of space, the field generated by each molecule will still be indistinguishable, outside of the source molecule, from that of a point dipole. If, however, we consider other shapes, e.g. needle shaped molecules, then the point dipole approximation becomes very poor, as far as the field in nearby regions is concerned. The second and less recognized place where the point dipole assumption matters, is at the surface of the sphere. If the molecules occupy an appreciable portion of space this sphere will cut right through the molecules. That may not be an unrepairable defect in Lorentz's argument, but at the least it requires recognition and a subsidiary argument. In view of the importance of the Lorentz field, $\mathbf{E} + 4\pi \mathbf{P}/3$, to our subject, we will provide an alternative derivation in a later section.

Maxwell, in his Treatise on Electricity and Magnetism,[31] whose first edition appeared in 1873, derives the same relationship as the other authors discussed in this section. Maxwell considers spherical inclusions of different (say higher resistivity), and then goes on to consider the resistivity of a large sphere containing many such inclusions. He allows the *unperturbed* current, that flows far away, to be incident on each little sphere. This ignores:

1) The fact that the large spherical region is a region of higher resistivity, and current flow through it is less than it is far away.

2) The fact that each little sphere causes a disturbance in the current flow which increases the current on neighboring little spheres.

These two corrections, presumably, cancel. That is Lorentz's key point when he argues that the net field due to the dipoles in the large sphere vanish at the center of the sphere. Even Lorentz's calculation, however, is still inadequate to establish the validity of Maxwell's reasoning since it applies only to the one small sphere at the center of the big sphere. Maxwell recognized that his derivation is valid only when the volume fraction of inclusions is small. Later authors, however, have on occasion claimed more on behalf of Maxwell than is warranted. Maxwell refers to "other ways" for obtaining his result and very likely had a better argument than the one he quoted in print. Lord Rayleigh, in an 1892 paper,[32] solved the partial differential equation for a conductor with periodic inhomogeneities, to find their effect on the resistance and to test the internal field expression, $\mathbf{E} + 4\pi \mathbf{P}/3$.

J. C. Maxwell Garnett[33] in 1904, in a paper on glasses containing small metal spheres, once again derived the Clausius-Mossotti-Lorenz-Lorentz relation. Garnett's derivation is based on Maxwell's equations for propagating waves, not just on static behavior. Thus Garnett's name has become attached to this approximation when invoked in the optical regime.[34,35] We shall, however, call it the Clausius-Mossotti relation, or CM approximation.

The appearance of the name J. C. Maxwell Garnett in connection with this paper, and a 1906 sequel, and nowhere else in physics is puzzling. We, therefore, append a few biographical remarks. William Garnett was J. C. Maxwell's first demonstrator after Maxwell arrived at the Cavendish Laboratory, and was coauthor of Maxwell's biography. His son, J. C. Maxwell Garnett, was born in 1880, the year after J. C. Maxwell's death. J. C. M. Garnett had a varied career. He was a barrister, principal of the College of Technology in Manchester, and authored papers on psychometry. After 1920 he became very concerned with the League of Nations and international relations. SCIENCE ABSTRACTS, PHYSICS lists no other papers in subsequent years, aside from the two we have mentioned.

There are many alternative derivations of the CM approximation, we shall present one subsequently. Another recent one is due to Barker.[36]

In the three decades following Garnett's paper a number of further publications appeared, but - at least in this author's prejudiced opinion - with little further in the way of real conceptual advances. The modern theory of inhomogeneous media can be dated to a 1935 paper by Bruggeman,[37] followed by a series of elaborations.[38] Bruggeman's papers provided two effective media theories, which we shall derive and discuss in detail. The better known of these, which treats the two constituents on a completely symmetrical basis, was rediscovered by a number of later investigators.

Dirk Anton George Bruggeman was born in Heerenveen, in the Netherlands, toward the end of the last century. He studied at the University of Groningen and then became a high school teacher of physics, mathematics and biology in the Hague. R. Casimir, father of the physicist H. B. G. Casimir, was the rector and founder of that school. H. B. G. Casimir was, in fact, one of Bruggeman's students. Bruggeman received his Ph.D. at Utrecht in 1930, where he was a student of Kramers. His Ph.D. thesis on elastic constants of crystal aggregates is presumably related to his subsequent papers on that subject.[39] He continued to do research while teaching and this led to the other cited papers.

The papers which followed Bruggeman's, for a number of years, came largely from Holland, and are reviewed in Böttcher's book.[40]

Solid state physics, after the advent of quantum mechanics, was preoccupied for some decades with the crystalline state. The theory of electrical conductivity was particularly successful in the case of metals, which in their pure state are homogeneous. Lattice defects do represent a source of spatial inhomogeneity and they provide scattering which contributes to the resistivity. But in that case the need for concern with spatial variation was easily eliminated. Instead of evaluating the resistivity of a sample with a particular distribution of defects, an ensemble was considered in which the defects took all possible positions, and the resistivity averaged over this ensemble. Thus the obvious notion that in the presence of spatial inhomogeneity the electric field is larger (and the current smaller) in the places where transport is hardest, was not really a part of the conventional solid state wisdom. The recent growth of interest in heterogeneity has many roots, but certainly the evolving concern with disordered materials was a major contributor and we will discuss that field briefly.

Interest in disordered materials in the fifties and sixties arose from a number of sources. The need for technological asessment of amorphous semiconductor device proposals[41] was an important component in this. Theoretical interest in electrons in disordered structures, however, preceded that. In the thirties disorder had been viewed as a source of scattering, acting on electrons which were presumed to be in a conduction band. This band was derived from the disordered structure by averaging out the disorder. Only in some very special ways were there attempts[42,43] which tried to go beyond this. By the early to middle fifties, however, it became generally clear that one could systematically ask and answer questions about the distribution of carriers in space and energy, in disordered structures. In 1953, Dyson[44] examined the vibrational spectrum of a disordered linear chain. The three years, 1953-55, also saw the emergence of three papers[45-47] which for the first time gave a systematic investigation of the electronic structure of disordered systems. These papers anticipate, sometimes in a very primitive fashion, concepts which were rediscovered later. Thus Parmenter[47] provides a very clear definition of what is now called the Coherent Potential Approximation, though he does not actually employ it. All of these papers[44-47] discuss band edge smearing. Reference 46 alludes to the localization of electrons to neighborhoods in parts of space which are particularly favorable for that energy. If the early electronic papers[45-47] are to be faulted, it is for the fact that despite their heavy concentration on one dimensional models they did not recognize that all states in a disordered one dimensional structure are essentially localized. That recognition did not come until 1960.[48,49] The study of disordered electronic structures has become a major industry in the last decade. We can here cite only a few items.[50-54]

Disordered structures are, almost by definition, inhomogeneous. Aside, however, from a general emphasis on inhomogeneity a more direct connection exists between the study of disordered materials and conductivity in inhomogeneous systems. An early, but very explicit attempt to establish this connection was given by deGennes, et al.[55] who were concerned with disordered binary alloys. If the electron on an A atom can only jump to an adjacent A atom what must the concentration of A atoms be to permit long range conductive motion? We will meet this problem again in Section 8, in a more detailed discussion of percolation theory. Alternatively we can view electrons in a disordered system as carriers in a stochastically corrugated potential. If the carrier energy is too low the carriers are confined to isolated potential pockets. Carriers with enough energy, however, can move in a large enough portion of space to be able to move from one pocket to another one, and can provide D.C. conductivity. This is again an illustration of the percolation threshold to be discussed later. This latter viewpoint of disordered structures was given in early and explicit description by Ziman,[56] and has since then been elaborated in great detail by many investigators.

Disordered structures can occur in a variety of forms, e.g. as liquids, as amorphous or glassy solids,[52] or as crystalline alloys with substitutional disorder. In most of these systems electronic transport is controlled by microscopic processes and not just by fluctuations of the conductivity over regions large enough to be described macroscopically. In some liquids, however, the latter possibility has been studied and proposed as an explanation for observed metal-nonmetal transitions. Metal-ammonia solutions have been studied in particular detail by Cohen and Jortner.[57] Expanded liquid mecury[57-59] as well as semiconducting liquids, e.g. Tellurium[60] and Tellurium-Thallium mixtures[61] have been studied from this viewpoint. The electrical conductivity of binary mixtures of phenol-water mixtures, doped with KCl to make them conducting, has been studied near the critical point,[62] where it is natural to expect long range fluctuations to become important.

2. THE INTERNAL FIELD

When a group of polarizable entities is subject to an external field it is conventional to take the effective field acting on a polarizable entity as $\mathbf{E} + \gamma \mathbf{P}$, rather than the space average field, which is $\mathbf{E}$. In the case of periodic crystals γ can, of course, be found by direct lattice computation. More random distributions may lead to easier calculations, but pose greater conceptual difficulties. If the distribution of polarizable entities were really random, the effective field would have to be $\mathbf{E}$. The key point - which unfortunately is carefully hidden in many approaches to the effective field - the polarizable entities do not overlap and are outside each other. The field inside the polarizable entities, as a result of the depolarization field, is depressed below the surrounding field. Thus the average field *outside* the polarizable entities, which is effective in acting on an additional entity, exceeds the space average field. This approach to the internal field, which we shall call the excluded volume approach, was discussed by Bragg and Pippard[63] who pointed out the close relationship of their work to that of Wiener.[22] M. Pollak[64-66] has also developed this viewpoint, independently, and with considerable generality. I have also invoked this approach in the past.[67] (In view of these several independent discoveries of the approach it seems likely that it has also been invoked by others, whose work has not come to my attention.) We shall, at first, limit ourselves to spherically symmetric polarizable entities. In the following analysis we will also view these polarizable spheres as immersed in a vacuum, i.e. in a medium with $\epsilon = 1$.

The equation $\mathbf{E} = \mathbf{D} - 4\pi \mathbf{P}$ tells us that the field $\mathbf{D}$ which has its sources on the capacitor plates is supplemented by the field of the polarized entities, and the space average of this supplement is $-4\pi\mathbf{P} = -4\pi n\mathbf{p}$ where n is the number of polarizable entities per unit volume and $\mathbf{p}$ the induced dipole moment of each. Thus the space integral of the field contributed by each dipole must be $-4\pi\mathbf{p}$. (That field is, of course, evaluated under the boundary conditions that the electrode charge which gave rise to $\mathbf{D}$ remains unchanged by the dipole field.) The fact that the integrated dipole field is $-4\pi\mathbf{p}$ can also be shown easily and independently by considering a uniform array of superposed dipoles, translated with respect to each other in the directions parallel to the capacitor plates. The resulting field then varies only in the direction perpendicular to the plates and does not extend beyond the space occupied by the polarizable entities. This field can easily be integrated to yield the quoted result. It is now easily shown by a further elementary integration that out of the total field integral of $-4\pi\mathbf{p}$, one third arises from inside the polarized volume. The remaining 2/3 must come from outside the volume. The integration leading to these results requires spherical symmetry, but does not require a uniformly polarizable sphere. The polarization can arise preferentially, for example, from the inner core of the sphere.

Now consider a given polarizable entity subject to the field $\mathbf{D}$ from the condenser plates, and also to the field of the other polarizable entities. The entity under consideration is guaranteed to be in the space outside of the other polarizable entities. The integral of the field, arising from a given polarized region, over the space outside of its own volume, is $-8\pi p/3$. Its contribution to the space average field, outside of its own volume is therefore $-8\pi p/3(V-v)$, where V is the volume of the dielectric and v the volume of the dipole source. Now $v \ll V$ and hence we can replace this by $-8\pi p/3V$. Thus the polarizable entity under consideration, exposed to the field of N other dipoles in the volume V is exposed to an average field $-8\pi Np/3V$ or $-8\pi np/3$.

Some supplementary considerations: The molecule (or polarizable entity) under consideration, C, is exposed to the field of another region A, and outside the volume of A. C is also outside the volume of dipole B, for example. That however does not give it any preferred locations in the field of dipole A. Another subsidiary point concerns the finite radius of the molecule under consideration. If the effective field is the field at its center then the effective field cannot really be the average of all points outside of the source dipole volume, but the excluded volume must be a larger sphere, of radius equal to the sum of two molecular radii. The dipole field integrated over the outside of this larger sphere is, however, still $-8\pi\mathbf{p}/3$.

The effective field is therefore

$$E_{eff} = D - 8\pi np/3 = D - \frac{2}{3} 4\pi P = (E + 4\pi P) - \frac{2}{3} 4\pi P = E + 4\pi P/3, \quad (2.1)$$

which is the usual Lorentz expression.

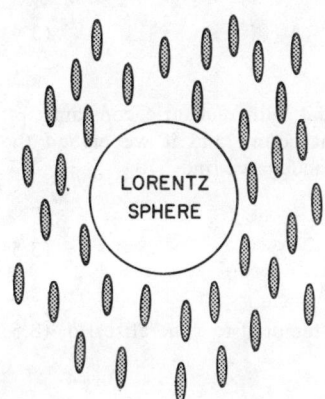

Fig. 1. Dielectric consisting of aligned prolate ellipsoids, with usual Lorentz sphere.

Cohen, et al.,[68] discuss the usual Lorentz sphere derivation of Eq.(2.1), when it is applied to needle-like ellipsoids, as shown in Fig. 1, aligned with the field. This yields an $E_{eff} = E + (4/3)\pi P$ which certainly differs from E. It is, on the other hand obvious that for such elongated structures the electric field is relatively uniform throughout the array, and E_{eff} must be almost identical with E. Ref. 68 also points out that the Lorentz sphere gives equally unreasonable results if the entities are flat disks, perpendicular to the field, but we will not repeat that argument here. We shall, however, return to this general point in the subsequent section.

3. THE CLAUSIUS-MOSSOTTI APPROXIMATION

Let us assume our polarizable entities have a polarizability α, i.e. $p = \alpha E_{eff}$. Then

$$\varepsilon - 1 = 4\pi n \alpha E_{eff}/E = 4\pi n \alpha (E + \frac{4\pi P}{3})/E = 4\pi n \alpha \frac{\varepsilon + 2}{3}. \quad (3.1)$$

Thus

$$(\varepsilon - 1)/(\varepsilon + 2) = (4\pi n\alpha)/3. \quad (3.2)$$

How do we compute α in Eq. (3.2)? Note that the internal field which leads to the left-hand side of Eq. (3.2) already takes into account the actual total polarization of other molecules. Thus in computing α we cannot allow for further polarization of the surroundings. That would be double-counting. Thus α is the polarizability when the molecule or entity is in a vacuum. If we assume that the polarizable entity is a sphere of dielectric constant ε_1 and radius a then elementary electrostatics yields

$$\alpha = (\varepsilon_1-1)a^3/(\varepsilon_1+2). \tag{3.3}$$

Putting together Eqs. (3.2) and (3.3) gives

$$\frac{\varepsilon-1}{\varepsilon+2} = \eta_1 \frac{\varepsilon_1-1}{\varepsilon_1+2}, \tag{3.4}$$

where η_1 is the fraction of space occupied by material with dielectric constant ε_1. Clearly it is only the relative dielectric constants that count, and if we embed the spheres in material of dielectric constant ε_0 instead of vacuum we find

$$\frac{\varepsilon-\varepsilon_0}{\varepsilon+2\varepsilon_0} = \eta_1 \frac{\varepsilon_1-\varepsilon_0}{\varepsilon_1+2\varepsilon_0} \tag{3.5}$$

If we have several polarizable constituents, we may be tempted to generalize Eq. (3.5) to

$$\frac{\varepsilon-\varepsilon_0}{\varepsilon+2\varepsilon_0} = \sum_i \eta_i \frac{\varepsilon_i-\varepsilon_0}{\varepsilon_i+2\varepsilon_0} \tag{3.6}$$

This equation has been given by Böttcher[40] It is indeed as justified as the equation

$$\frac{\varepsilon-1}{\varepsilon+2} = \frac{4\pi}{3} \sum_k N_k \alpha_k \tag{3.7}$$

used commonly in dielectric theory,[69] where N_k and α_k are, respectively, the number density and polarizability of the kth species. Eq. (3.6), however, does not seem quite as well founded as Eq. (3.5). Eqs. (3.6) and (3.7) assume that the *same* effective field, $E + 4\pi P/3$, acts on all the polarizable entities present. Our derivation, however, of this internal field made it clear that **P** represented the *actual* polarization in the surrounding medium, allowing for the effects of the polarizable entity, under consideration, on its surrounding region. This polarization will in turn be correlated with the polarizability of the entity under consideration. There will be, on the average, a greater polarization surrounding a highly polarizable entity than a poorly polarizable one.

Consider the result given by Eq. (3.5). First of all, if we have a mixture of two materials, A and B, Eq. (3.5) gives us different results, depending on whether we view the system as A embedded in B, or the other way around. Fig. 2 shows how the conductivity in a two phase system changes, if the conductivities of the two components are drastically different. The highest lying curves in both parts of Fig. 2 represent Eq. (3.5). The left hand side shows how conductivity of the good conductor drops as insulating inclusions are added. The right-hand side of Fig. 2 shows how the resistivity of the poorly conducting material drops (going in from the right hand edge) as portions of it are replaced by the good conductor. We notice immediately that these approximations yield no percolation threshold, i.e. a conductor retains conductivity, until it is completely replaced by insulator, whereas a highly resistive material remains highly resistive until it is completely replaced by good conductor. As will be discussed later, this is not the way most systems behave. Is it wrong? This can best be answered by reference to the results of Hashin and Shtrikman[70] enlarging on the work of Brown[71] They used a variational technique to demonstrate that the two CM expressions are maximal and minimal bounds for a macroscopically homogeneous and *isotropic* medium. (This isotropy requirement rules out the case in which the mixture consists of striations parallel or perpendicular to the current flow.) They have furthermore shown that systems exist which do take on these maximal and minimal values. They have thus shown that without some detailed geometrical information about the distribution of the two phases one cannot do better than point to the whole range between the two CM expressions. In other words, the behavior shown in Fig. 1 for the CM expressions is possible. The occurrence of a percolation type transition, at some intermediate composition, as shown for the Br.s. approximation in Fig. 2, is not inevitable. Another way of saying the same thing: If the minority component in a two phase mixture occurs as a skin completely surrounding and separating the other material, it can have a controlling effect on the behavior.

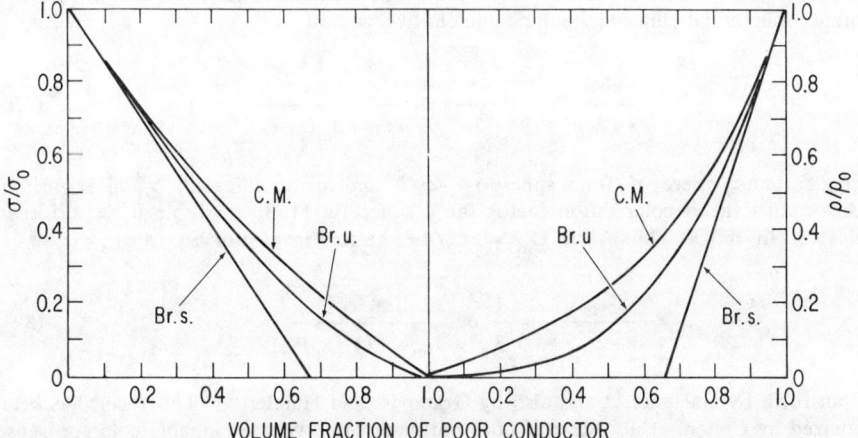

Fig. 2. Effective conductivity and resistivity for a system with components of drastically different conductivity. Left side shows drop in conductance as poorly conducting component increases. Right side shows rise in resistance as poorly conducting component increases. C.M. is Clausius-Mossotti approximation. Br.u. and Br.s. represent Bruggeman's unsymmetrical and symmetrical effective medium theories. In the left half C.M. and Br.u. are calculated on the assumption that poor conductor is introduced into a continuous background of good conductor. The right hand side assumes that the poor conductor is the continuous material.

There are shortcomings of the CM approximation which are characteristic of any molecular field theory. The molecular field does not do justice to the variability in local environments. This will be discussed in later sections.

Note that Eq. (3.5), derived for material 1 embedded in material 0, nevertheless gives the correct answer ($\varepsilon = \varepsilon_1$) even if $\eta = 1.0$. The derivative, however, $d\varepsilon/d\eta$, at $\eta = 1.0$ is quite wrong unless material 0, when present in small amounts, still remains as a skin surrounding material 1. Eq. (3.6), however, applied to two components (1,2) introduced into embedding material 0, gives incorrect answers, not just incorrect derivatives when $\eta_1 + \eta_2 = 1.0$. Even if all of material 0 has been displaced the result still depends on ε_0. This is closely related to observations made by Kerner,[72] in the discussion of a different approach, and was also noted by Wiener.[22]

Cohen, et al.[68] have proposed a form of Eq. (3.5) for the case where the inclusions are ellipsoids, rather than spheres. The ellipsoids are assumed to be aligned with each other and embedded in an isotropic medium. Three effective dielectric constants can then be found, one for each of the principal axes of the ellipsoids. Let n_m be the depolarization factor for the axis under consideration.

In the ordinary Lorentz derivation of Eq. (3.5) the structure of the l.h.s. arises from the geometry of the hypothetical sphere cut out of the material, the structure of the r.h.s. term comes from the shape of the polarizable entities. If we invoke this Lorentz sphere and ellipsoidal polarizable entities we find

$$\frac{\varepsilon-\varepsilon_0}{\varepsilon+2\varepsilon_0} = \frac{4\pi}{3} \delta \frac{\varepsilon_1-\varepsilon_0}{4\pi\varepsilon_0+n_m(\varepsilon_1-\varepsilon_0)} \qquad (3.8)$$

in c.g.s. units, where n for a sphere is $4\pi/3$, and for an ellipsoid: $\sum_1^3 n_i = 4\pi$. In m.k.s. units the depolarization factor for a sphere is $1/3$, and $\sum_1^3 L_i = 1$ for an ellipsoid. In m.k.s. units, with $L_m = n_m/4\pi$ we find instead of Eq. (3.8)

$$\frac{\varepsilon-\varepsilon_0}{\varepsilon+2\varepsilon_0} = \frac{1}{3} \delta \frac{\varepsilon_1-\varepsilon_0}{L_m\varepsilon_1+(1-L_m)\varepsilon_0} \qquad (3.9)$$

as put forth by Galeener,[73] and also by Granqvist and Hunderi.[74] This result has been criticized by Cohen, et al.[68] since it does not meet some very reasonable common sense criteria which we have discussed in an earlier section. Cohen, et al. instead propose a Lorentz cavity with the same shape as the particles. This leads to a result (in mks units)

$$\frac{\varepsilon-\varepsilon_0}{L_m\varepsilon+(1-L_m)\varepsilon_0} = \delta \frac{\varepsilon_1-\varepsilon_0}{L_m\varepsilon_1+(1-L_m)\varepsilon_0}. \qquad (3.10)$$

How is this ellipsoidal cavity justified? One way would be to forsake Lorentz for Clausius.[27] Clausius invoked a cavity which jacketed the polarizable entity tightly. This author is, however, a little uneasy about defending that approach. The alternative is to follow the excluded volume derivation that we have given for the internal field. It is immediately apparent, even without writing down any details, that now the same depolarization factor appears in the calculation of the effective field as in the response of the ellipsoidal region to which that field applies. It is, after all, the lowering of the field in the surrounding polarized volumes which raises the field at the site under consideration above the space average. In fact, a straightforward extension of our derivation of Eq. (3.5) leads to Eq. (3.10), as shown in Ref. 63.

What if the ellipsoids are not aligned? That creates considerable complexity above and beyond the preceding discussion, at two levels. First of all, unless the complete alignment is replaced by a completely random orientation, we have to contend with spatial correlations in alignment. Furthermore we face the same questions raised in connnection with Eq. (3.6). The polarization surrounding a given ellipsoid can be correlated with its alignment.

4. LORENTZ CATASTROPHE

Eq. (3.2) can be written in the form

$$\varepsilon = (1 + \frac{1}{3} 8\pi n \alpha)/(1 - \frac{1}{3} 4\pi n \alpha). \qquad (4.1)$$

Eq. (4.1), with its possibility of a vanishing denominator has on occasion[75] been proposed as an explanation for the onset of spontaneous polarization, i.e. ferroelectricity. Such discussions emphasize the fact that for certain crystalline arrangements the Lorentz factor can become much larger than $4\pi/3$. It is clear, however, that as long as we fill part of space with non-overlapping polarizable media the polarizability of the resulting structure has an upper bound given by the case in which *all* of space has been filled by the highly polarizable medium. In other words α, in Eq. (4.1), cannot be large enough to cause a vanishing denominator. This is a point which has long been understood.[76] Then why does ferroelectricity occur?

Ferroelectrics differ in two principal ways from our array of polarizable regions. First of all in a crystal the local field can induce relative motion of ions. This brings electronic shells into regions where they see a crystalline field which differs from that at their original site. This is a source of polarization not allowed in the case of stationary polarizable entities. Furthermore in a crystal we do not deal with independently polarizable units, each with its own separate restoring mechanism opposing the action of the field. When ions are moved together, in a long-range coordinated way, their restoring action on each other can be much weaker than if we try to move them one at a time, with the other ions fixed. It has thus become customary to view the onset of ferroelectricity (more specifically "displacive" ferroelectricity, in contrast to ferroelectricity which represents an ordinary order-disorder transition) as the instability of a long-wavelength mode of a vibration associated with lattice polarization,[77,78] i.e. with instability of an optical mode.

5. BRUGGEMAN'S UNSYMMETRICAL EFFECTIVE MEDIUM THEORY

From Eq. (3.5), or from elementary considerations, we can easily show that if a small volume fraction, $d\eta$, is cut out of a uniform material of conductivity σ, in the form of spheres then the resulting change in conductivity is

$$d\sigma = -\frac{3}{2} d\eta \sigma. \qquad (5.1)$$

Now let us continue to remove successive portions of the material, and at each stage invoke Eq. (5.1). We use the effective conductivity of the remaining material on the r.h.s. of Eq. (5.1). Let us also assume that the successive removals are uncorrelated. Thus in contrast to the CM approximation, the removal volumes can overlap. If a fraction α of the original conducting medium has been removed and a fraction $d\eta$ of the resulting mixture is then removed

$$(1 - \alpha)d\eta = d\alpha. \qquad (5.2)$$

Combining Eqs. (5.1) and (5.2) yields

$$d\sigma/\sigma = -\frac{3}{2} d\alpha/(1 - \alpha), \qquad (5.3)$$

which can be integrated to yield

$$\sigma = (1 - \alpha)^{3/2} \sigma_0, \qquad (5.4)$$

where σ_o is the original unperturbed conductivity when $\alpha = 0$.

Instead of removing the spheres entirely we could replace them with material of a different conductivity σ_1. In that case the above chain of reasoning leads us to the result:

$$(\sigma - \sigma_1)^3/\sigma = (1 - \alpha)^3 (\sigma_0 - \sigma_1)^3/\sigma_0. \qquad (5.5)$$

Fig. 2 shows the solution of Eq. (5.5) for the cases $\sigma_1 \ll \sigma_0$ and $\sigma_1 \gg \sigma_0$. As in the case of the CM approximation, no percolation threshold turns up. Consider the case $\sigma_1 = 0$, leading to Eq. (5.4). At each stage Eq. (5.4) predicts an effective conductivity. Removing further material forces the current to detour around the new cavity, but this is always possible under the assumption that the newly formed cavity is surrounded by a homogeneous medium with an effective conductivity. In actual fact, however, we'll be removing material from an inhomogeneous medium, and at times will be removing the last conducting link in a path.

Let us turn aside for a second to consider resistor networks to which we return later in more detail. Fig. 3 taken from Leath[79] shows a two-dimensional resistor network from which sites to which the resistors connect have been removed at random. It is clear that the removal of a site can have a very variable effect, depending on its environment. A site taken out of a region of occupied surrounding sites has a modest effect. Removing a site which is already disconnected has no effect at all. Removing a site which opens up a continuous link has a strong effect. Clearly the considerations leading to Eq. (5.5) do not do justice to this sort of variability. Fig. 3, while specifically drawn for resistor networks, has an obvious implication for inhomogeneous continuous media. Note that Eq. (5.5) differs from the CM approximation because Eq. (5.5) allows for *overlapping* spherical regions. It is thus, in a rough sense, more realistic as a consequence of the greater variety of shapes for the inclusions.

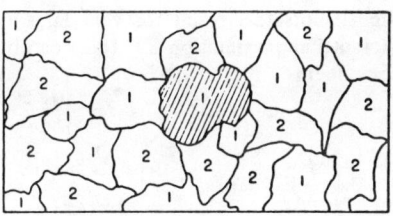

Fig. 3. Randomly generated cluster of 193 occupied, connected sites (•), surrounded by 169 vacant, boundary sites (o).

Fig. 4. The shaded crystal of type 1 is surrounded by crystals of both types, which are imagined to be replaced by a single medium of uniform conductivity.

6. BRUGGEMAN'S SYMMETRICAL EFFECTIVE MEDIUM THEORY

Fig. 4 illustrates a two component medium. Let us consider the cross-hatched volume, take it as spherical, and assume it is embedded in a uniform medium with an effective conductivity σ_m. If the field far from the inclusion is $\mathbf{E}_0$ then elementary considerations lead to a dipole moment associated with the volume under consideration

$$\mathbf{p} = \frac{3}{4\pi} V \frac{\sigma_1 - \sigma_m}{\sigma_1 + 2\sigma_m} \mathbf{E}_0, \tag{6.1}$$

where V is the volume of the region. This polarization produces a deviation from $\mathbf{E}_0$. The space integral of the deviation, as discussed previously, is $-4\pi\mathbf{p}$. Thus if the average deviation from $\mathbf{E}_0$ is to vanish, the total polarization summed over the two types of inclusion must vanish. Thus

$$\delta_1 \frac{\sigma_1 - \sigma_m}{\sigma_1 + 2\sigma_m} + \delta_2 \frac{\sigma_2 - \sigma_m}{\sigma_2 + 2\sigma_m} = 0. \tag{6.2}$$

This is a quadratic equation in σ_m whose positive solution is

$$\sigma_m = \frac{1}{4} \left(\gamma + (\gamma^2 + 8\sigma_1 \sigma_2)^{1/2} \right), \tag{6.3}$$

where

$$\gamma = (3\delta_2 - 1)\sigma_2 + (3\delta_1 - 1)\sigma_1. \tag{6.4}$$

Fig. 2 plots this result for very large conductivity ratios. In contrast to the Clausius-Mossotti approximation Eq. (6.2) can be generalized without difficulty to any number of components

$$\sum \delta_i (\sigma_i - \sigma_m)/(\sigma_i + 2\sigma_m) = 0. \tag{6.5}$$

This approximation, which treats the components on a symmetrical basis, has become the most commonly invoked approximation in this field. It agrees with the most obvious CM approximation at each end of a two component mixture range, i.e. with the CM approximation in which the predominant material is continuous. Fig. 2 shows a percolation threshold. Enough conductor put in an insulator makes it conducting, enough insulator put into a conductor blocks current flow. The accuracy with which this percolation threshold is described will be discussed in the subsequent section on resistor networks. The existence of a percolation threshold in this theory, and the particular form of (σ/σ_o) and (ρ/ρ_o) shown in Fig. 2 were emphasized in an unpublished comment[80] which helped to generate some of the current interest in effective medium theory and its relation to percolation theory. Actually, however, the behavior shown in Fig. 2 for Bruggeman's symmetrical approximation was appreciated as early as 1946, by Polder and van Santen.[81]

We have derived Bruggeman's symmetrical approximation by requiring the average value of the electric field deviation to vanish. As has been pointed out by Davidson and Tinkham[82] it can also be obtained by requiring the deviation in current, from the average, to vanish.

Eq. (6.5) is readily generalizable in further ways. The spherical regions can be replaced by ellipsoids.[81] The conductivity of the spherical regions need not be isotropic, but can be tensorial.[83] While we have focussed on the D.C. case, the approximation is clearly applicable to higher frequencies.[83,84]

Let us make a slight detour to comment on the quality of Bruggeman's symmetrical approximation as a function of the dimensionality of the system. Bruggeman already gave the two dimensional result,[37] and his theory is trivially extendable to the case of a linearly extended one dimensional conductor. All three cases (1,2 and 3D) can be combined to yield

$$\delta_1 \frac{\sigma_1-\sigma_m}{\sigma_1+(n-1)\sigma_m} + \delta_2 \frac{\sigma_2-\sigma_m}{\sigma_2+(n-1)\sigma_m} = 0, \qquad (6.6)$$

where n is the dimensionality of the system. The one dimensional equation yields an exact answer. In that case there is no opportunity for spatial current fluctuations and the voltage drop across each region can be calculated exactly. As the number of dimensions increases the opportunity for current flow around (or toward) a region of differing conductivity increases. This is reflected in Eq. (6.6) by the increasing coefficient of σ_m in the denominator. The environment of a given region is, however, not really a fixed effective medium but is variable. That is why this theory is an approximation. Clearly as the path around the component in question becomes more important, the fluctuations in this path also become more important. Thus we would expect this symmetrical effective medium theory to be better in two dimensions than in three. We shall return to this point in Sec. 8.

7. FURTHER EXTENSIONS OF CONTINUOUS MEDIA THEORIES

The viewpoints expounded in the preceding sections can be extended and refined in many ways, and the literature contains a great variety of approximations and expressions. A useful and compact tabulation of some of the more likely ones has been given by Tinga, et al.[85] who in addition to the results of others include their own work on mixtures of coated ellipsoids. A more theoretically oriented review has been given by Beran.[86] The applicability of multiple scattering theory techniques to transport in inhomogeneous systems was recognized at least as early as 1958,[87] and has since then become a very popular viewpoint[50,83,88]. It is not clear, however, to this author that progress is to be made by paying more attention to the shapes of the perturbation theory diagrams than to the shapes of the particles. In fact our inability to give simple and compact characterizations of the statistics of complex geometrical distributions seems to be the chief stumbling process to further progress. Spheres may well be a good first approximation if the fluctuations arise purely randomly. But many textures arise in a more complex way, e.g. the dendritic structures often found as a result of phase separation processes. Thin films can be particularly complex and can be variable in thickness, not just in composition,[74] and can leave us with questions not just about particle shape, but whether we are dealing with a two or three dimensional conduction process.

Hori and Yonezawa[88] have developed an approximation based on cumulant expansions. Böttcher and Bordewijk[89] have reviewed mixture theories with emphasis on mixed dielectric powders. A broad and elementary review covering resistance and other properties has been given by Hale.[90] van Beek[91] has assembled a comprehensive tabulation of approximations with emphasis on Maxwell-Wagner polarization resulting from different relaxation times for different constituents.

Phenomenologically adjusted expressions have also been proposed on a number of occasions. One recent example[82] gives a simple expression which has the right value and derivative at each end of a two-phase mixture range, and also gives a reasonable percolation threshold.

In much of this paper we emphasize the case where the inhomogeneities in resistivity are pronounced, since this is the most severe test of an approximation. Nevertheless the case of very modest fluctuations does arise, e.g. in the treatment of resistance changes arising from the equilibrium temperature fluctuations.[3] Perturbation theories have been discussed in a definitive way by Herring.[92] In the case of isotropic fluctuations in a conductivity which is isotropic at each point he finds

$$\sigma_{eff} = <\sigma>[1 - \frac{1}{3}(<(\sigma-<\sigma>)^2>/<\sigma>^2)]. \quad (7.1)$$

$$\rho_{eff} = <\rho>[1 - \frac{2}{3}(<(\rho-<\rho>)^2>/<\rho>^2)]. \quad (7.2)$$

$<>$ denotes spatial averages. The effective conductivity would be equal to the space average conductivity if the material were lamellar, with the interfaces parallel to the direction of current flow. In a random mixture, however, the current is not equally free to flow preferentially along the high conductivity channels, but must detour to reach the high conductivity regions. Thus $\sigma_{eff} < <\sigma>$. For the resistivity we would have $\rho_{eff} = <\rho>$ if the interfaces were perpendicular to the direction of current flow. Since, however, in a random mixture the current can detour and seek out low resistivity paths, $\rho_{eff} < <\rho>$. (Note that $\sigma_{eff} = \rho_{eff}^{-1}$, but $<\sigma> \neq <\rho>^{-1}$.)

8. RESISTOR NETWORKS AND PERCOLATION

We have already alluded to the concept of a percolation threshold, as illustrated, for example, by Bruggeman's symmetrical theory, in Fig. 2. The formation (or breakup) of a continuous conducting network as the amount of the conducting component is varied is most easily investigated in a periodic array of resistors whose presence or absence is stochastically controlled. The first questions that were investigated in this context related only to the connectivity of the resulting network, and not to the magnitude of the resistivity. Shante and Kirkpatrick[93] have summarized this work, known as percolation theory. The study of the onset of conduction in continuous media is more difficult, though arguments exist[94-98] to the effect that in a mixture of conducting and nonconducting material about 15 percent of conducting material is needed to create continuous conducting pathways. Refs. 94 and 95 invoke plausibility arguments based on the percolation theory of discrete lattices, whereas Refs. 96, 97, and 98 are based on a computational analysis of more realistic models of continuous materials. Clearly this percolation threshold depends on the geometry of the constituents, and as was pointed out in connection with the CM approximation in Sec. 3, the threshold can be anywhere from zero to one-hundred percent. There is some experimental[99] evidence to support a small value, in the range of 15 to 17 percent conducting material, for the percolation threshold. This small value can be contrasted with the behavior of spherical particles in a cubic lattice. Such spheres will only be in contact if they occupy 0.5236 of the available space.[100] It can also be contrasted to the percolation threshold given by

Bruggeman's symmetrical theory, shown in Fig. 2. This corresponds to 1/3 conducting material, considerably in excess of the 15 to 17 percent. We will return to a discussion of this discrepancy later in this section.

As stated, the treatment of resistor networks in which the values of the resistors (or their presence or absence) are chosen stochastically, but in which the resistor positions are laid out in a periodic array, has been much more comprehensive and successful than the continuous case. These networks can easily be modelled on a computer, are treated by percolation theory,[93] and can also be treated by minor modifications of the methods discussed in Secs. 3-6. In particular Bruggeman's symmetrical effective medium theory is readily extended to these networks[101] in the case where the "bonds," i.e. the individual resistors, are chosen stochastically. We simply view each possible choice of resistor as embedded in an otherwise uniform network of "effective" resistors. The resistor under consideration will then have an excess or deficit voltage, compared to the effective resistors far from it. The effective resistor value is then chosen so that the average voltage deviation vanishes.

Figs. 5 and 6 show Kirkpatrick's computer results for the relative conductivity of three and two dimensional networks, as resistors are removed at random. The dashed lines represent the percolation probability, i.e. the pobability that a resistor present in the network will be part of, or connected to, a continuous current flow path (though possibly on a dead end branch of such a path) through the specimen. The solid lines represent the effective medium results. The thresholds for the onset of conduction are 25 percent and 50 percent respectively for the two cases. We see that effective medium theory works well, except near the threshold region. A good many papers have studied the detailed behavior in this threshold region, reminiscent of critical phenomena at phase transitions, and we will here cite only a few recent studies.[102-105] The critical behavior can be seen in Fig. 5, but does not show clearly on the scale used in Fig. 6. Effective medium theory fails to predict this critical behavior, and as seen in connection with Fig. 5 does not necessarily give a good estimate of the threshold location. Near threshold connecting networks are first formed, as illustrated in Fig. 3. (In that case the sites to which resistors are connected, rather than the resistors, have been selected independently.) Effective medium theory, as any molecular field theory, takes each resistor (or site) to be in the same environment. Near threshold, as already pointed out in Sec. 5, this is a poor assumption. Isolated occupied sites, sites which are along a single strand path, sites surrounded by other occupied sites, and sites along a dead end branch, are all in very different environments.

If instead of removing the resistors at random we remove the sites at random, we have a problem which is a little more difficult to treat. Watson and Leath[106-108] have provided an effective medium approach to this problem, which we shall not discuss in detail. Figs. 7, adapted from Ref. 106, and 8, adapted from Ref. 109, illustrate this approximation for a two dimensional square lattice. In contrast, however, to the case in which the bonds are removed at random, there is not a single obvious and unique direct extension of Bruggeman's symmetrical theory applicable to this case.[110]

The appearance of conductivity at the percolation threshold depends on the formation of thin filamentary connections. Clearly if we are not dealing with an idealized computer model but with real materials then such details as the formation process, or the surface or interface energies between phases, will have a strong effect on

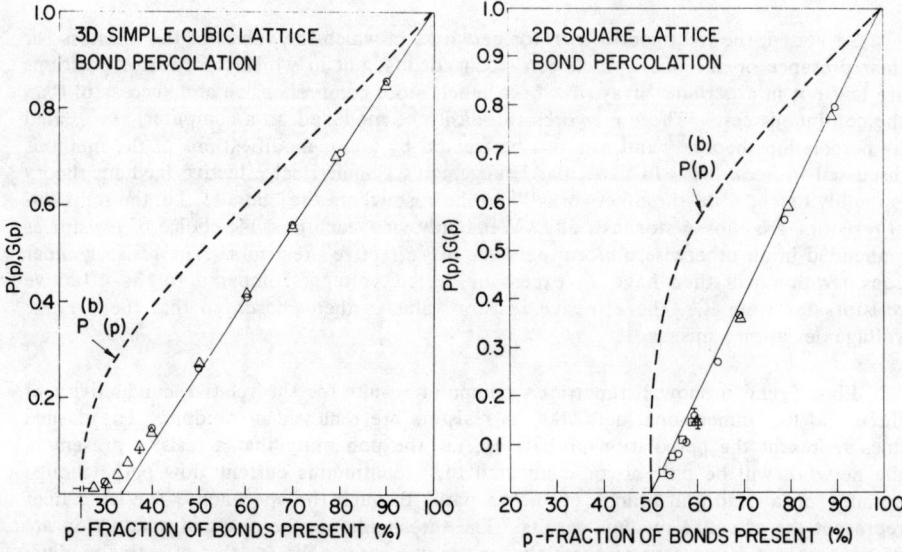

Fig. 5. Percolation probability, P(p) (dashed line), and conductance, G(p) (data points), for bond percolation on 3D simple cubic network. G(p) is normalized to unity at p=1. The networks studied ranged in size from $15 \times 15 \times 15$ to $25 \times 25 \times 25$. The solid line indicates the prediction of the effective medium theory. The arrow indicates the position of the percolation threshold p_c.

Fig. 6. Results for bond percolation on 50×50 site 2D square networks. Labelling conventions same as in Fig. 5. The threshold concentration for this lattice is $p_c = 1/2$, an exact result.

the shape of the bridges which cause the onset of conductivity. Thus in real two-phase systems[111] the percolation thresholds can be very different from any of the theoretical predictions.

We have seen that the percolation threshold predicted by the symmetrical effective medium theory occurs at higher percentages of conducting material than predictions made on a more careful basis, amounting to a discrepancy of about a factor of 2 in the case of volume inhomogeneities. We shall here analyze the reason for this discrepancy. We have already, in connection with Fig. 3 pointed out that we cannot expect good

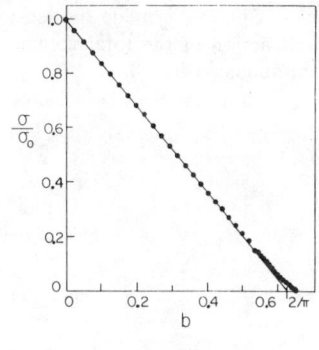

Fig. 7. Relative conductivity of a two dimensional square lattice as sites are removed. The circles are experimental points derived from a screen mesh. The horizontal axis gives the fraction of bonds removed, as the sites are cut out. The solid line is the theory of Ref. 106, $\sigma(b) = 1 - \pi b/2$.

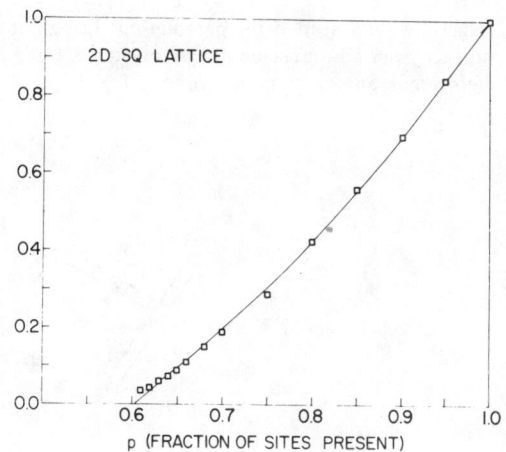

Fig. 8. Same data as in Fig. 7, but plotted against fraction of sites present, and showing deviation from computer simulation near threshold.

answers near the percolation threshold, but have not discussed the sign of the discrepancy.

Eq. (6.2) is based on spherical conducting (or polarizable) entities. Spherical shapes provide a minimum of surface area for a given volume of conducting material and a fixed number of particles. Surface area can be taken as a crude indicator of the difficulty the conducting particles have in making contact with each other. This is seen particularly easily in two dimensions. Fig. 9 makes it clear that randomly oriented elongated sticks can make conducting contact with a relatively small volume content. (Some readers will be familiar with an old game, "Pick up Sticks," based on this principle.)

In three dimensions we can compare plate or disk shaped entities, with sticks or spaghetti, and with spheres. Disks or plates stacked at random (i.e. without the influence of gravity) will constitute the three dimensional equivalent of Fig. 9. Their thickness will be almost immaterial, and they can provide a continuing conducting linkage at a very small volume fraction. Spaghetti do not block each other quite as readily, they can poke past each other. This intuitive judgement is also consistent with the fact that for a given particle density and a given volume fraction they provide less surface area.* Nevertheless it is clear that their enlongated shape does help provide

*The same ratio of long axis/short axis is also assumed.

exactly what's needed for percolation: Long continuous links. Spheres provide the least surface area and must obviously occupy a more appreciable fraction of the total volume before guaranteeing enough contact for the existence of continuous pathways.

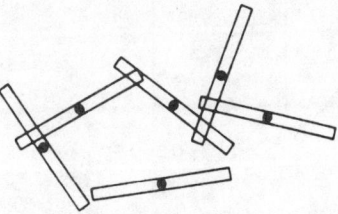

Fig. 9. Elongated structures can provide contact more easily than spheres with the same volume. From Ref. 112, where it is invoked in a somewhat different context.

We thus see that our assumption of spherical conducting entities is the most conservative possible assumption. As Fig. 3 makes clear, once we put occupied sites next to each other we get linearly extended shapes, which deviate from spherical, and therefore are characterized by more favorable depolarization factors. This concept has been invoked, and studied more quantitatively, by Granqvist and Hunderi[113] who have used the symmetrical effective medium theory for randomly oriented ellipsoids to show that lower percolation thresholds can be achieved that way.

In principle one could - at least in the case of a resistive network - use an effective medium theory in which the elementary entities were all the possible local arrangements of resistors in some given sized neighborhood, giving each possible arrangement its proper statistical weight. Such approaches, however, have been generally unsuccessful,[50] and at their best would converge very slowly. Furthermore the number of different possible local cluster arrangements which have to be taken into account rises very rapidly with cluster size.[108] It's also clear that in such an approach, which permits a complex geometry for the basic constituents, it is not just the relative conductivity of the inclusions and the surrounding neighborhood that counts, but also the exact degree to which conducting pathways match at the interfaces.

To the extent that we have discussed resistive networks we have emphasized periodic geometries and the two component case (resistor present or absent). In semiconductors, particularly in amorphous semiconductors, we face a number of variable range hopping problems. A carrier can tunnel from one site to a nearby one with a probability varying exponentially with the range. As pointed out by Miller and Abrahams,[114] and discussed very clearly by Ambegaokar, et al.,[115] this hopping can be represented as a resistive network with connecting conductances between sites proportional to the hopping probability. It was recognized by Ambegaokar, et al.[115] and by Pollak[116] that such a network, with a tremendous range for the resistances, is essentially a percolation problem. (This is true whether the geometry is stochastic or periodic.) All those resistances which are below some limiting value, which is chosen to include enough resistors to just allow a continuous network, matter. The remaining resistors have too little conductivity to affect the effective conductance of the medium appreciably, since they are in parallel with paths of much lower resistivity. A considerable literature has grown up in this area, and we cite only two additional recent items.[117,118] The subject of hopping conductivity includes many other even more complex aspects, which are still further from our subject.

9. SPATIAL INHOMOGENEITY NEAR LATTICE DEFECTS

If inhomogeneities appear on a sufficiently small scale we must go beyond the description in terms of macroscopic constitutive relations and thus inevitably become more specialized. In this section we will consider localized scatterers in metals, i.e. scattering by lattice defects small compared to the mean free path. While the case of electrons in metals has been studied in detail, we wish to stress, at the very beginning, that closely related questions can arise in other types of transport, e.g. in semiconductors, or in heat flow via phonon transport. The latter has, in fact, been studied by Erdös from a viewpoint closely related to the one to be presented here.[119-121] As discussed in Sec. 1 it has become customary to ignore these spatial variations in fields and in current. As we shall point out subsequently this does introduce errors into the resistivity calculation, but these are of a somewhat secondary nature.

The pile up of localized charge, in the presence of current flow, near a defect is of real importance in electromigration theory where we are concerned with the force these charges exert on the localized defect, and thus on their contribution to the defect motion. Only in the electromigration theory papers have these charges been explicitly acknowledged. Even in electromigration theory, however, there is no agreement on the resulting force and a number of somewhat divergent viewpoints have been expressed.[122] We shall not, here, concern ourselves with the electromigration force, but only with the closely related spatial inhomogeneities. To emphasize the basic concepts involved we shall concentrate on the one dimensional case.

Lattice defects act as perturbations in two ways: They introduce carrier scattering and they also change the carrier density. Let us consider the scattering first, and for the moment ignore the changes in carrier density. Consider a one-dimensional degenerate Fermi gas, which in addition to uniformly distributed lattice vibration scattering, also has a thin barrier at x=0, with a reflection probability r. We will assume that in the presence of transport there is an electric potential, $V(x)$, which specifies the deviation of the bottom of the conduction band from its equilibrium value. $\psi_+(x)$ will denote the energy level up to which carriers moving to the right are present, $\psi_-(x)$ similarly for carriers moving to the left. In equilibrium $\psi_+ = \psi_-$ and both are independent of x. This equilibrium value of ψ will be taken to be origin for our ψ scale. Thus ψ_+ and ψ_- measure deviations from this value. The difference between ψ_+ and ψ_- gives us current flow and

$$j = \frac{1}{2}(\psi_+ - \psi_-)v\, dn/dU. \qquad (9.1)$$

Here j is particle current, v the velocity at the Fermi surface and dn/dU the total density of states at the Fermi surface. Thus $(1/2)dn/dU$ is the density of states with a positive velocity.

Let us, at first, consider the system without the localized barrier. Carriers are scattered from right to left, due to the lattice vibrations, and this will be assumed to be given by

$$(n_+ - n_-)/2\tau = \left(\frac{1}{2\tau}\right)\frac{1}{2}(\psi_+ - \psi_-)dn/dU = j/2\tau v. \qquad (9.2)$$

The factor 2 in the l. h. s. denominator of Eq. (9.2) arises from the fact that we are taking the excess of n_+ over n_- rather than over the population average. In the steady state the rate of scattering of electrons out of the positive velocity class, as specified by Eq. (9.2) must be balanced by the rate at which electrons with a positive velocity pile up, as a result of a non-uniform current. The latter rate is given by $-dj_+/dx$, or

$$(\partial\rho_+/\partial t)_{flow} = -\partial j_+/\partial x = -\frac{\partial}{\partial x}\left(\frac{1}{2}\psi_+ v\, dn/dU\right), \qquad (9.3)$$

whereas Eq. (9.2) states

$$(\partial\rho_+/\partial t)_{scatt.} = -j/2\tau v. \qquad (9.4)$$

Setting the sum of Eqs. (9.3) and (9.4) equal to zero, as required for time independence, yields

$$\frac{\partial}{\partial x}\left(\frac{1}{2}\psi_+ v\, dn/dU\right) = -j/2\tau v. \qquad (9.5)$$

Similarly for electrons moving to the left

$$\frac{\partial}{\partial x}\left(\frac{1}{2}\psi_- v\, dn/dU\right) = -j/2\tau v. \qquad (9.6)$$

(Eqs. (9.5) and (9.6) are equivalent to Eqs. (5a) and (5b) of Ref. 123. Unfortunately, in Ref.123, the signs were written incorrectly.) Eq. (9.5) can be integrated to give

$$\psi_+(x_2) - \psi_+(x_1) = -j(x_2-x_1)\frac{dU}{dn}/\tau v^2. \qquad (9.7)$$

The expression for the drop in ψ_- is identical to the one for ψ_+.

Now let us discuss the electrical potential V. This is determined by Poisson's equation

$$\nabla^2 V = -4\pi\rho/\varepsilon. \tag{9.8}$$

For simplicity we take ε, the dielectric constant of the cores, to be spatially uniform. Let us now define a new quantity ψ, which averages ψ_+ and ψ_-, to get a measure of the total charge present. The charge disturbance is then measured by the extent to which the top of the occupied state range (ψ) has separated from the band bottom (-eV). Thus

$$\nabla^2 V = 4\pi e \varepsilon^{-1} (\psi + eV) dn/dU, \tag{9.9}$$

$$\frac{d^2V}{dx^2} - \frac{1}{\lambda^2} V = \frac{1}{\lambda^2}(-\psi/e). \tag{9.10}$$

with $\lambda^{-2} = 4\pi e^2 \varepsilon^{-1} (dn/dU)$. λ is the usual Thomas-Fermi screening length. Eq. (9.10) then tells us that V follows the long-range variation of $(-\psi/e)$, but V can deviate from $(-\psi/e)$ over distances of the order of λ. Combining this with Eq. (9.7), and using $i = -ej$, where i is electrical current yields a conductivity

$$\sigma = e^2 v^2 \tau \, dn/dU \tag{9.11}$$

So far we have simply provided a cumbersome derivation of the conventional result given in Eq. (9.11). We have done so because the methods presented here are extendable to non-uniform systems. Assume that we have a conductor as discussed, but now with an additional localized barrier characterized by the reflection probability r. On either side of the barrier we can still have solutions of the form given by Eq. (9.7). Now we must match these solutions in a way which gives us continuity of current across the barrier. To discuss this we must introduce additional notation. Let the subscripts + and − denote, as before, electrons moving, respectively, to the right and to the left. The additional subscripts R and L will identify positions to the right and left of the barrier. If the particle flow incident on the barrier from the left equals that incident from the right no net current goes across the barrier. Thus the current flow across the barrier is proportional to this difference in incident carrier flow:

$$\frac{1}{2} \psi_{+L} v \frac{dn}{dU} - \frac{1}{2} \psi_{-R} v \frac{dn}{dU}. \tag{9.12}$$

Note, however, that only (1-r) of the incident flow is transmitted. Thus the actual electron flow across the barrier is

$$j = \frac{1}{2}(1-r) v \frac{dn}{dU}(\psi_{+L} - \psi_{-R}). \tag{9.13}$$

ψ_{-R} in Eq. (9.13) can be replaced, through the use of Eq. (9.1), by terms involving ψ_{+R} and j. This gives us:

$$(\psi_{+L} - \psi_{+R}) = \frac{r}{1-r}\frac{2}{v}\frac{dU}{dn} j. \tag{9.14}$$

Since Eq. (9.1) holds on each side of the barrier we can use it to replace the ψ_+ terms in Eq. (9.14) by ψ_- terms, yielding

$$(\psi_{-L} - \psi_{-R}) = \frac{r}{1-r}\frac{2}{v}\frac{dU}{dn} j. \tag{9.15}$$

Thus we find the same discontinuity in ψ, for both directions of motion, at the barrier. Eq. (9.10) tells us that V need not be similarly discontinuous. The discontinuity in ψ is reflected by a change in $(-eV)$ which is spread out over a region whose length is of order λ. Far away from the barrier, however, the barrier produces a change in $(-eV)$ equal to that in ψ_+ or ψ_-. Using $i = -ej$ we find

$$\Delta V = \frac{i}{e^2}\frac{2r}{1-r}\frac{1}{v}\frac{dU}{dn} = \frac{i}{e^2}\frac{2r}{1-r}\frac{dp}{dn}, \tag{9.16}$$

where p is the momentum at the Fermi surface. If we take our one dimensional model very literally, and assume that aside from the barrier we have a periodic medium and two degenerate spin states for each spatial wave-function Eq. (9.16) yields a resistance

$$R = \Delta V/i = \hbar \pi r/(1-r)e^2. \tag{9.17}$$

Note that the barrier resistance varies as $r/(1-r)$. As r is allowed to approach unity, and the barrier becomes impenetrable, the resistance becomes infinite. The drop in ψ is localized to the barrier site. The corresponding electric field is not quite as well localized as that, as shown in Fig. 10. The screening length measures the spread of the exponentially decaying field. The voltage drop is thus very localized compared to the typical mean free path, which in most metals is much larger than the screening length. That shouldn't really be a surprise. Tunneling spectroscopy, in which we displace the band stucture on one side of a tunneling barrier relative to that on the other, can only work if the voltage drop is localized to the immediate vicinity of the barrier. An alternative way of discussing this localization: The incident current piles up at the barrier until the resulting field is big enough to let the incident current pass. It is this latter view which is most readily generalized to point scatterers in three dimensions. A dipole moment gets established near each scattering element. The dipole sources are contained within a region which exceeds the actual range of the scattering potential by about λ. Thus the residual resistivity field, i.e. the field due to current flow past the defect scattering consists of a set of dipole fields. These exhibit a strong field concentration within the scattering volume. Further details are given in earlier papers.[124,125] This viewpoint was viewed with considerable (and unpublished) skepticism when it first appeared.[124] In recent years it has become accepted by most of the participants in the electromigration theory debates.[122]

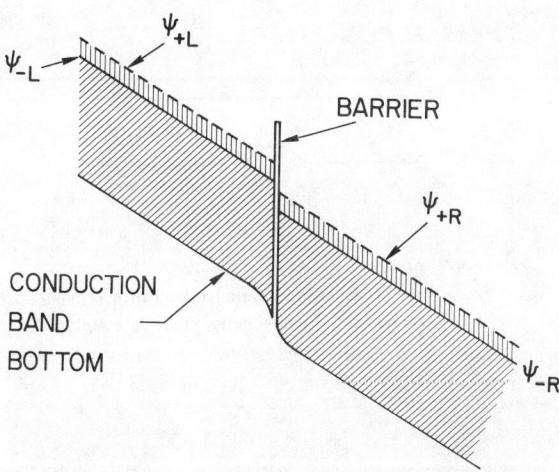

Fig. 10. The quasi-Fermi levels are discontinuous at the barrier. The potential drop, however, reaches exponentially, with the range of the screening length, into the regions on each side of the barrier.

Electromigration theory has, however, stressed other contributions to the local variations in electric field, which our semi-classical discussion has suppressed. The most important of these, first discussed by Bosvieux and Friedel,[126] consists of the interference oscillations between the waves incident on the barrier and the resulting reflected waves. These terms give rise to an oscillatory charge density which is first order in the scattering potential. Since resistance is second order (or higher) in the scattering potential it is clear that the resulting oscillatory field must vanish after integration over all space. The Bosvieux-Friedel oscillations are important for electromigration, which probes the local field, but not for the calculation of resistivity.

As pointed out before, an inhomogeneity can also modify the number of carriers and this is illustrated in Fig. 11. Within a semi-classical one-dimensional treatment a perturbation, as shown in Fig. 11 will only change the carrier density and introduce no scattering. This has been treated[123] by the method invoked in this section, generalizing it to allow a number of the parameters, such as n(x) and v(x), to be space dependent. Consider the case of a localized perturbation, small compared to the mean free path. If we were very naive we'd simply assume that the result given in Eq. (9.11) controls the ratio i/E in each part of space, and integrate the resulting field. While incorrect as a description of the spatial field, that procedure actually does give the correct result[123] for the total voltage drop. Furthermore, while the field variation E(x) is more spread out than this procedure would suggest, the extra spreading is on the scale of the screening length λ, not on the scale of the mean free path.

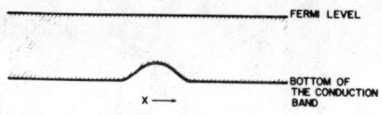

Fig. 11. One-dimensional model of a repulsive lattice inhomogeneity, creating a region of lowered carrier density

The treatment of the general three dimensional case including both carrier density changes and localized scattering and their interaction, has not yet been done as completely as one would like to see. A semi-classical and very approximate discussion exists,[127] and a more rigorous quantum mechanical discussion,[128] but the latter does not come to grips with the actual magnitude of the complex expression involved. Nevertheless, it is clear[127] that the addition (or subtraction) of carriers in a location where they are exposed to both the defect field and lattice scattering can produce somewhat complex deviations from Matthiessen's rule. These have been neglected in the extensive existing literature on such deviations.

While the localized dipole formation resulting from the localized scattering has become accepted by a number of electromigration theorists, the effects associated with Fig.11, called "carrier density modulation," seem to be less well understood. We shall, therefore, introduce an additional viewpoint to help the unconvinced reader. Near an attractive impurity potential, even if it is completely screened by bound states, the conduction band states are still perturbed in such a way that the electrons move faster, near the impurity. The higher velocities at the Fermi surface leads to increased current fluctuations. Through fluctuation-dissipation theory this, in turn, implies a higher conductivity.

While we have confined this discussion to introductory concepts, it seems appropriate here to allude to an unsettled question. How do defects interact? The literature provides two alternative viewpoints. Woo and I[67] have suggested that each localized scatterer causes a perturbation in the current flow and thus changes the current incident on other nearby defects, and this results in corrections very similar to those found in the discussions of internal fields in dielectrics. A contrasting view offered by Sorbello[129] views each lattice defect as subject to an unperturbed current flow and the resulting *field* will then be sensed by other neighboring defects. The extent to which these two discussions are complementary or contradictory still needs elucidation.

This section has been concerned with inhomogeneities which are small compared to the mean free path. The other extreme, where inhomogeneities are large compared to the mean free path, is the central subject of this conference. In that case there will be a simple distortion of the lines of current flow. In an appendix we provide a comparison of the two cases. To permit a comparison we shall consider a material characterized by a mean free time τ_0 which has inclusions characterized by a different mean free time τ'. Note that it is only the intensity of electron scattering which will differ between the two materials. We will then compare the effect of large inclusions with the effect of the same amount of included material dispersed into isolated atoms.

10. GALVANOMAGNETIC EFFECTS

Early concern with the Hall effect in inhomogeneous media arose from interest in powders with poorly conducting interface layers[130] and from the non-uniformity provided by dislocations.[131,132] A correct and somewhat general attempt, however, to match the boundary conditions for continuity of normal current and tangential field at a cavity interface, in the presence of a magnetic field, did not appear until 1956, when Juretschke et al.[133] considered cylindrical and spherical cavities. The spherical cavities were assumed to be far enough apart to be non-interacting. The cylindrical cavities included the three possible cases in which the cylinders are parallel to either the magnetic field, or to the current, or else perpendicular to both. The cylinders were allowed to have arbitrary cross-sections (not necessarily circular), and could have any arrangement and occupy any fraction of the original material, as long as a conducting path remained. At the time Ref. 133 appeared, the subject was still considered to be a somewhat incidental curiosity. Three lengthy unpublished notes, written by various subsets of the authors of Ref. 133 were condensed, for the purposes of formal and joint publication, to yield a final paper substantially less than one printed page in length. The first real appreciation of the full scope of the subject came in 1960 with a paper by Herring,[92] which in contrast to most other papers on this subject is largely (but by no means entirely) a perturbation theory, assuming modest fluctuations. Among many other points Herring showed that non-uniformities can give rise to a magnetoresistance which does not saturate with increasing magnetic field, as would be expected for uniform free-electron like materials. We shall return briefly to this subject later in this section. A more detailed treatment is given by A. B. Pippard.[12]

The Hall effect in inhomogeneous materials is of interest because Hall measurements are used to measure carrier densities, and many materials are only available in porous form, at less than theoretical density.[134-136] Furthermore, as pointed out in Sec. 1, and discussed in detail by Cohen,[137] there are many systems which have pronounced spontaneous fluctuations in carrier density.

The Hall effect and magnetoresistance can be treated most easily in the case of inhomogeneities far enough apart to be noninteracting. de Wit[138] extended the theory of Juretschke, et al.,[133] (which was limited to non-conducting spherical and cylindrical cavities) to spherical inclusions of arbitrary conductivity and Hall mobility. de Wit's theory is still limited to weak magnetic fields and does not treat magnetoresistance. A much more general formulation, allowing ellipsoidal inclusions and strong fields was provided by Stroud and Pan.[139]

More concentrated and statistically randomly distributed inhomogeneities have also been considered. A number of authors have applied Bruggeman's symmetrical approximation to the high field magnetoresistance in metals,[83,140,141] considering cavities, or else polycrystalline material consisting of a mixture of free-electron like materials and crystallites with open orbitals. This latter system has also been the subject of a detailed recent experimental study by Martin et al.[142] An effective medium theory for the Hall effect, based upon Bruggeman's symmetrical approximation, has been provided by Cohen and Jortner[143] and has been compared[144] to computer simulation of inhomogeneous materials. Kirkpatrick[145] has used the cylindrical cavity results of Ref. 133, and physical plausibilty considerations, to yield an expression for the rapid increase in Hall coefficient as a percolation threshold is approached from the conducting side.

The recent statistical discussions[83,140,141,143] have emphasized Bruggeman's symmetrical theory. We would like to point out that there are other approaches. Without attempting to defend or judge its validity we will discuss Bruggeman's unsymmetrical approximation for the low field Hall effect. Ref. 133 tells us that if we introduce a small volume fraction, $d\eta$, of spherical cavities into an isotropic material, then the effective Hall constant increases:

$$dR = \frac{3}{4} R d\eta. \quad (10.1)$$

In accordance with the reasoning used in Sec. 5 continue to cut cavities, at random, into the existing porous material. If $d\alpha$ is the volume fraction of actual solid material removed then the fraction of mixed material taken out is

$$d\eta = d\alpha/(1-\alpha). \quad (10.2)$$

Substituting Eq. (10.2) into Eq. (10.1) and integrating yields

$$R_m = R_o/(1-\alpha)^{3/4}. \quad (10.3)$$

R_o is the original Hall constant, for $\alpha = 0$.

The anomalous magnetoresistance of free-electron type metals has been discussed on many occasions.[92,139]. The diagonal components of the magnetoresistance tensor in a homogeneous material in which the electrons follow closed orbits in k space under a magnetic field are, according to theory, independent of magnetic field, at high magnetic fields.[146] In actual fact, however, this does not seem to be observed and the transverse magnetoresistance increases with magnetic field.[147] Spatial inhomogeneity has been one of several explanations offered for this effect. We will briefly discuss that, relying on the very physical picture provided by Sampsell and Garland.[147] Before proceeding, we must point out that the magnetoresistance and magnetoconductance are tensors, and the diagonal components of the magnetoconductance are not just reciprocals of those of the magnetoresistance.[146] The magnetoconductance in homogeneous materials with closed orbits, at high fields, has a component in the direction of the magnetic field which is independent of field, the other two components become small with increasing magnetic field. We shall invoke that in the subsequent discussion.

In the vicinity of a cavity the current flow must be detoured around the cavity. In the absence of a magnetic field the current flow displaced from the cavity is compensated by regions of higher current density along the upper and lower sides of the cavity, as shown in the lower left hand corner of Fig. 12, taken from Ref. 147. Consider now the case where the axis of a cylindrical cavity, the direction of current flow, and the magnetic field are all perpendicular. We shall also limit ourselves to the high field case, $\omega_c \tau \gg 1$. (ω_c is the cyclotron frequency, τ the carrier scattering time.) In this regime the electrons execute many cyclotron orbits between scattering events, and can move perpendicular to the magnetic field **B** only with difficulty. The difficulty increases with increasing **B**. The carriers are unimpeded by the magnetic field in their motion parallel to the field. This leads us to the results shown in the upper portions of Fig. 12. The arriving charges deposit on the cavity surfaces facing the incoming and departing current flow, and this generates a dipole field. The component of the electric field parallel to the direction of **B** is effective, current can easily be taken in this direction. The component of the dipole field parallel to the average current direction, however, is relatively ineffective and we cannot easily get the increased current densities needed above and below the cavity. Thus, to get the extra required total current above and below the cavity, a large region of distortion is utilized, since it is relatively easy to move the carriers in the direction of **B**. The vertical range of distortion shown in Fig. 12, which keeps increasing with **B** is, of course, accompanied by increasing dipole fields at the cavity and by extra dissipation, and therefore by greater magnetoresistance. Consider a number of cylindrical cavities of the form shown in Fig. 12, displaced with respect to each other in the direction of **B**, but close enough so that each cavity interferes with the attempt to detour current around a nearby cavity. In that case one would expect particularly strong effects on the magnetoresistance.

Fig. 12 and the associated discussion, as well as all of those in the literature, are limited to inhomogeneities large compared to the mean free path. As shown, however, in Sec. 9, inhomogeneities small compared to the mean free path can act like large scale inhomogeneities in a number of generally unexpected ways. This leads to the obvious and still unanswered question: Are large scale inhomogeneities really needed to explain

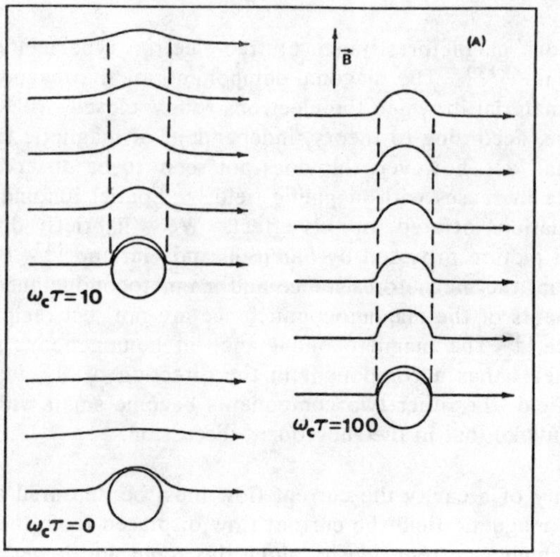

Fig. 12. Projections of current lines (injected uniformly at the far left) in the region near a cylindrical void, for different values of $\omega_c \tau$. The dashed lines indicate regions of high current density. The plane of the projection contains the magnetic field axis and the direction of current flow and is perpendicular to the cylinder axis.

the lack of saturation (with **B**) in the magnetoresistance, or are atomic sized defects adequate?

11. SMALL SAMPLES AND FIELD EFFECT TRANSISTORS

Most of our discussion has followed the typical approach to this field in which the effective resistance of a large sample is under discussion. Resorting to the jargon of statistical mechanics: We have discussed the thermodynamic limit. In biological systems, however, and in computers, we often deal with highly miniaturized components. Computer development[148,149] has been largely an attempt to push bits/cm^2 and bits/cm.^{3} (This author has, in fact, tried to go beyond discussions based on the scaling of known devices and has attempted to move toward an understanding of the ultimate physical limitations imposed by physics on the size and scale of the computational process.)[150] We try to make components smaller since this permits us to make more of them in one set of process steps, and thus at a lower cost per device. As a result a given type of system is likely to have more components, each of them smaller. Reliability for the overall system will, typically, require each component to be good. (There are exceptions, e.g. in memory we can easily utilize redundancy.) As the number of components increases, and their size decreases the components become more delicate in several ways. First of all as the component size decreases atoms will have to move over smaller distances to cause deterioration. Furthermore as the devices become smaller the importance of fluctuations in their internal structure becomes more significant, and it is this latter point which relates to the subject of this paper. In view of the increasing number of components per system, and the greater role of fluctuations in determining the device performance we will want to know something about the statistical distribution of properties. Unfortunately, studies of this sort are in their infancy, and part of our purpose here is to point to an area of genuine technological importance.

Fluctuations and inhomogeneities in devices can arise from the particular method of preparation, or can have a more fundamental statistical character. An illustration of the former is given in a study by Kirkpatrick and Mayadas[151] of aluminum interconnections in integrated circuits, containing small high resistivity precipitates. If the precipitate size is comparable to the width of the interconnections, fluctuations can become very important.

A more fundamental sort of fluctuation arises from the stochastic nature of spatial donor and acceptor distribution in semiconductors. For many years a standard lunch-table conversation topic in semiconductor device laboratories has pointed to the fact that as we make these devices smaller we can no longer think of the n region in a pnp transistor as a region of fixed donor density, but more as a thin and somewhat irregular array of donors, past which carriers thread their way. Unfortunately, no analytic way to come to grips with these questions has been available. One limited exception: Shockley's theory[152] of breakdown in p-n junctions which allowed for fluctuations and calculated breakdown at the weakest point. In addition Herring[92] has discussed the role of donor and acceptor fluctuations in *bulk* semiconductor properties. Recently, Keyes[153,154] has treated the role of fluctuations in insulated gate field effect transistors. Keyes has idealized the problem in several ways, as we shall point out subsequently. It is, thus, only a start toward an important range of problems. Nevertheless it is a significant step, considering the previous lack of formulation.

A field effect transistor[155] is shown schematically in Fig. 13. Positive charge brought on to the metallic gate electrode produces a negative image charge on the silicon surface and this introduces an n-type surface layer on a p-type substrate, called a "channel." This permits current flow between the n-type source and drain regions. In the absence of the positive gate charge we have a p-type surface connected to the n-type source and drain regions. Current then cannot flow easily since one of the two p-n junctions involved will be reverse biased. The voltage on the gate required to deplete holes from a Si surface layer and to bend the bands enough to permit electrons to appear represents the onset of conduction and is called the threshold voltage. This voltage depends on the density of acceptors in the p-type silicon. Keyes[153,154] points out that the spatial fluctuations in acceptor density lead to fluctuations in the local threshold voltage, and hence not all parts of the transistor turn on together. We thus deal with a percolation threshold. Keyes' theory is limited to the case when the source-drain voltage is small. This is one of the idealizations; in real circuits the source-drain voltage is at its maximum value when the transistor is first turned on. Furthermore, Keyes' theory is based on the theory of a long channel[155] in which all the conducting electrons at the Si surface have to find their image charges on the gate, rather than on the drain junction, for example. Unfortunately, it is in the highly miniaturized short channel devices that the effects discussed by Keyes become most significant.

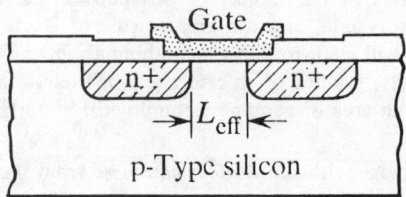

Fig. 13. Field effect transistor. A conducting gate lies over an insulating oxide. Charges on the gate control the conductivity along the silicon surface, underneath the oxide.

We should point out that fluctuations in doping are not the only source of variability in field effect transistor threshold. Ionic charges can be embedded in the insulator and these can have an effect on both the carrier mobility and on the threshold. In contrast to Keyes' papers there is a highly developed literature, stressing the role of fixed charges in the insulator. Their effect on threshold voltage is basic to all of the technologically oriented literature.[155] The effect of insulator charges on the mobility was first discussed by Stern and Howard.[156] The effect of their fluctuations on the threshold voltage was discussed by Nicollian and Goetzberger.[157] Of the extensive later literature related to these fluctuations we can here cite only a few items.[158-162] We will not attempt to decide when Keyes' mechanism is important and when, instead, the oxide charges matter. We shall, however, limit ourselves to Keyes' theory for two reasons:

1) Keyes discusses a source of fluctuation which is more basic and inevitable. Other devices (e.g. junction transistors) confront us with very similar problems.

2) Keyes explicitly concerns himself with a small sample and the variation between transistors. We have already stressed the importance of that in integrated circuitry.

Keyes avoids the complexity of a real continuum problem by dividing the Si transistor into cubes whose height equals the thickness of the depletion layer at the surface of the Si just before the onset of conduction (i.e. the depth of the region from which p-type carriers have been eliminated). He then concerns himself with an array of cubes along the Si surface, one cube high, each with its own threshold, depending on the number of acceptors in that cube. This is a modification of Shockley's approach[152] in treating p-n junction breakdown, appropriate to this problem. It is a physically plausible assumption, and represents an approximate solution to the question: As positive gate charge is introduced and the gate charge produces a depletion of mobile holes at the Si surface, how rapidly along the Si surface can the depth of the depletion layer and the resulting potential at the Si-SiO$_2$ interface vary? It is this potential, after all, which determines the onset of n-type conduction.

Keyes does not discuss the relationship of the cube size to the screening length in the material, i.e. to the ease with which transport fields, parallel to the Si surface can change with distance, along the Si surface. Keyes is concerned with the point at which the conduction process is initiated, through the presence of n type carriers in a set of contacting cubes leading from source to drain. He does not do actual source drain current calculations, for more positive gate voltages. One can thus argue that the exact spatial transport field involved in the source-drain conduction process is not considered by Keyes, and he thus need not worry about the screening length. Actually, however, a stronger point can be made, very similar to that already invoked in Sec. 9.

Let us assume that the cubes are large compared to the mean free path. This is a much more restrictive assumption than we used in Sec. 9, necessitated here by the fact that the array of cubes is two dimensional. (In one dimension electron exchange between two adjacent regions is easily calculated, particularly for a degenerate electron gas. This is true even in the presence of rapid spatial variations. With a more complex two or three dimensional velocity distribution this no longer holds.) Let ψ be the local quasi-Fermi level of the electrons, i.e. the location of the Fermi-level needed to account for the local electron density. Then the current flow will be $\sigma \nabla \psi / e$. σ is the local conductivity, appropriate to the number of carriers present at the point in question. This expression includes both diffusion currents and field-driven currents, as discussed by Swanson,[163] for example. We can thus calculate the drop in ψ (rather than in electric potential, V) in the usual way, according to macroscopic theory. As discussed, however, in Sec. 9, the long range variation in $(-\psi/\varepsilon)$ and that in V must be the same, otherwise huge space charges are built up. Thus the cubes need not be large compared to the screening length. Screening in a field transistor geometry is a complex problem, but it is clear that the screening length will be largest near threshold where there is a minimum number of carriers near the surface.

We will not actually try to repeat Keyes' somewhat complex calculations here, but only to extract a particular important result regarding the variability of the percolation threshold in small samples. This is shown in Fig. 14, which shows results for 3 x 3, 6 x 6, 10 x 10, and 18 x 18 arrays. The vertical coordinate shows the fraction of conductive elements. The horizontal scale shows the fraction of arrays with a continuous conducting path between source and drain. Fig. 14 makes it very clear that small arrays exhibit a good deal more variation in behavior than large ones. Fig. 14 is, of course, not dependent on the detailed physical mechanism causing threshold fluctuations, as long as that mechanism treats adjacent cubes as stochastically independent regions, and as long as the mechanism is uniform in its action over the whole array.

Keyes' assumption that the cubes are "on" or "off" is, of course, another idealization. n-type carriers are always present, but near threshold their number is highly variable depending on the potential at the Si-SiO$_2$ interface for the cube in question. When the conductivity varies by many orders of magnitude we have a problem of the type discussed in Sec. 8, related to variable range hopping in semiconductors. A refined approach to the field effect transistor problem which would attempt to calculate actual conductivities near threshold could then proceed in the manner suggested by Ambegaokar and Halperin.[115] and by Pollak.[116] We start with the best conducting cube and

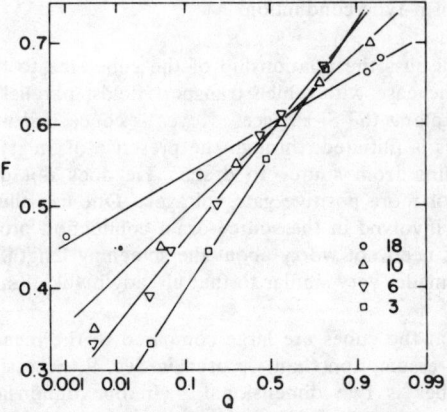

Fig. 14. The probability of finding a conductive array, Q, as a function of the fraction of conductive elements in the array, F. The 4 curves represent differing sizes for the two dimensional array.

successively add the cube of next lower conductivity until a continuous path between source and drain is established.

ACKNOWLEDGMENTS

I am indebted to a number of colleagues for preprints, for copies of original figures, and for explanations. I owe a particular apology to those who at considerable effort supplied me with input which as a result of a lack of space, time, and intelligence was not put to use.

N. G. van Kampen of Utrecht has kindly done the detective work needed to learn something about Bruggeman. A. B. Pippard of the Cavendish Laboratory and D. A. Fiske of the Education Department of the City of Manchester have supplied me with biographies of Garnett. P. Mirone of the University of Modena sent information about Mossotti and a copy of Mossotti's 1850 paper. N. Lipari of my laboratory translated that for me. C. J. F. Böttcher and P. Bordewijk of Leiden provided galley proofs from volume 2 of their book. The library of my own laboratory helped me to track down a good many papers, some from obscure and old publications, with minimal delay. B. Abeles brought the paper by Bragg and Pippard to my attention.

Finally, I must acknowledge the particularly strong help and influence of my colleague, Scott Kirkpatrick, with whom I've had countless discussions.

APPENDIX

Section 9 was concerned with inhomogeneities small compared to the mean free path. The other extreme, where inhomogeneities are large compared to the mean free path, is the central subject of this conference. In that case there will be a simple distortion of the lines of current flow. This appendix gives a comparison of the two cases. To permit a comparison we shall consider a material characterized by a mean free time τ_0 which has inclusions characterized by a different mean free time τ'. Note that it is only the intensity of electron scattering which will differ between the two materials. All other properties of electrons will be taken to be the same and we will assume that an electron continues undeflected when it crosses the boundary between the matrix and the inclusion. This is probably a fair representation of the case where the inclusion is an isotope of the matrix. We have, however, chosen the model primarily because it permits us to vary the state of dispersion and nothing else. We shall compare two cases in both of which a matrix with mean free time τ_0 contains a small percentage, x, by volume, of the other material. In one case the inclusions will be broken up into spheres large compared to the mean free path; in the other case the inclusions will be in the form of spheres which are small compared to the mean free path.

In the case of the macroscopic inclusions a current flow **j** will result in charges at the interfaces. These charges correspond to a polarization per unit volume

$$\mathbf{P} = \frac{3x}{4\pi} \frac{\mathbf{j}}{\sigma_0} \frac{\tau'-\tau_0}{\tau'+2\tau_0}, \tag{A.1}$$

where σ_0 is the conductivity of the matrix. We shall now derive an expression similar to (A.1) for the case of the localized obstacle.

In the case of localized obstacles the dipole moment produced by the incident current is given[124] by $\mathbf{p} = \mathbf{i}_\infty \, 3\pi \, S_0 \, \hbar/4e^2k^2$ where $\mathbf{i}_\infty$ is the undisturbed current flow, far from the obstacle. S_0 is the scattering cross-section, as used in transport theory, weighting collisions by $(1-\cos\theta)$ with θ the deflection angle. k is the Fermi-surface wave vector of the supposedly isotropic electron gas. The scattering cross section of a volume V (small compared to the mean free path) is V/ℓ' where ℓ' is the mean free path in the inclusion. In the absence of the inclusion there would be a scattering cross section V/ℓ_0, where ℓ_0 is the mean free path in the matrix. The extra scattering cross section responsible for the dipole formation is then

$$S_0 = \frac{V}{\ell'} - \frac{V}{\ell_0} = \frac{4}{3} \pi a^3 \frac{1}{v} \left(\frac{1}{\tau'} - \frac{1}{\tau_0} \right), \tag{A.2}$$

where a is the radius of the inclusion and v the electron velocity. If (A.2) is substituted in the expression for **p** we find a dipole moment per scatterer

$$\mathbf{p} = \frac{\pi^2 a^3 \hbar}{k^2 e^2 v} \left(\frac{1}{\tau'} - \frac{1}{\tau_0} \right) \mathbf{i}. \tag{A.3}$$

With the use of $\sigma/\ell = e^2k^2/3\pi^2\hbar$ this becomes

$$\mathbf{p} = \frac{a^3}{3}\left(\frac{1}{\sigma_0} - \frac{1}{\sigma'}\right)\mathbf{i}. \qquad (A.4)$$

n such spheres per unit volume will constitute a fraction $x = 4\pi na^3/3$ of the unit volume and will give a polarization n**p** per unit volume. This polarization is then given by

$$\mathbf{P} = n\mathbf{p} = \frac{3x}{4\pi}\frac{\mathbf{j}}{\sigma_0}\frac{\tau' - \tau_0}{3\tau'}. \qquad (A.5)$$

(A.1) and (A.5) are identical except for the factors containing τ' and τ_0. The two factors are compared in Fig. 15.

For $\tau' = \tau_0$, corresponding to a homogeneous solid, the polarization vanishes. At this point the two curves are tangent. In fact, it follows directly from the transport equation, if $|\tau' - \tau_0| \ll \tau_0$, and if a first order perturbation approximation is used, that the polarization is independent of the state of dispersion. For values of τ' which differ appreciably from τ_0 the state of dispersion does matter. Consider first the case $\tau'/\tau_0 < 1$ in which the inclusions scatter more actively than the surrounding matrix. If the inclusions are well dispersed into volumes each of which is small compared to the mean free path, then the average current j_∞ is incident on each scatterer. If, however, the inclusions consist of larger volumes, then the current flow incident upon each atom in the inclusions will be less than j_∞, and therefore the difference between τ_0 and τ' will not be as effective in generating polarization, as in the dispersed case. In the case $\tau'/\tau_0 > 1$ the current in the larger inclusions exceeds j_∞ and therefore the difference in scattering ability is more effectively used in the case where the inclusions are large.

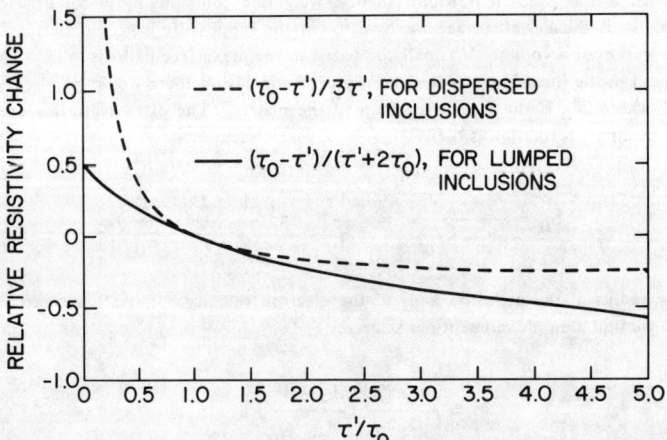

Fig. 15. The resistivity change for a small percentage of material characterized by τ', in a matrix characterized by τ_0. The resistivity change is not shown but the dimensionless quantity $\sigma_0 \Delta\rho/3x$ is plotted instead. x is the volume fraction of the inclusions.

REFERENCES

1. J. G. Morley, Physics Reports **28**, 245 (1976).
2. J. C. R. Turner, Chem. Eng. Sci. **31**, 487 (1976).
3. R. F. Voss, and J. Clarke, Phys. Rev. Lett. **36**, 42 (1976).
4. W. J. Skocpol, M. R. Beasley, and M. Tinkham, J. Appl. Phys. **45**, 4054 (1974).
5. H. Bush, Ann. Phys. (Leipz.) **64**, 401 (1921).
6. R. Landauer, Comments Solid State Phys. **4**, 139 (1972).
7. G. E. Pike, W. J. Camp, C. H. Seager, and G. L. McVay, Phys. Rev. **B10**, 4909 (1974).
8. R. M. Bielefeld, G. E. Pike, R. T. Johnson, Jr., Phys. Rev. **B15**, 5912 (1977).
9. J. E. Gubernatis and J. A. Krumhansl, J. Appl. Phys. **46**, 1875 (1975).
10. R. J. S. Brown and J. Korringa, Geophysics **40**, 608 (1975).
11. J. Korringa, J. Math. Phys. **4**, 509 (1973).
12. A. B. Pippard, this volume.
13. J. Lambe and R. C. Jaklevic, Phys. Rev. Lett. **22**, 1371 (1969).
14. B. Abeles and P. Sheng, this volume.
15. J. S. Helman and B. Abeles, Phys. Rev. Lett. **37**, 1429 (1976).
16. B. Abeles, Phys. Rev. **B15**, 2628 (1977).
17. J. A. Krumhansl in *Amorphous Magnetism,* H. O. Hooper and A. M. deGraaf, eds. (Plenum Press, N.Y., 1973) p. 15.
18. J. Larmor, *Mathematical and Physical Papers,* Vol. II (Cambridge 1929)p. 45 and following pages. [Equivalent to J. Larmor, Philos. Trans. R. Soc. Lond. A, **128**, 238 (1897)].
19. J. H. Van Vleck, *The Theory of Electric and Magnetic Susceptibilities,* (Oxford, 1932)p. 5.
20. C. J. F. Böttcher, *Theory of Electric Polarization* 2nd ed., Vol. I (Elsevier, Amsterdam, 1973)pp. 2, 169.
21. W. Fuller Brown, in *Handbuch der Physik,* Vol. 17, S. Flügge, ed. (Springer, Heidelberg 1956)p. 50.
22. O. Wiener, Abhandlungen der Mathematish-Physischen Klasse der Koenigl. Saechsischen Gesellschaft der Wissenschaften **32**, 509 (1912).
23. Dictionary of Scientific Biography, C. C. Gillispie, ed. Vol. I (Scribner's N. Y., 1970)p. 347.
24. O. F. Mossotti, Memorie di Matematica e di Fisica della Società Italiana delle Scienze Residente in Modena, Vol. 24, pt 2, (1850)p. 49-74. I have a copy of this paper but the copy does not provide most of the information contained in the citation, and I am dependent on other sources for the exact citation.
25. Dictionary of Scientific Biography, C. C. Gillispie, ed. Vol. IX, (Scribner's, New York, 1974)p. 547.
26. L. Lorenz, Wiedemannsche Annalen **11**, 70 (1880).
27. R. Clausius, *Die mechanische Behandlung der Electricität* (Vieweg, Braunschweig, 1879)p. 62.
28. H. A. Lorentz, *The Theory of Electrons,* (B. G. Teubner, Leipzig, 1909; Reprint: Dover, N.Y. 1952).
29. Dictionary of Scientific Biography, C. C. Gillispie, ed. Vol. VIII (Scribner's, New York, 1973)p. 501.
30. Dictionary of Scientific Biography, C. C. Gillispie, ed. Vol. VIII (Scribner's, New York, 1973)p. 488.
31. J. C. Maxwell, *A Treatise on Electricity and Magnetism,* Vol. 1 (Reprint: Dover, New York, 1954) Sec. 314, p. 440.
32. J. W. S. Rayleigh, Philos. Mag. **34**, 481 (1892).
33. J. C. M. Garnett, Philos. Trans. R. Soc. Lond. **203**, 385 (1904).
34. J. I. Gittleman and B. Abeles, Phys. Rev. **B15** 3273 (1977).

35. B. Abeles and J. I. Gittleman, Appl. Opt. **15,** 2328 (1976).
36. A. S. Barker, Jr., Phys. Rev. **B7,** 2507 (1973).
37. D. A. G. Bruggeman, Ann. Physik (Leipz.) **24,** 636 (1935).
38. D. A. G. Bruggeman, Ann. Physik (Leipz.) **24,** 665 (1935); **25,** 645 (1936); **29,** 160 (1937); Phys. Z. **37,** 906 (1936).
39. D. A. G. Bruggeman, Naturwissenschaften **19,** 814 (1931); Z. Phys. **92,** 561 (1934).
40. C. J. F. Böttcher, *Theory of Electric Polarisation,* 1st ed. (Elsevier, Amsterdam, 1952) Sec. 64, p. 415.
41. *Fundamentals of Amorphous Semiconductors,* (National Academy of Sciences, Washington, D.C. 1972), Chapters VI and VII.
42. T. Muto, Sci. Papers Inst. Phys. Chem. Research (Tokyo) **30,** 99 (1936); ibid **34,** 377 (1938).
43. J. M. Luttinger, Philips Res. Rep. **6,** 303 (1951).
44. F. J. Dyson, Phys. Rev. **92,** 1331 (1953).
45. H. M. James and A. S. Ginzbarg, J. Phys. Chem. **57,** 840 (1953).
46. R. Landauer and J. C. Helland, J. Chem. Phys. **22,** 1655 (1954).
47. R. H. Parmenter, Phys. Rev. **97,** 587 (1955).
48. R. Landauer, "One Dimensional Quantum Transport Theory." Talk at the *Conference on Statistical Mechanics and Irreversibility* (Queen Mary College, Dec. 1960).
49. N. F. Mott and W. D. Twose, Adv. Phys. **10,** 107 (1961).
50. R. J. Elliott, J. A. Krumhansl, and P. L. Leath, Rev. Mod. Phys. **46,** 465 (1974).
51. H. Ehrenreich, and L. M. Schwartz, in *Solid State Physics,* Vol. 31 (Academic Press, N. Y., 1976) p. 149.
52. N. F. Mott, and E. A. Davis, *Electronic Processes in Non-Crystalline Materials,* (Clarendon, Oxford, 1971).
53. C. Papatriantafillou, and E. N. Economou, and T. P. Eggarter, Phys. Rev. **B13,** 910 (1976).
54. C. Papatriantafillou, E. N. Economou, Phys. Rev. **B13,** 920 (1976).
55. P. G. deGennes, P. Lafore, and J. P. Millot, J. Phys. Chem. Solids **11,** 105 (1959).
56. J. M. Ziman, J. Phys. C **1,** 1532 (1968).
57. M. H. Cohen, and J. Jortner, J. Phys. Chem. **79,** 2900 (1976). See also papers by other authors in this same volume reporting the proceedings of a conference on this subject, starting at p. 2789.
58. M. H. Cohen, I. Webman, and J. Jortner, J. Chem. Phys. **64,** 2013 (1976).
59. J. Jortner and M. H. Cohen, Phys. Rev. **13,** 1548 (1976).
60. M. H. Cohen and J. Jortner, Phys. Rev. Lett. **30,** 699 (1973).
61. R. J. Hodgkinson, J. Phys. C **9,** 1467 (1970).
62. C. H. Shaw and W. I. Goldburg, J. Chem. Phys. **65,** 4906 (1976).
63. W. L. Bragg and A. B. Pippard, Acta. Cryst. **6,** 865 (1953).
64. M. Pollak and M. Knotek, J. Non-Cryst. Solids **4,** 459 (1970).
65. M. Pollak, Proc. R. Soc. Lond. **A325,** 383 (1971).
66. H. A. Pohl and M. Pollak, J. Chem. Phys. **66,** 4031 (1977).
67. R. Landauer and J. W. F. Woo, Phys. Rev. **B5,** 1189 (1972).
68. R. W. Cohen, G. D. Cody, M. D. Coutts, and B. Abeles, Phys. Rev. **B8,** 3689(1973).
69. C. J. F. Böttcher, Theory of Electric Polarization, 2nd ed., Vol. 1 (Elsevier, Amsterdam 1973) p. 168.
70. Z. Hashin and S. Shtrikman, J. Appl. Phys. **33,** 3125 (1962).
71. W. F. Brown, Jr., J. Chem. Phys. **23,** 1514 (1955).
72. E. H. Kerner, Proc. Phys. Soc. Lond. **B69,** 802 (1956).
73. F. L. Galeener, Phys. Rev. Lett. **27,** 421 (1971).

74. C. G. Granqvist and O. Hunderi, *Optical properties of ultrafine gold particles*, Phys. Rev. B, to be published.
75. See, for example, the discussion of Slater's theory of ferroelectricity in $BaTiO_3$ in F. Jona and G. Shirane, *Ferroelectric Crystals* (Macmillan, New York 1962)p. 190.
76. C. J. F. Böttcher, Theory of Electric Polarization, 2nd ed., Vol. I (Elsevier, Amsterdam, 1973)p. 171.
77. W. Cochran in *Structural Phase Transitions and Soft Modes* ed. by E. J. Samuelsen, E. Andersen and J. Feder, (Universitets Forlaget, Oslo, 1971).
78. J. D. Axe and G. Shirane, Phys. Today **26**, No. 9, 32 (Sept. 1973).
79. P. L. Leath, Phys. Rev. **B14**, 5046 (1976).
80. R. Landauer, *Poor Man's Percolation Theory*, an unpublished note of Jan. 29, 1971, prepared for the National Academy of Sciences Ad Hoc Committee on the Fundamentals of Amorphous Semiconductors.
81. D. Polder and J. H. van Santen, Physica (Utr.) **12**, 257 (1946).
82. A. Davidson and M. Tinkham, Phys. Rev. **B13**, 3261 (1976).
83. D. Stroud, Phys. Rev. **B12**, 3368 (1975).
84. B. E. Springett, Phys. Rev. Lett. **31**, 1463 (1973).
85. W. R. Tinga, W. A. G. Voss, and D. F. Blossey, J. Appl. Phys. **44**, 3897 (1973).
86. M. J. Beran, Phys. Status Solidi A **6**, 365 (1971).
87. R. Landauer in *Proceedings of the International Conference on the Electronic Properties of Metals at Low Temperatures*, August 1958, Geneva, N. Y. This volume received a limited informal distribution. The particular point cited here was repeated in Ref. 125.
88. M. Hori and F. Yonezawa, J. Phys. **C10**, 229 (1977) and references therein to earlier papers by these authors.
89. C. J. F. Böttcher and P. Bordewijk, *Theory of Electric Polarization*, 2nd ed., Vol. 2 (Elsevier, Amsterdam, in galley proof) Sec. 98, p.476.
90. D. K. Hale, J. Mater. Sci. **11**, 2105 (1976).
91. L. K. H. van Beek, Prog. in Dielectrics **7**, 69 (1967).
92. C. Herring, J. Appl. Phys. **31**, 1939 (1960).
93. V. K. S. Shante and S. Kirkpatrick, Adv. Phys. **20**, 325 (1971).
94. R. Zallen and H. Scher, Phys. Rev. **B4**, 4471 (1971).
95. H. Scher and R. Zallen, J. Chem. Phys. **53**, 3759 (1970).
96. A. S. Skal, B. I. Shklovskii, and A. L. Efros, Zh. Eksp. Teor. Fiz. Pis'ma Red. **17**, 522 (1973) [JETP Lett. **17**, 377 (1973)].
97. I. Webman, J. Jortner, and M. H. Cohen, Phys. Rev. **B11**, 2885 (1975).
98. I. Webman, J. Jortner, and M. H. Cohen, Phys. Rev. **B14**, 4737 (1976).
99. I. Webman, J. Jortner, and M. H. Cohen, Phys. Rev. **B15**, 5712 (1977).
100. R. F. Meredith and C. W. Tobias, J. Appl. Phys. **31**, 1270 (1960).
101. S. Kirkpatrick, Rev. Mod. Phys. **45**, 574 (1973).
102. P. L. Leath, Phys. Rev. Lett. **36**, 921 (1976).
103. J. P. Straley, Phys. Rev. **B15**, 5733 (1977).
104. M. E. Levinshtein, J. Phys. **C10**, 1895 (1977).
105. J. P. Straley, J. Phys. **C10**, 1903 (1977).
106. B. P. Watson and P. L. Leath, Phys. Rev. **B9**, 4893 (1974).
107. G. E. Pike and C. H. Seager, Phys. Status Solidi B **75**, 289 (1976).
108. J. Bernasconi and H. J. Wiesmann, Phys. Rev. **B13**, 1131 (1976).
109. A. B. Harris and S. Kirkpatrick, Phys. Rev. **B16**, 542 (1977).
110. Y. Yuge, J. Stat. Phys. **16**, 339 (1977).
111. B. Abeles, H. L. Pinch and J. I. Gittleman, Phys. Rev. Lett. **35**, 247 (1975).
112. G. E. Pike and C. H. Seager, Phys. Rev. **B10**, 1421 (1974).

113. C. G. Granqvist and O. Hunderi, "Conductivity of inhomogeneous materials. Effective-medium theory with dipole-dipole interaction," Phys. Rev. B, to be published.
114. A. Miller and E. Abrahams, Phys. Rev. **120,** 745 (1960).
115. V. Ambegaokar, B. I. Halperin, and J. S. Langer, Phys. Rev. **B4,** 2612 (1971). See also B. I. Shklovskii and A. L. Efros, Zh. Eksp. Teor. Fiz., **60,** 867 (1971) [Sov. Phys. JETP **33,** 468 (1971)].
116. M. Pollak, J. Non-Cryst. Solids **11,** 1 (1972).
117. R. M. Hill, Phys. Status Solidi A **34** 601 (1976).
118. H.Böttger and V. V. Bryksin, Phys. Status Solidi B **78,** 9 (1976).
119. S. B. Haley and P. Erdös, J. Phys. Chem. Solids **33,** 477 (1972).
120. Y. P. Joshi and D. P. Sing, Physica (Utr.) **81A** 475 (1975).
121. P. Erdös, Phys. Rev. **139,** 1249 (1965).
122. R. Landauer, "Geometry and Boundary Conditions in the Das-Peierls Electromigration Theorem," cites a number of the more recent contributions to this debate. Phys. Rev. B, to be published.
123. R. Landauer, J. Phys. C **8,** 761 (1975).
124. R. Landauer, IBM J. Res. Develop. **1,** 223 (1957).
125. R. Landauer, Z. Phys. **B21,** 247 (1975).
126. C. Bosvieux and J. Friedel, J. Phys. Chem. Solids **34,** 937 (1973).
127. R. Landauer, Phys. Rev. **B14,** 1474 (1976).
128. B. Bell, "A Microscopic Theory of the Driving Force in Electromigration," unpublished.
129. R. S. Sorbello, J. Phys. Chem. Solids **34,** 937 (1973).
130. J. Volger, Phys. Rev. **79,** 1023 (1950).
131. W. T. Read, Philos. Mag. **46,** 111 (1955).
132. R. Landauer, Phys. Rev. **94,** 1386 (1954).
133. H. J. Juretschke, R. Landauer, and J. A. Swanson, J. Appl. Phys. **27,** 838 (1956).
134. E. Goldin and H. J. Juretschke, Trans. Met. Soc. AIME **212,** 357 (1958).
135. H. J. Juretschke and R. Steinitz, J. Phys. Chem. Solids **4,** 118 (1958).
136. H. H. Wieder, Thin Solid Films **31,** 123 (1976).
137. M. H. Cohen, this volume.
138. H. J. de Wit, J. Appl. Phys. **43,** 908 (1972).
139. D. Stroud and F. P. Pan, Phys. Rev. **B13,** 1434 (1976). See also a paper by these authors in this volume.
140. H. Stachowiak, Physica (Utr.) **45,** 481 (1970).
141. K. D. Schotte and D. Jacob, Phys. Stat. Solidi A **34,** 593 (1976).
142. P. M. Martin, J. B. Sampsell, and J. C. Garland, Phys. Rev. **B15,** 5598 (1977).
143. M. H. Cohen and J. Jortner, Phys. Rev. Lett. **30,** 696 (1973).
144. I. Webman, J. Jortner and M. H. Cohen, Phys. Rev. **B15,** 1936 (1977); R. D. Swenumson and J. C. Thompson, Phys. Rev. **B14,** 5142 (1976).
145. S. Kirkpatrick in *The Properties of Liquid Metals,* S. Takeuchi ed. (Taylor & Francis, London, 1973)p. 351.
146. A. C. Smith, J. F. Janak, and R. B. Adler, *Electronic Conduction in Solids.* (McGraw Hill, New York, 1967) sec. 9.6, p. 222.
147. J. B. Sampsell and J. C. Garland, Phys. Rev. **B13,** 583 (1976).
148. R. W. Keyes, Proc. IEEE **63,** 740 (1975).
149. R. Landauer, in: *Optical Information Processing,* Y. E. Nesterikhin, G. W. Stroke, and W. E. Kock, eds. (Plenum Press, New York, 1976).
150. R. Landauer, Ber. Bunsenges. Phys. Chem. **80,** 1048 (1976).
151. E. S. Kirkpatrick and A. F. Mayadas, J. Appl. Phys. **44,** 4370 (1973).
152. W. Shockley, Solid-State Electron. **2,** 35 (1961).

153. R. W. Keyes, IEEE J. Solid State Circuits **10,** 245 (1975).
154. R. W. Keyes, Appl. Phys. **8,** 251 (1975).
155. D. L. Critchlow, R. H. Dennard, and S. A Schuster, IBM J. Res. Develop., **17,** 430 (1973).
156. F. Stern and W. E. Howard, Phys. Rev. **163,** 816 (1967). See also S. Kawaji and Y. Kawaguchi in *Proceedings of the International Conference on the Physics of Semiconductors, Kyoto, 1966,* J. Phys. Soc. Jap. Suppl. **21,** 336 (1966).
157. E. H. Nicollian and A. Goetzberger, Bell Syst. Tech. J. **46,** 1055 (1967).
158. J. R. Brews, J. Appl. Phys. **46,** 2181 (1975).
159. J. R. Brews, J. Appl. Phys. **46,** 2193 (1975).
160. N. Mott, M. Pepper, S. Pollitt, R. H. Wallis, and C. J. Adkins, Proc. R. Soc. Lond. A **345,** 169 (1975).
161. E. Arnold, Surf. Sci., **58,** 60 (1976).
162. F. Stern, Phys. Rev. **B9,** 2762 (1974).
163. J. A. Swanson, IBM J. Res. Develop. **1,** 39 (1957).

DISCUSSION

D. L. MITCHELL (N.S.F.): In one of your earlier papers you have discussed the electrical resistance of disordered one-dimensional lattices. (Ed. Note: See R. Landauer, Phil. Mag. 21, 863 (1970)). As the length of the disordered lattice increased to infinity the sample became either infinitely conductive or infinitely resistive.

LANDUAER: You are referring to a calculation which is inherently quantum mechanical. This subject has become very sophisticated and is one which I have not touched on in this oral version of the talk.

MITCHELL: Do you see any obvious connection between the quantum mechanical description of the transport properties of, say, a one dimensional disordered system and the macroscopic description of transport in an inhomogenoeus medium.

LANDAUER: I really do not think so. Beginning in about 1960, people realized that the resistivity of a one dimensional array of scatterers does not behave classically. For instance, the electrical resistance of an ensemble does not increase linearly with the length of the ensemble but rather as an exponential function of the length. Furthermore, the variation of the resistance -- depending on the particular choice of ensemble member -- over the ensemble is tremendous. However, all of these effects are essentially quantum mechanical. In some regions of the material there is no transport at all, with electronic wavefunctions decaying exponentially. In other regions, the material may be completely transparent. To my mind these properties do not bear any relationship to the effective medium theory or to the other classical theories we have been considering here.

R. CHANG (Rockwell International): Are we in this conference addressing the subject of point defects, that is, those defects which may be only a few atoms in size? Although a material with such defects is certainly inhomogeneous, it seems to me to be quite a different subject.

LANDAUER: You are correct in making a statistical characterization of this conference. With a few exceptions this conference is largely concerned with inhomogeneities at least as big as a mean free path. I feel however -- and this is a minority opinion -- that people have not thought carefully enough about transport and optical effects in the microscopic regime. Our education in solid state physics has taught us that when impurities are small compared to a mean free path, fields and currents may be treated as spatially uniform; under these circumstances one has merely to balance things in momentum space in order to obtain an answer which, with some qualifications, is pretty good. However, this approach is physically wrong, and it is this point which I want to emphasize. It is fashionable for us to view transport effects in momentum space rather than in real space, and in most situations this point of view causes no difficulties. In some instances, however, it is essential to talk about what happens in real space. As I have emphasized previously, in electromigration one is probing a local field and a local current, so that one must be particularly careful about real space effects. In addition, microscopic spatial inhomogeneities may have an important influence on some galvanomagnetic effects. For example, if we consider the high field Hall effect, the moment we cut a hole into an otherwise homogeneous conducting material we observe large circulating currents. I suspect -- although I do not claim to have the answer -- that there may be a vestige of these currents, perhaps not so minor a vestige, when the hole becomes microscopic in size.

W. T. DOYLE (Dartmouth College): I would like to comment on the Sampsell and Garland result you showed of a column of distorted current flowing above a cylindrical hole in a conducting medium (Ed. Note: Figure 12). There is an analogous effect in hydrodynamics which is known as a Taylor column and which pertains to rotating fluids. If one takes a rotating fluid such as an enclosed bucket of water and places an object such as a ping pong ball in the center of the water, then Taylor showed fifty years ago that when the ping pong ball is moved all the water above and below it moves with the ball.

A. B. PIPPARD (Cavendish Laboratory): I would like to comment that the difference between large and small inhomogeneities is shown up very strongly in light scattering theory, where objects smaller than a wavelength are easy to deal with. If one persists in using microscopic theory when the bodies are larger, then it becomes necessary to incorporate the first ten thousand normal modes of the bodies into the analysis. And when this formidable task is accomplished one ends up with the macroscopic result, a result not all that different from Descartes's theory of the rainbow. In other words, there is a smooth continuity between large and small but this connection does not mean that the same treatment should apply to both regimes. In fact, one can probably differentiate between the two regimes on the basis of which method is easiest to begin to apply.

M. H. COHEN (Univ. of Chicago): Concerning the point of view you have discussed, one can show in quite general terms that there are extinction theorems which hold for all of these problems. One finds that the external sources, those responsible for the external fields usually appearing in transport problems, are completely screened out as one moves far into the conductor. In a long wire one can obviously not have a finite electric field arising from sources outside the wire. In fact, the field which one must put into the transport equation has its sources entirely within the conductor. It is the internal dipoles you have been discussing that generate the internal field and, so long as one does not attempt to probe the detailed structure of these sources, your point of view is the correct one.

ANALYTICAL PROPERTIES OF THE COMPLEX EFFECTIVE DIELECTRIC
CONSTANT OF A COMPOSITE MEDIUM WITH APPLICATIONS TO THE
DERIVATION OF RIGOROUS BOUNDS AND TO PERCOLATION PROBLEMS

David J. Bergman
Department of Physics and Astronomy
Tel-Aviv University, Ramat-Aviv, Israel.

ABSTRACT

The complex effective dielectric constant of a composite medium ε_e is a function of the complex dielectric constants of the homogeneous components ε_i which depends also on the microscopic geometry. A characteristic geometric function is introduced to describe this dependence for the case of ε_e and other, similar material constants, and its general analytical properties are derived and discussed. A useful representation is found for this function and we show how it may be used to derive rigorous bounds on ε_e in various situations. The characteristic function is also used to discuss the behavior of ε_e in a disordered conductor-dielectric mixture near the percolation threshold of the conductor. It is found that Re ε_e diverges as the threshold is approached from either side. At the threshold itself, both Reε_e and Imε_e are shown to acquire a peculiar frequency dependence.

I. INTRODUCTION

The problem of calculating the average or effective bulk dielectric constant ε_e of a composite or heterogeneous material is discussed in a mathematical framework that separates the dependence of ε_e on the ε_i of the various components from its dependence on the microscopic geometry.

A characteristic geometric function is introduced which gives the dependence on the microscopic geometry not only of ε_e but of other similar effective constants as well, e.g., the magnetic permeability, the thermal or electrical conductivity, and the diffusivity. This function has some general analytical properties which are derived and discussed in Section II. In Section III these properties are used to derive some rigorous bounds on the characteristic function, and hence on ε_e, for various cases.

In Section IV we discuss the case of a two-phase disordered composite near the percolation threshold for one of the phases. The analytical properties of the characteristic function allow us to characterize the percolation threshold in a simple way and to derive some interesting results: Considering a composite made of a pure conductor and a pure dielectric, we find that Reε_e diverges as the percolation threshold for conductivity is approached from either side. At the threshold itself, both Reε_e and Imε_e (i.e., the conductivity) exhibit a peculiar frequency dependence, even though the pure components have a frequency independent dielectric constant and a frequency independent conductivity, respectively.

ISSN: 0094-243X/78/046/$1.50 Copyright 1978 American Insitute of Physics

II. THE CHARACTERISTIC GEOMETRIC FUNCTION AND ITS MATHEMATICAL PROPERTIES.

Mathematically, the crucial aspect of the problem lies in the fact that there are two physically relevant vector fields, a curl-free field $\vec{E}$ and a divergence-free field $\vec{D}$, which are linearly related with the help of the dielectric constant ε. Clearly, the same kind of mathematical structure exists in the case of electrical conduction, where $\vec{E},\vec{D},\varepsilon$ are replaced by $\vec{E},\vec{J}$ (the electric current), σ (the conductivity). The process of molecular diffusion also has the same structure, the appropriate quantities now being $\vec{\nabla}n$, where n is the density of the diffusing species, the diffusion current $\vec{J}$, and the diffusion coefficient or diffusivity D. In what follows, we will discuss the problem in terms of $\vec{E},\vec{D},\varepsilon$.

If we are interested in the electrical properties of a heterogeneous material only on a length scale which is large compared to the scale of heterogeneity, we can treat the material as though it were homogeneous and ascribe to it an effective dielectric constant ε_e. This is defined so as to give the correct value for the total electrostatic energy stored when this material serves as the filler in a parallel plate condenser

$$\varepsilon_e E_o^2 \equiv \frac{1}{V} \int \varepsilon(\vec{r}) E^2(\vec{r}) \, dV, \tag{1}$$

or the alternative definition

$$\frac{D_o^2}{\varepsilon_e} \equiv \frac{1}{V} \int \frac{D^2(\vec{r})}{\varepsilon(\vec{r})} \, dV \tag{2}$$

Here $\vec{E}(\vec{r})$ is the actual spatially fluctuating electric field in the heterogeneous filler, while $\vec{E}_o$ is the uniform field that would exist in a homogeneous filler with the same potential applied to the plates. Similarly, $\vec{D}(\vec{r})$ is the actual fluctuating displacement while $\vec{D}_o$ is the uniform value it would have for a homogeneous filler with the same total charge deposited on the conducting plates. The local dielectric constant $\varepsilon(\vec{r})$ has a fixed value ε_i in each phase i, hence we can write it as

$$\varepsilon(\vec{r}) = \theta_\varepsilon(\vec{r}) \equiv \sum_{i=1}^{k} \theta_i \varepsilon_i, \tag{3}$$

where $\theta_i(\vec{r})$ is equal to 1 if $\vec{r}$ is inside phase i material, and to zero otherwise.

In order to calculate ε_e by means of Eq. (1), we must first obtain the field $\vec{E}(\vec{r})$ by solving the partial differential equation

$$\vec{\nabla} \cdot (\varepsilon \vec{E}) = 0 \tag{4}$$

with appropriate boundary conditions. Since this equation, as well as Eq. (1), are homogeneous in all the ε_i and in ε_e, we can simplify matters by dividing all ε's by one of the ε_i, e.g. ε_k. In this way we find

$$h_i \equiv \frac{\varepsilon_i}{\varepsilon_k} , \quad i = 1 \ldots k-1$$

$$h_k \equiv 1 \tag{5}$$

$$m \equiv \frac{\varepsilon_e}{\varepsilon_k} = \frac{1}{V} \int \theta_h (\vec{\nabla}\psi)^2 \, dV ,$$

where

$$\theta_h \equiv \frac{\varepsilon(\vec{r})}{\varepsilon_k} = \sum_1^k \theta_i h_i \tag{6}$$

$$\vec{E} = |E_o| \vec{\nabla}\psi .$$

The boundary value problem which determines $\vec{E}$ can be stated in terms of the scalar potential field ψ

$$\left\{ \begin{array}{l} \vec{\nabla} \cdot (\theta_h \vec{\nabla}\psi) = 0 \\[6pt] \psi = 0, L \text{ at the two condenser plates whose separation is L} \\[6pt] \frac{\partial \psi}{\partial n} = 0 \text{ at the condenser walls} \end{array} \right\} \tag{7}$$

The function $m(h_1 \ldots h_{k-1})$ includes all the geometric information that is needed to calculate ε_e, or any of the other mathematically similar quantities such as effective magnetic permeability, effective electrical or thermal conductivity, effective diffusivity. It therefore deserves to be called the characteristic geometric function of the composite.

An alternative formulation is obtained if we try to calculate ε_e by means of the other definition, i.e., Eq. (2). Introducing a scalar field Φ by the following definition

$$\vec{D} \equiv |D_o| \theta_h \vec{\nabla}\Phi , \tag{8}$$

we now get

$$\frac{1}{m} = \frac{1}{V} \int \theta_h (\vec{\nabla}\Phi)^2 \, dV. \tag{9}$$

The field Φ satisfies a boundary value problem that is different from Eq. (7), namely,

$$\left\{\begin{array}{l} \vec{\nabla}\cdot(\theta_h \vec{\nabla}\Phi) = 0 \\[6pt] \Phi = 0, \text{const. at the two condenser plates} \\[6pt] \frac{\partial \Phi}{\partial n} = 0 \quad \text{at the condenser walls} \\[6pt] \frac{1}{S} \int \theta_h \frac{\partial \Phi}{\partial n} dS = 1 \end{array}\right\} \qquad (10)$$

The last integral is over the entire area of a single condenser plate, or any plane parallel to it.

In discussing the case of a lossy medium, or a composite made of a mixture of a dielectric and a conductor, we may characterize the low frequency behavior by complex dielectric constants. This leads to complex values of h_i, and the fields ψ and Φ will in general be complex too. The effective complex dielectric constant is then defined by

$$\varepsilon_e E_o^2 \equiv \frac{1}{V} \int \varepsilon(\vec{r}) |E(\vec{r})|^2 dV. \qquad (11)$$

It can easily be checked that in this case $\text{Re}\varepsilon_e$ again determines the electrostatic energy stored in the composite medium, while $\text{Im}\varepsilon_e$ now determines the rate of dissipation, both processes being referred to a fictitious homogeneous medium. Eq. (11) looks different from Eq. (1), and would lead to

$$m \equiv \frac{\varepsilon_e}{\varepsilon_k} = \frac{1}{V} \int \theta_h |\vec{\nabla}\psi|^2 dV , \qquad (12)$$

which looks different from Eq. (5) when ψ is complex. Despite this apparent difference, we can show that (12) and (5) in fact yield the same complex function $m(h_1 \ldots h_{k-1})$. In order to see this, we first transform Eq. (12) to a surface integral

$$m = \frac{1}{V} \oint \psi^* \theta_h \vec{\nabla}\psi \cdot d\vec{s} . \qquad (13)$$

Since the integrand here is nonzero only at the condenser plates, where ψ is real, we may replace ψ^* by ψ. Transforming back to a volume integral we regain Eq. (5).

We now turn to examine the properties of $m(h_i)$ in detail. We consider a specific sample of heterogeneous material with a finite volume and a well defined, though perhaps not well known, microscopic geometry. The boundary value problems (7) and (10) usually have a solution for arbitrary complex values of h_i. Whenever a solution exists for (7), we can easily show that m has a derivative given by

$$\frac{\partial m}{\partial h_i} = \frac{1}{V} \int \theta_i (\vec{\nabla}\psi)^2 \, dV \; . \tag{14}$$

Similarly, whenever a solution exists for (10), we can show that the function

$$\tilde{m}(\tilde{h}_1 \ldots \tilde{h}_{k-1}) \equiv \frac{1}{m}$$

$$\tilde{h}_i \equiv \frac{1}{h_i} \tag{15}$$

has a derivative given by

$$\frac{\partial \tilde{m}}{\partial \tilde{h}_i} = \frac{1}{V} \int \theta_i (\theta_h \vec{\nabla}\Phi)^2 \, dV. \tag{16}$$

Thus, m (or $\tilde{m}$) is analytic at any point where the boundary value problem (7) (or(10)) has a solution, and that is then a regular point. In the usual case, when both (7) and (10) have solutions, it is easy to see that Φ must be just a constant multiple of ψ. This relationship will fail to hold at points where the solution of (10) leads to $\Phi=0$ on both of the condenser plates. In that case we get from Eq. (9) (by transforming it to a surface integral) that $\tilde{m}=0$, and hence m=∞. Thus, (7) does not have a solution then. (Note however that Φ is a solution to the homogeneous counterpart of (7), i.e., the boundary value problem where ψ must vanish on both condenser plates). Moreover, as such a point is approached through regular points of ψ, it becomes more and more difficult to satisfy the nonzero boundary condition of (7). This forces ψ to have large fluctuations, and this makes m increase without limit. Clearly, such a non-regular point of m is a pole. A similar discussion shows that $\tilde{m}$ has a pole whenever the solution of (7) leads to a vanishing of the total charge on the condenser plates

$$\int \theta_h \frac{\partial \psi}{\partial n} \, dS = 0 \; . \tag{17}$$

In that case, (10) has no solution but its homogeneous counterpart, i.e., the boundary value problem where the total charge on each plate vanishes as in (17), is solved by ψ, the solution of (7).

The only other possitibility for a non-regular point of m and $\tilde{m}$ would be if at some point both (7) and (10) failed to have a solution. While we are unable at this time to offer a definite proof, we will conjecture that this case never occurs and proceed with the discussion under that assumption. In this case, when poles are the only singularities in m and $\tilde{m}$, including points at infinity, both m and $\tilde{m}$ are rational function of their variables.

Restricting ourselves first to a two-phase material, where there is just one h variable and where

$$\theta_h = h\theta_1 + \theta_2 , \tag{18}$$

we can show that the poles of m(h) all lie on the negative real axis and that they are all simple poles: At such a pole, we use the solution of (10) (or of the homogeneous counterpart of (7)) to form the following integral

$$0 = \int \Phi^* \vec{v} \cdot (\theta_h \vec{\nabla}\Phi) \, dV = -\int (h\theta_1 + \theta_2) |\vec{\nabla}\Phi|^2 \, dV . \tag{19}$$

Clearly, the final integral can only vanish if h is real and negative. For such values of h, Eq. (16) shows that $d\tilde{m}/d\tilde{h}$ is positive, and hence $\tilde{m}$ has a simple zero and m has a simple pole.

A similar result for the poles of $\tilde{m}$ and the zeros of m follows by using the solution of (7) (or of the homogeneous counterpart of (10)) to write an equation similar to (19). This shows that h must again be real and negative. Eq. (14) then shows that $dm/dh > 0$. Therefore, the zeros of m and the poles of $\tilde{m}$ are real, negative, and simple.

When there are more than one h-variables (i.e., when there are more than two phases in the composite), the situation is more complicated. The functions m and $\tilde{m}$ usually have (k-2)-dimensional complex manifolds of poles and zeros, where k is the number of phases and k-1 is the number of h-variables. If we restrict ourselves to real values of h_i, we can still show by a slight generalization of the previous discussion that a zero or a pole can only occur if some of the h_i are negative. We can also show that for real values of h_i, the poles are all simple.

A useful way to discuss the properties of m or $\tilde{m}$ in this case is to consider these functions along a line in h-space. In particular if we restrict ourselves to a line of the form

$$h_i(h_1) = a_i + b_i h_1 + \sum_\alpha \frac{c_i^{(\alpha)}}{h_i^{(\alpha)} - h_1} , \quad i > 1, \tag{20}$$

where the coefficients $a_i, b_i, c_i^{(\alpha)}, h_i^{(\alpha)}$ are all real and

$$b_i, c_i^{(\alpha)} > 0, \tag{21}$$

we find that $\text{Im} h_i$ has the same sign as $\text{Im} h_1$. Consequently, we can again show that the poles and zeros of $m(h_i(h_1))$ and of $\tilde{m}(\tilde{h}_i(\tilde{h}_1))$ are all real. Furthermore, since for real h_1 we find

$$\frac{dh_i}{dh_1} > 0 , \quad \frac{d\tilde{h}_i}{d\tilde{h}_1} > 0, \tag{22}$$

we again conclude that these poles and zeros are all simple.

For some purposes it is convenient to discuss the properties of ε_e by means of slightly different variables and functions[1]

$$u_i \equiv 1 - h_i \quad , \quad v_i \equiv 1-\tilde{h}_i$$
$$f(u_i) \equiv 1-m \quad , \quad \phi(v_i) \equiv 1-\tilde{m} \quad . \tag{23}$$

Clearly, f and ϕ are just as deserving as m for the title of characteristic geometric function. For the case of a single u variable, all the analytic properties of f and ϕ, including the fact that $f(0) = \phi(0) = 0$, can be summarized by writing the following representation for f(u)

$$f(u) = \sum_\alpha \frac{B_\alpha}{\frac{1}{u} - \frac{1}{u_\alpha}} \quad , \quad B_\alpha > 0, \quad 1 < u_\alpha \leq \infty \quad , \tag{24}$$

and a similar representation for $\phi(v)$. When there are several u variables, we can again consider the function $f(u_i)$ along a line in u-space given by

$$u_i(u_1) = \sum_\alpha \frac{b_i^{(\alpha)}}{\frac{1}{u_1} - \frac{1}{u_i^{(\alpha)}}} \quad , \quad i > 1, \tag{25}$$

where $u_i^{(\alpha)}, b_i^{(\alpha)}$ are real, and $b_i^{(\alpha)} > 0$. It is easy to check that $f(u_i(u_1))$ again has only real, simple poles, although they are no longer restricted to lie above $u_1 = 1$. It is also clear that $df(u_i(u_1))/du_1 > 0$. Consequently we can again represent $f(u_i(u_1))$ by an expression like Eq. (24), except that the poles may now range over the entire real axis. The same result holds for $\phi(v_i(v_1))$.

To conclude this section, we note that $f(u_i)$ and $\phi(v_i)$ have some remarkable convexity properties for real values of their variables in the interval $(-\infty, +1)$. It can be shown by a direct calculation[1] that the even order differentials $d^{2n}f$ (or $d^{2n}\phi$), viewed as polynomials in du_i (or dv_i), are always positive definite, while all the odd order differentials $d^{2n+1}f$ (or $d^{2n+1}\phi$) are positive for positive du_i (or dv_i).

In the case of a two-phase system this means that derivatives of $f(u)$ and $\phi(v)$ in any order are always positive for $-\infty < u,v < 1$. These convexity properties have already been used to derive various types of rigorous bounds on ε_e.[1]

III. RIGOROUS BOUNDS FOR ε_e

Various rigorous bounds for real values of ε_e have been derived in the literature. The simplest of these is the well known result

$$\left(\sum_i \frac{P_i}{\varepsilon_i}\right)^{-1} \leqslant \varepsilon_e \leqslant \sum_i P_i \varepsilon_i \, , \tag{26}$$

where P_i is the volume fraction of the phase i. Better bounds were derived by Hashin and Shtrikman[2] for the case when the composite is random and isotropic, and these were later shown to hold also for the case of a composite with a cubic point symmetry.[3] When information is available on other average bulk coefficients of the same composite (e.g., magnetic permeability, thermal or electrical conductivity, diffusivity), this can be used to derive improved bounds on ε_e.[4] This has been done both by the use of variational principles,[4,3] as well as by focusing attention on the convexity properties of $f(u_i)$ and $\phi(v_i)$.[1] We will now show how the analytical properties of f and ϕ, and in particular the representation of Eq. (24), can be applied to the derivation of such bounds.

The bounds which can be obtained depend on the information which is available. Suppose that we have a two-phase medium where the volume fractions P_1, P_2 are known, and where the macroscopic point symmetry is either isotropic or cubic. Under these assumptions, we know the first two derivatives of $f(u)$ at the origin[1]

$$f'(0) = P_1$$
$$\tfrac{1}{2}f''(0) = \tfrac{1}{3}P_1 P_2. \tag{27}$$

It is convenient to introduce the following notation

$$S \equiv \frac{1}{u}$$
$$S_\alpha \equiv \frac{1}{u_\alpha} \tag{28}$$
$$F(S) \equiv f(u) = \sum_\alpha \frac{B_\alpha}{S - S_\alpha} \, ,$$

where

$$B_\alpha > 0 \, , \quad 1 > S_o > S_1 > \ldots > S_n \geqslant 0. \tag{29}$$

From (27) and (28) we get the following sum rules

$$\sum_\alpha B_\alpha = P_1$$

$$\sum_\alpha B_\alpha S_\alpha = \tfrac{1}{3} P_1 P_2,$$

(30)

which must be satisfied. In order to determine bounds for $F(S)$, we calculate its linear variation for a fixed value of S subject to the constraints of (30). We do this by allowing S_α and B_α to have small increments δS_α and δB_α, but we use (30) to eliminate δB_0 and δS_0 from the result for δF. In this way we get

$$\delta F(S) = \sum_{\alpha \neq 0} \delta B_\alpha \left(\frac{S_\alpha - S_0}{S - S_0}\right)^2 \frac{1}{S - S_\alpha} +$$

$$+ \sum_{\alpha \neq 0} B_\alpha \delta S_\alpha \frac{(S_0 - S_\alpha)(S_0 + S_\alpha - 2S)}{(S - S_\alpha)^2 (S - S_0)^2}$$

(31)

We must now seek to maximize or minimise $F(S)$ by independently varying all the B_α and S_α, $\alpha \neq 0$. Suppose that $S > 1$ ($S < 0$), then the coefficient of δB_α in (31) is always positive (negative) and by taking $\delta B_\alpha < 0$ we always get $\delta F < 0$ ($\delta F > 0$). A lower (upper) bound on F is therefore obtained by taking $B_\alpha = 0$ for all $\alpha \neq 0$. This also eliminates S_α for $\alpha \neq 0$, and we are left with B_0 and S_0, which must be determined by the sum rules of (30), i.e.,

$$B_0 = P_1$$

$$B_0 S_0 = \tfrac{1}{3} P_1 P_2$$

(32)

We thus get the following bound for F

$$\frac{P_1}{S - \tfrac{1}{3} P_2} \lessgtr F(S) \quad \text{for} \quad \left\{\begin{array}{l} S > 1 \\ S < 0 \end{array}\right\}.$$

(33)

The other bound can be found by exchanging the roles of the two phases. This will bring about the following changes

$$S \to 1 - S$$

$$F(S) \to \frac{1 - S F(S)}{1 - S},$$

(34)

and we now find, following (33), that

$$\frac{P_2}{1-S-\frac{1}{3}P_1} \gtrless \frac{1-S}{1-S} F(S) \quad \text{for} \begin{Bmatrix} S > 1 \\ S < 0 \end{Bmatrix} . \tag{35}$$

This leads to the other bound on $F(S)$. The bounds of (33) and (35) exactly reproduce the bounds of Hashin and Shtrikman[2] for a two-phase composite. This is of course no great achievement since Eqs. (27), which were used in the derivation, were obtained to begin with from these bounds.[1]

Suppose now that we know the value of one of the material constants described by $f(u)$. This means that we know F at some point denoted by $S_+ (=1/u_+)$. What can we say about values of F at other points? To Eqs. (30) we must now add the further constraint

$$\sum_\alpha \frac{B_\alpha}{S_+ - S_\alpha} = F(S_+) . \tag{36}$$

This can be used to eliminate one more variable from the expression for δF, namely δB_1, leaving us with

$$\delta F(S) = \sum_{\alpha>1} \delta B_\alpha \left(\frac{S_\alpha - S_o}{S-S_o}\right)^2 \left(\frac{1}{S-S_\alpha} - \frac{S_+ - S_1}{(S_+ - S_\alpha)(S-S_1)}\right)$$

$$+ \sum_{\alpha>1} B_\alpha \delta S_\alpha \frac{S_o - S_\alpha}{(S-S_o)^2} \left(\frac{S_o + S_\alpha - 2S}{(S-S_\alpha)^2} - \frac{(S_+ - S_1)(S_o + S_\alpha - 2S_+)}{(S_+ - S_\alpha)^2 (S-S_1)}\right)$$

$$+ B_1 \delta S_1 \left(\frac{S_o - S_1}{(S-S_o)(S-S_1)}\right)^2 \frac{S_+ - S}{S_+ - S_1} \tag{37}$$

Focusing attention on the last term, we suppose that $S_+ > S > 1$. Then the coefficient of δS_1 is positive, and $\delta S_1 < 0$ leads to $\delta F < 0$. A lower bound is obtained by taking $S_1 = 0$ and eliminating all $B_\alpha, S_\alpha, \alpha > 1$. The values of B_o, B_1, S_o are found by solving Eqs. (30) and (36).

The other bound is found in this case by exchanging the roles of the two phases, and by considering the other type of function - ϕ rather than f: We define

$$\hat{\phi}(u) \equiv 1 - \frac{\varepsilon_1}{\varepsilon_e} = \hat{F}(S) = 1 + \frac{1-S}{S(1-F(S))} \tag{38}$$

The function $\hat{\phi}(u)$ or $\hat{F}(S)$ has the same analytical properties as $f(u)$ or $F(S)$, except that the derivatives of $\hat{\phi}(u)$ at the origin are now given by[1]

$$\hat{\phi}'(0) = P_2$$

$$\frac{1}{2}\hat{\phi}''(0) = \frac{2}{3}P_1P_2 \tag{39}$$

Thus, we can use the representation of (28) for $\hat{F}(S)$, and the whole discussion leading to a bound for $\hat{F}$ is the same as it was for F. Only the constraint equations are different:

$$\sum_\alpha B_\alpha = P_2$$

$$\sum_\alpha B_\alpha S_\alpha = \frac{2}{3} P_1 P_2 \tag{40}$$

$$\sum_\alpha \frac{B_\alpha}{S_+ - S_\alpha} = \hat{F}(S_+) .$$

In this way, for the case under discussion (i.e., $S_+ > S > 1$) we get a lower bound for $\hat{F}(S)$ and this leads to an upper bound for $F(S)$. It is straightforward to check that in this way an upper and a lower bound are always obtained for any real values of S, S_+, as long as these values are outside the "non-physical" segment $(0,1)$. The bounds found in this way for ε_e coincide with bounds that were recently derived in a very tedious and not completely rigorous manner by the use of a variational principle.[5]

In a composite made of more than two phases the key to the construction of bounds for ε_e is the choice of a proper trajectory in the space of u_i. This trajectory should pass through all the points where we have information on f as well as through the point where we wish to derive bounds, and it should have the form of Eq. (25). We demonstrate this for one case of a three-phase isotropic or cubic composite where we know the volume fractions P_1, P_2, P_3 as well as one value of the function f, i.e., $f(u_1^+, u_2^+)$.

We first determine a line of the form

$$u_2 = \frac{au_1}{1 - \frac{b}{a} u_1} \tag{41}$$

that passes through u_1^+, u_2^+ as well as through the point u_1, u_2 that we are interested in. Assuming that we get

$$0 < b < a, \tag{42}$$

the function $f(u_1, u_2(u_1))$ must have all of its poles on the real u_1 axis in the interval $(1, \infty)$. This function is thus similar to $f(u)$ and we can construct bounds for it in a similar way. The expansion of $f(u_1, u_2)$ near the origin is[1]

$$f(u_1,u_2) = P_1 u_1 + P_2 u_2 + \frac{1}{3}[P_1(1-P_1)u_1^2 + P_2(1-P_2)u_2^2$$
$$-2P_1 P_2\, u_1 u_2] + \text{higher order terms.} \qquad (43)$$

Defining

$$F(S) \equiv f(S^{-1}, u_2(S^{-1})) \qquad (44)$$

and expanding $F(S)$ as in (28), we get the following sum rules by considering the first two derivatives of $f(u_1, u_2(u_1))$:

$$\sum_\alpha B_\alpha = P_1 + aP_2$$
$$\sum_\alpha B_\alpha S_\alpha = bP_2 + \frac{1}{3}[P_1(1-P_1) + a^2 P_2(1-P_2) - 2aP_1 P_2]. \qquad (45)$$

These equations replace Eqs. (30), and to them we must now add an equation like (36).

Applying this procedure to the following example

$$P_1 = \frac{5}{18} \qquad P_2 = \frac{6}{18} \qquad P_3 = \frac{7}{18}$$

$$\varepsilon_1^+ = 1 \qquad \varepsilon_2^+ = 6 \qquad \varepsilon_3^+ = 9 \qquad \varepsilon_e^+ = 4\tfrac{1}{2}$$

$$\varepsilon_1 = 1 \qquad \varepsilon_2 = 11\tfrac{1}{2} \qquad \varepsilon_3 = 18, \qquad (46)$$

we find the following upper bound for ε_e

$$\varepsilon_e < 7.846. \qquad (47)$$

This should be compared to the Hashin-Shtrikman bound for this example[2]

$$\varepsilon_e < 10.026, \qquad (48)$$

and to a previously derived improved bound that also includes the information about ε_e^+[1] (the Hashin-Shtrikman bounds ignore this information)

$$\varepsilon_e < 8.014. \qquad (48a)$$

Obviously, a lower bound can also be found by this method.

IV. THE PERCOLATION THRESHOLD IN A RANDOM CONTINUOUS TWO-PHASE MEDIUM

Consider a two-phase medium where phase 2 is a conductor while phase 1 is an insulator. Instead of $\varepsilon_1, \varepsilon_2, \varepsilon_e$ we now consider conductivities $\sigma_1, \sigma_2, \sigma_e$. Clearly, the presence or absence of percolation in phase 2 depends on the value of m(h) at h = 0: When m(0) > 0 phase 2 percolates, while when m(0) = 0 phase 2 does not percolate.

In order to discuss the percolation threshold, we must have a statistical model for the composite which depends on some continuously varying parameter P. E.g., P could be the volume fraction of the percolating phase P_2. Instead of m(h) for a particular sample of composite material with a well defined microscopic geometry, we must now deal with an m(h) that is averaged over an appropriate statistical ensemble. The percolation threshold is the critical value P_c such that for $P > P_c$ there is percolation while for $P < P_c$ there is not.

In order to give a reasonable description for any specific composite, the statistical ensemble that describes the structure must lead to a narrow distribution of the physically relevant values of m(h). Thus, the averaging procedure should produce very little smearing of m for real, positive values of h. On the other hand, wherever m(h) has a pole, even a small amount of smearing will change the pole into a cut. Thus, we expect that the averaged functions m(h) and $\tilde{m}(\tilde{h})$ have a cut or cuts on the negative real axis, but remain analytic everywhere else. We also expect that the distribution is narrow enough to preserve multiplicative relationships between these functions, such as $m\tilde{m} = 1$. In particular, we will need to use the relationship

$$\tilde{M}(h) \cdot m(h) = h, \qquad (49)$$

where $\tilde{M}$ is the same type of function as $\tilde{m}$, but with the roles of the two phases reversed, i.e.,

$$\tilde{M} \equiv \frac{\varepsilon_1}{\varepsilon_e} = \frac{\sigma_1}{\sigma_e}. \qquad (50)$$

We note from Eq. (49) that for $P > P_c$, i.e., above the percolation threshold of phase 2, $\tilde{M}(0) = 0$. Taking the derivative of that equation at h = 0, we then find

$$m(0)\tilde{M}'(0) = 1 \qquad (51)$$

If we let P approach P_c, $m(0) \to 0$, and therefore $\tilde{M}'(0) \to \infty$. Since $\tilde{M}(0)$ is always bounded, in fact,

$$0 \leq \tilde{M}(0) \leq 1, \qquad (52)$$

this proves that at the percolation threshold the uppermost edge of the cut in $\tilde{M}(h)$ is right at the origin. From Eq. (49) it then follows that the same must be true of $m(h)$. At the same time, it is easy to convince oneself that away from P_c the upper edge of the cut is below the origin. (E.g., for $P < P_c$ a singularity in σ_e can only result if there are channels stretching across the entire sample along which the negative resistance $\sim 1/\sigma_1$ exactly cancels the positive resistance $\sim 1/\sigma_2$. This requires a finite negative value of σ_1, because every channel contains a finite amount of phase 1 material.) The picture which emerges is therefore that as P_c is approached from either side, the upper edge of the cut approaches $h = 0$.

For small, positive values of $P-P_c$ we will write the following expression for the upper edge of the cut $h_c(P)$

$$h_c(P) \cong -C (P-P_c)^\gamma, \quad C, \gamma > 0. \tag{53}$$

For values of h close to $h_c(P)$, we then write the following form for $m(h)$

$$m(h) \cong A (P-P_c)^\alpha + B (h-h_c)^\beta, \quad A, B, \alpha > 0, \quad 0 < \beta < 1. \tag{54}$$

This form will now be used to evaluate the effective complex dielectric constant κ_e (in this section we use κ to denote a complex dielectric constant, while ϵ and σ denote real quantities) for a conductor - dielectric mixture near the percolation threshold of the conductor. The complex dielectric constants κ_1, κ_2 of the pure phases and the value of h are given by

$$\kappa_1 = \epsilon_1$$

$$\kappa_2 = \frac{4\pi\sigma_2}{i\omega} \tag{55}$$

$$h = \frac{\kappa_1}{\kappa_2} = \frac{i\omega\epsilon_1}{4\pi\sigma_2},$$

and we assume that $|h| \ll 1$. We evaluate κ_e by expanding $m(h)$ around $h = 0$

$$\kappa_e = \kappa_2 m(h) \cong \frac{4\pi\sigma_2}{i\omega} m(0) + \epsilon_1 m'(0)$$

$$= \frac{\epsilon_1 \beta B}{(C(P-P_c)^\gamma)^{1-\beta}} + \frac{4\pi\sigma_2}{i\omega} [A(P-P_c)^\alpha + B(C(P-P_c)^\gamma)^\beta]. \tag{56}$$

For small, positive values of P_c-P, the upper edge of the cut will be written as

$$h_c(P) \cong -C'(P_c-P)^{\gamma'}, \qquad C', \gamma' > 0. \tag{57}$$

In this regime, it is more convenient to use the function $\tilde{M}(h)$ (since $m(h)$ must satisfy $m(0) = 0$ for all $P < P_c$), for which we now write

$$\tilde{M}(h) \cong A'(P_c-P)^{\alpha'} + B^{-1}(h-h_c)^{1-\beta}, \quad A'\alpha' > 0. \tag{58}$$

The reason why we have B^{-1} and $1-\beta$ appearing in this expression rather than a new pair of coefficients, is because for $P = P_c$ the two expressions (54) and (58) must obey Eq. (49). We now evaluate κ_e by an expansion similar to (56).

$$\frac{1}{\kappa_e} = \frac{\tilde{M}(h)}{\kappa_1} = \frac{1}{\varepsilon_1}\tilde{M}(0) + \frac{i\omega}{4\pi\sigma_2}\tilde{M}'(0)$$

$$\cong \frac{1}{\varepsilon_1}[A'(P_c-P)^{\alpha'} + B^{-1}(C'(P_c-P)^{\gamma'})^{1-\beta}] +$$

$$+ \frac{i\omega}{4\pi\sigma_2}\frac{(1-\beta)B^{-1}}{(C'(P_c-P)^{\gamma'})^{\beta}} \tag{59}$$

From (56) and (59) we find that

$$\varepsilon_e \cong \frac{\varepsilon_1 B\beta}{(C(P-P_c)^{\gamma})^{1-\beta}} \qquad \text{for } P > P_c, \tag{60}$$

$$\varepsilon_e \cong \frac{\varepsilon_1}{A'(P_c-P)^{\alpha'} + B^{-1}(C'(P_c-P)^{\gamma'})^{1-\beta}} \qquad \text{for } P<P_c. \tag{61}$$

Thus, ε_e diverges as P_c is approached from either side.

At $P = P_c$, the previously derived expressions for κ_e fail, because one cannot make a Taylor expansion around a singular branch point. Returning to Eq. (54) (or (58)) we see that in this case we get

$$\kappa_e = \kappa_2 B h^\beta = B \kappa_1^\beta \kappa_2^{1-\beta} = B \varepsilon_1^\beta \left(\frac{4\pi\sigma_2}{i\omega}\right)^{1-\beta}, \tag{62}$$

so that σ_e and ε_e satisfy

$$\sigma_e \sim \omega^\beta, \qquad \varepsilon_e \sim \omega^{\beta-1} \tag{63}$$

FOOTNOTES

1. D.J. Bergman, Phys. Rev. B14, 4304 - 4312 (1976).
2. Z. Hashin and S. Shtrikman, J. Appl. Phys. 33, 3125-3131 (1962).
3. D.J. Bergman, Phys. Rev. B14, 1531-1542 (1976).
4. S. Prager, J. Chem. Phys. 50, 4305-4312 (1969).
5. See the bounds derived in Section III,F of Ref. 3.

DISCUSSION

R. L. MILLS (Ohio State Univ.): When you speak of a finite number of poles, you are referring to a finite sample with a finite number of inclusions. Am I correct in assuming that your argument goes through smoothly for an infinite sample in which the poles become a cut?

BERGMAN: Although some people would argue about the smoothness of the transition, essentially you are correct.

J. P. STRALEY (Univ. of Kentucky): You have three exponents in your expression for m(h), whereas there are only two in scaling theory. I observe that in the limit h=0 there is a confluent singularity effect with two divergent p terms. Which one of them wins?

BERGMAN: Which one will predominate depends on the values of the exponents. In the kind of general treatment I have discussed, we do not attempt to calculate the exponents explicitly. However, it would be interesting if the persons who calculate these exponents would note that the only singularity of m(h) is a cut on the real axis. If you have an expansion for the function around any point (with a reasonable number of terms), you can find not only how this point behaves with changes of T, you can also derive the critical exponent and the critical behavior in the neighborhood of the point; you do not have to worry about other singularities because there are none.

S. KIRKPATRICK (IBM): One of the appealing features of the effective medium theory is the ease with which one extends the treatment for two values of conductance to a distribution of values. It seems rather difficult to do that with what you have shown us. Am I correct?

BERGMAN: I think you are correct. It is easy to extend the theory to a finite number of different types of grains, but the theory of a continuous distribution is much more difficult.

THE ELECTRONIC PROPERTIES OF INHOMOGENEOUS MATERIALS; METAL-NONMETAL TRANSITIONS*

Morrel H. Cohen
James Franck Institute and Department of Physics
University of Chicago
Chicago, Illinois 60637

Joshua Jortner
Chemistry Department
Tel-Aviv University
Tel-Aviv, Israel

Itzhak Webman
Chemistry Department
Louisiana State University
Baton Rouge, Louisiana 70803

ABSTRACT

We survey inhomogeneous materials, both microscopic and macroscopic. We elucidate the conditions under which the transport properties of a microscopically inhomogeneous material become local so that the problem of determining macroscopic properties becomes the same as for macroscopically inhomogeneous materials. We review percolation briefly, survey the basic theoretical tools for dealing with inhomogeneous materials, and describe a few typical theoretical results. Finally, we review the results of our studies of metal-nonmetal transitions in microscopically inhomogeneous materials.

INTRODUCTION

A wide variety of materials shows metal-nonmetal transitions not as phase changes, but as continuous changes of electronic properties. We have atributed the smearing out of these transitions to microscopic inhomogeneity and have developed a quantitative theory of such transitions.[1-11] We review here the basis for that work, covering briefly much of the theory of inhomogeneous materials in the process and go on to several applications of the theory.

In section II, we begin with a broad survey of macroscopically and microscopically inhomogeneous materials. We characterize fluctuations in microscopically inhomogeneous materials and discuss behavior near a critical point. We then elucidate the circumstances in which the properties of microscopically inhomogeneous materials become similar to those of macroscopic systems. In section III we give a brief review of percolation.[12-16] In section IV, we

---

*Supported in part by NSF Grant DMR76-29218 and by the NSF Materials Research Laboratory Program at the University of Chicago.

discuss the basic theoretical tools available for studying the properties of inhomogeneous materials and illustrate them with several results. In section V, we conclude with a discussion of several continuous metal-nonmetal transitions in microscopically inhomogeneous materials.

II. INHOMOGENEOUS MATERIALS

A material can be macroscopically or microscopically inhomogeneous. It is macroscopically inhomogeneous when the inhomogeneity is on a scale much larger than any of the important microscopic lengths of the material. Such inhomogeneity is usually static. Examples of one-phase, macroscopically inhomogeneous materials are anisotropic substances in polycrystalline form and an ordered material in a domain structure. Inhomogeneous two-phase materials can be formed by precipitation, by spinodal decomposition, or as composites. The same processes can lead to multiphase inhomogeneous materials. In all of the macroscopically inhomogeneous materials the geometry and topology of the inhomogeneities can be widely variable.

The material is microscopically inhomogeneous when the scale of inhomogeneity is comparable to an important microscopic length. Such microscopic inhomogeneities can be either static or dynamic. Any of the above macroscopic inhomogeneities has microscopic analogues. Those would be nonequilibrium inhomogeneities frozen-in on a microscopic distance scale. However, there can also be equilibrium inhomogeneous states, e.g. the vortex arrays which occur in rotating superfluid helium or in type II superconductors in a sufficiently large magnetic field. Other examples are discommensurations as can occur in dislocation arrays and incommensurate charge- and spin-density wave states. Clustering can occur in multicomponent systems. Disordered versions of all of the above can occur.

Well defined microscopic entities occur in the examples cited above, but that need not be the case. The inhomogeneities may arise entirely from random fluctuations. In ordered systems, the fluctuations would be dynamic but, in the cases of interest to us, slow on an electronic time scale.

In disordered systems, the fluctuations can be static or can have a dynamic component as well. In single-component materials, the fluctuating quantities can be the number density, as in expanded liquid Hg, or the bonding configuration as in liquid Te, Se and their alloys. In multicomponent systems, the fluctuating quantity can be the composition. Generally, the fluctuating quantity is an order parameter, magnetization in magnetic systems, polarization, or the orientation of the directrix in liquid crystals.

The fluctuations can simply be those present in thermodynamic equilibrium, which in simple situations, will have a Gaussian probability distribution. Let X be the locally fluctuating quantity. P(X), its probability distribution, is given by

$$P(X) = \frac{e^{-(X-\bar{X})^2/2\sigma_X^2}}{(2\pi\sigma_X^2)^{1/2}}, \qquad (1)$$

where $\bar{X}$ is the thermodynamic mean value. The variance σ_X^2 is proportional to $\partial^2 F/\partial Y^2$, the curvature of the free energy F with respect to the thermodynamic variable Y conjugate to X, a generalized susceptibility $\mathcal{X}$. Pronounced fluctuations can occur whenever $\mathcal{X}$ is large. That, in turn can occur near a critical point, within smeared phase transitions, or when clustering or its generalization occurs.

We have implicitly assumed that the fluctuating quantity X is a local variable and can be regarded as a random function of position r within the material, as in Fig. 1. Such fluctuations are most readily characterized by their autocorrelation function

$$C_X(R) \equiv \langle X(\vec{r})X(\vec{r}+\vec{R}) \rangle - \langle X(\vec{r}) \rangle \langle X(\vec{r}+\vec{R}) \rangle, \qquad (2)$$

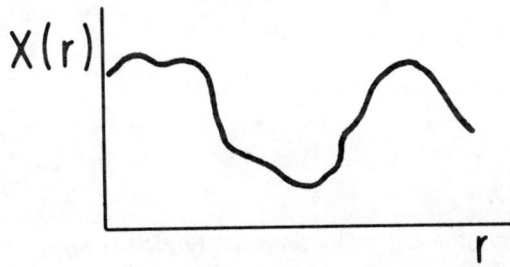

Fig. 1. Instantaneous position dependence of a fluctuating quantity $X(\vec{r})$ characterizing the local condition of a microscopically inhomogeneous material.

where the average can be regarded either as an ensemble average or a volume average over the position $\vec{r}$. In (2), we have

$$\langle X(\vec{r}) \rangle = \langle X(\vec{r}+\vec{R}) \rangle = \bar{X} \qquad (3)$$

Fig. 2 shows a typical separation dependence for $C_X(R)$. It vanishes asymptotically as $e^{-R/\xi}/R$, where ξ is the coherence length or, more precisely the Ornstein-Zernike fluctuation decay length. It is weakly dependent on distance within a radius b_X, the short correlation length. Two cases can be distinguished: 1.) $b_X \simeq \xi$, which may be the case for liquid metal-ammonia solutions,[3,4,17] and 2.) $b_X \neq \xi$, which appears to be the case for liquid Hg[1,7] and liquid Te.[5]

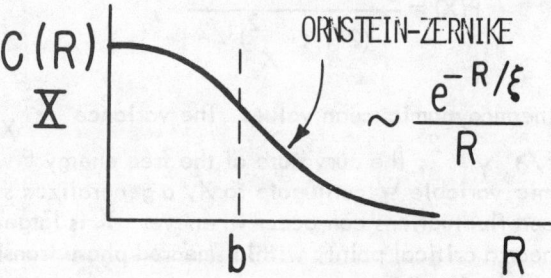

Fig. 2. A typical separation dependence of the autocorrelation function $C_X(R)$ defined in Eq. (2). b is the short correlation length and ξ_1 the Ornstein-Zernike fluctuation decay length.

The simplest approximation to $C_X(R)$ is that of a step function, as shown in Fig. 3,

$$C(R) = \langle X^2 \rangle_v - \langle X \rangle_v^2, \quad R < b_X$$
$$= 0, \quad R > b_X \quad (4)$$

$$V = 4\frac{\pi}{3} b_X^3 \quad (5)$$

$\langle \; \rangle_v$ means average over the volume V. The probability $P(X)$ is now proportional to $\exp\{-F_v(X)/kT\}$ where

$$F_v = \frac{1}{2} \chi^{-1} V (X - \bar{X})^2 \quad (6)$$

so that

$$\sigma_X^2 = kT \chi/V \quad (7)$$

is the appropriate value of σ_X^2 to insert in (1). The step function approximation to $C_X(R)$ is equivalent to supposing X to remain constant within each volume V of the material at some random value having the above probability distribution.

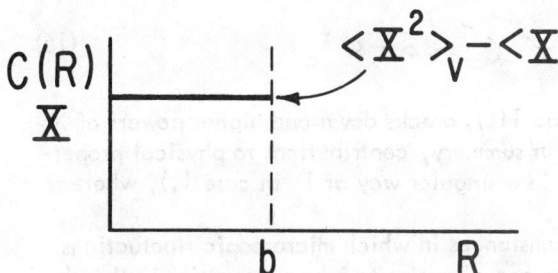

Fig. 3. Step function approximation to the correlation function $C_X(R)$.

The susceptibility $\mathcal{X}$ diverges at a critical point, raising the possibility that the fluctuations in X are divergent there. Suppose X is an order parameter which becomes nonzero below a critical temperature, T_c. Both ξ and $\mathcal{X}$ then diverge at T_c:

$$\xi \propto \epsilon^{-\nu} \tag{8}$$

$$\mathcal{X} \propto \epsilon^{-\gamma} \tag{9}$$

$$\epsilon = |(T - T_c)/T_c| \tag{10}$$

In case 1.), $b_X \cong \xi$, inserting (8.) and (9.) into (7.) gives

$$\sigma_X^2 \propto \epsilon^{3\nu - \gamma} \tag{11}$$

Below T_c, the order parameter $\overline{X}$ is

$$\overline{X} \propto \epsilon^{\beta} \tag{12}$$

Thermodynamic scaling relates the critical indices γ, β, and ν,

$$\gamma + 2\beta = 3\nu \tag{13}$$

in three dimensions[18] so that

$$\sigma_X^2 \propto \epsilon^{2\beta} \tag{14}$$

Thus, in case 1.), $\sigma_X^2 \to 0$ at the critical point, and the fluctuations drop out there but with a singular temperature derivative. On the other hand, in case 2.), $b_X \not\cong \xi$, b may be considered temperature independent near the critical point and

$$\sigma_X^2 \propto \epsilon^{-\gamma} \to \infty \text{ at } T_c \tag{15}$$

The Gaussian approximation, Eq. 11.), breaks down and higher powers of X must be kept in the exponent. In summary, contributions to physical properties from fluctuations disappear in a singular way at T_c in case 1.), whereas they are large at T in case 2.).

We now elucidate the circumstances in which microscopic fluctuations such as we have described above can give rise to inhomogeneities in the physical properties of the system leading to behavior similar to that expected from macroscopic inhomogeneities. We work in the independent-electron approximation but expect the points we make to be of general validity. The one-electron eigenfunction Ψ for energy E satisfies the Schrödinger equation

$$H\Psi = E\Psi \tag{16}$$

The Hamiltonian H consists of a kinetic energy and a potential energy $V(\vec{r})$

$$H = \frac{p^2}{2m} + V \tag{17}$$

Because of disorder, fluctuations, etc., $V(\vec{r})$ is a random function of position. We suppose that the position dependence of V on an atomic scale is determined by the local value of X. The wave function possesses an amplitude A and phase ϕ which are random functions of position,

$$\Psi = Ae^{i\phi} \tag{18}$$

Correlation functions $C_A(R)$ and $C_\phi(R)$ can be defined for A and ϕ, just as $C_X(R)$ was defined for X in Eq. (2). Short correlation lengths b_A and b_ϕ can also be defined, within which A is approximately constant and ϕ varies in an orderly fashion. For the kinds of systems we shall be dealing with, we can suppose that

$$b_A \cong b_X , \tag{19}$$

although this is not true in general. Outside b_ϕ, the phase becomes totally uncorrelated over a phase coherence length l_ϕ, which can be defined through

$$\left\langle e^{-i\phi(\vec{r})} e^{i\phi(\vec{r}+\vec{R})} \right\rangle \sim \frac{f(R)e^{-R/l_\phi}}{R} \tag{20}$$

where f(R) can be constant or an oscillatory function. The phase coherence length is closely related to the mean free path entering electronic transport when the latter is large compared to the deBroglie wavelength. As disorder increases, however, the phase coherence length becomes shorter than the de Broglie wavelength, and the concept of a mean free path loses its meaning. We shall be concerned with systems in which

$$b_\phi \gtrsim l_\phi \tag{21}$$

The lengths just introduced enable us to develop in a systematic way the concept of local electronic structure. Consider the Green's function

$$G(\vec{r},\vec{r}',E^-) = \lim_{\delta \to 0^+} \sum_j \frac{\psi_j(\vec{r})\psi_j^*(\vec{r}')}{E - E_i - i\delta} \tag{22}$$

where

$$H\psi_i = E_i \psi_i \tag{16'}$$

The total density of states N(E) can be expressed in terms of the Green's function via

$$N(E) = \frac{1}{\pi}\int d^3r \, \text{Im} G(\vec{r},\vec{r};E^-). \tag{23}$$

N(E) is seen to be a superposition of contributions from every point in the material,

$$n(\vec{r},E) = \frac{1}{\pi}\,\text{Im}\,G(\vec{r},\vec{r};E^-). \tag{24}$$

Taking a time or ensemble average of (22) gives us $\langle G(\vec{r},\vec{r}';E^-)\rangle$, which decays to zero with increasing $|\vec{r}-\vec{r}'|$ over a distance given by b_ϕ or smaller. $n(\vec{r},E)$, on the other hand, involves only amplitudes A, and varies appreciably only over distances greater than b_x. Thus $n(\vec{r},E)$ is a local density of states containing information about the local electronic structure, for example energy gaps or deep minima.

One can also use these concepts to establish the existence of local electronic response functions. Consider the linear response of a material to a static local electric field $\vec{E}(\vec{r})$,

$$\vec{j}(\vec{r}) = \int \overleftrightarrow{\sigma}(\vec{r},\vec{r}') \cdot \vec{E}(\vec{r}') \, d^3r', \tag{25}$$

where $\vec{j}(\vec{r})$ is the local current density and $\overleftrightarrow{\sigma}(\vec{r},\vec{r}')$ the nonlocal conductivity tensor. It is convenient to change variables in (25) by introducing

$$\overleftrightarrow{\sigma}'(\frac{\vec{r}+\vec{r}'}{2},\vec{r}-\vec{r}') \equiv \overleftrightarrow{\sigma}(\vec{r},\vec{r}')$$

and

$$\vec{\rho} = \vec{r} - \vec{r}'$$

Eq. (25) becomes

$$\vec{j}(\vec{r}) = \int \overleftrightarrow{\sigma}(\vec{r} + \frac{\vec{\rho}}{2}, \vec{\rho}) \cdot \vec{E}(\vec{r} + \vec{\rho}) d^3\rho \qquad (26)$$

Examination of the Kubo relations for $\overleftrightarrow{\sigma}(\vec{r}, \vec{r}')$ shows that the range of $\overleftrightarrow{\sigma}$ in $\vec{r} - \vec{r}'$ is given by b_ϕ or is shorter, whereas the variation of $\overleftrightarrow{\sigma}$ with $\frac{1}{2}(\vec{r} + \vec{r}')$ is on the scale of b_X. Insertion of these facts into the formal solution of the equation of continuity together with Eq. (26) for $\vec{E}(\vec{r})$ and $\vec{j}(\vec{r})$ shows that both vary on the scale of b_X. When $b_\phi \ll b_X$, we may neglect the dependence of E on $\vec{\rho}$ in (26) and obtain a local constitutive equation

$$\vec{j}(\vec{r}) = \overleftrightarrow{\sigma}(\vec{r}) \cdot \vec{E}(\vec{r}) \qquad (27)$$

where

$$\overleftrightarrow{\sigma}(\vec{r}) = \int \overleftrightarrow{\sigma}(\vec{r}, \vec{r}') d^3 r'. \qquad (28)$$

Now fluctuations in $n(\vec{r}, E)$ producing local energy gaps or density of states minima would be associated with locally small values of $\overleftrightarrow{\sigma}(\vec{r})$. Charge transport across these poorly conducting regions by tunneling over distances of order b_X could then be important and could yield important nonlocal behavior in $\overleftrightarrow{\sigma}(\vec{r}, \vec{r}')$. However, we shall confine our attention only to situations in which tunneling through barriers imposed by fluctuations is unimportant.

Thus when $b_\phi \ll b_A (\leq b_X)$ and tunneling is unimportant, the problem of determining the properties of a microscopically inhomogeneous medium becomes identical to the problem of determining the properties of a macroscopically inhomogeneous medium except for the difference in length scales. If we further make the step function approximation for $\overleftrightarrow{\sigma}(\vec{r})$, we have $\overleftrightarrow{\sigma}$ constant over volumes of radius b_X with a probability distribution $P(\overleftrightarrow{\sigma})$ of its values. Thus b_X and $P(\overleftrightarrow{\sigma})$ characterize the problem, as in the macroscopic case.

We now turn to a brief description of metal-nonmetal transitions in microscopically inhomogeneous materials. For simplicity, let us consider an isotropic local conductivity σ. It depends only on the value X of the quantity which characterizes the local state of the material

$$\sigma = \sigma(X). \qquad (29)$$

From the presumed known distribution P(X), the distribution $P(\sigma)$ may be determined. Suppose that when σ is less than some characteristic value σ^*, the material is locally nonmetallic. Conversely, when σ exceeds σ^*, the material is locally metallic. σ^* will correspond to some characteristic value X^* of X, $\sigma^* = \sigma(X^*)$, and will equal Mott's minimum metallic conductivity[19] when that concept is applicable. The material is thus a heterogeneous mixture of metallic and nonmetallic regions, the metallic volume fraction being

$$C = \int_{X^*}^{\infty} P(X)dX = \int_{\sigma^*}^{\infty} P(\sigma)d\sigma \tag{30}$$

A metal-nonmetal transition occurs when C decreases below a critical value, as we discuss in the next section. The details of the transition may depend on the nature of $P(\sigma)$. A unimodal distribution of σ is shown in Fig. 4a, and a bimodal distribution shown in Fig. 4b. The conductivity is not

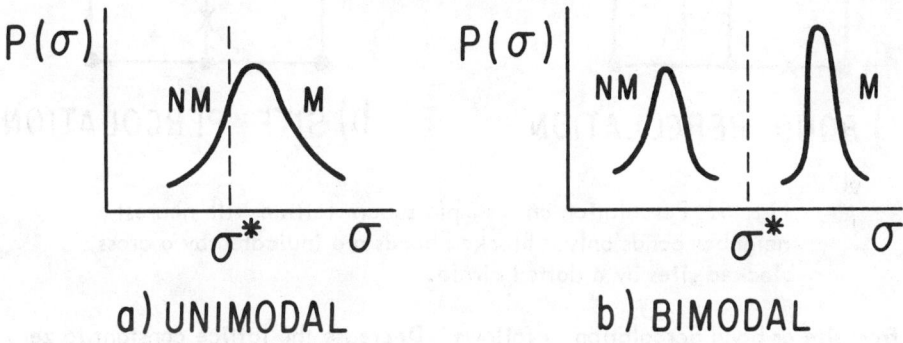

Fig. 4. Examples of the probability distribution $P(\sigma)$ of the local conductivity. Values of $\sigma > \sigma^*$ are metallic (M) and $< \sigma^*$ nonmetallic (NM).

the only property dramatically affected by such a transition. The optical properties, thermoelectric power, etc., change, and one can carry out similar considerations for any such property. The nonmetallic state can be quite varied, for example, a semiconductor as in liquid Hg or liquid Te, an insulator, or an electrolyte with an electronic component to the conductivity near the transition as in metal-ammonia solutions.

III. CONDUCTION AS PERCOLATION

Before proceeding to the calculation of the electronic properties of microscopically/macroscopically inhomogeneous materials and thence to metal-nonmetal transitions, it is useful to review percolation[12-16] briefly. Diffusion is random motion in an ordered medium. Percolation, on the other hand, is exemplified by regular motion through a random medium. Examples are flow of water through porous rock or of coffee through coffee grounds in a percolator.

Percolation can occur on lattices or through continuous media (continuum percolation[20,2,6]). In lattice percolation there are two cases commonly considered, bond percolation in which a fraction p_b of bonds between nearest neighbors are present or open (unblocked) and site percolation in which a fraction p_s of sites together with all attached bonds are present or open. These two cases are illustrated in Fig. 5. Continuum percolation can be generated

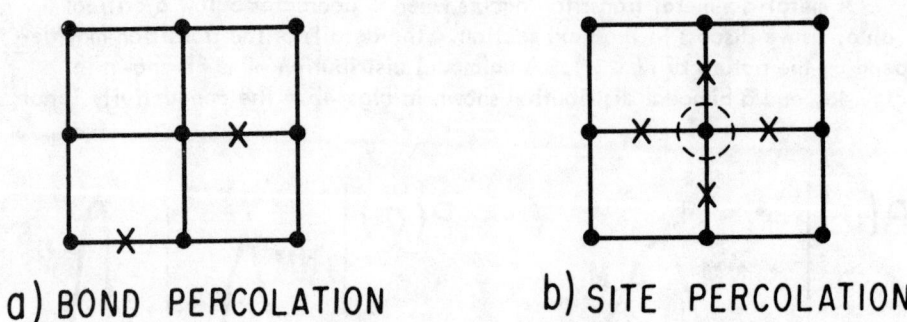

Fig. 5. Percolation on a simple square lattice with nearest neighbor bonds only. Blocked bonds are indicated by a cross, blocked sites by a dotted circle.

from site or bond percolation as follows. Decrease the lattice constant to zero relative to the distance within which lattice sites remain connected or bonded. Decrease the site occupation to zero. Let the allowed volume fraction C associated with open sites remain finite. C then replaces p in the percolation theory.

Sites are linked via unblocked bonds into clusters. As p increases the mean cluster size grows until, at some critical value p*, an infinite cluster emerges. The percolation probability P(p) is defined as the fraction of the total number of allowed sites linked into the infinite cluster. P(p) vanishes below p*, the percolation threshold, and increases rapidly above p*, ultimately reaching unity as shown in Fig. 6. p_b^* is $\frac{1}{4}$ for the simple cubic lattice and

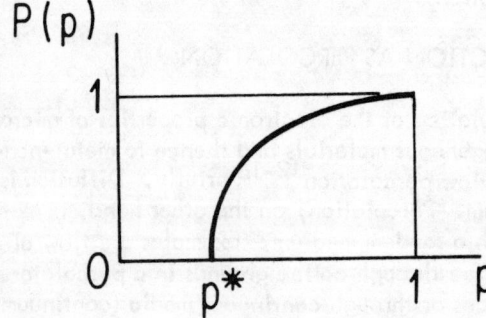

Fig. 6. Percolation probability P as a function of open bond or site fraction p. p* is the percolation threshold.

$\frac{1}{2}$ for the simple square. C* is 15% for continuum percolation in three dimensions.

Conduction frequently appears as a percolation process in inhomogeneous materials. Consider a two-valued distribution of conductivities, $\sigma = 0$ with probability $1 - C$ and σ_0 with probability C. Percolation theory tells us that for $C < C^*$, the conducting regions of the material are disjoint and isolated as in Fig. 7a. The macroscopic conductivity remains zero. When C

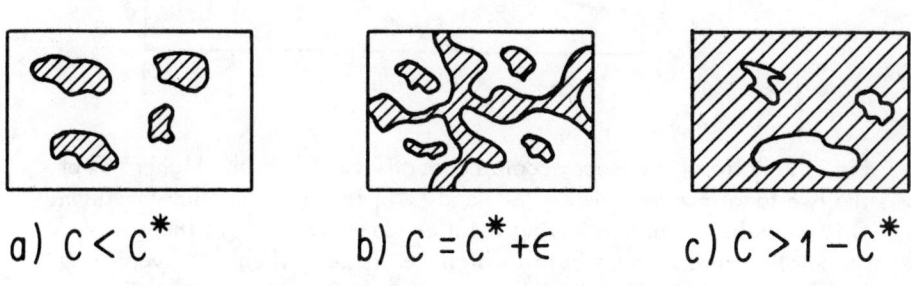

a) $C < C^*$ b) $C = C^* + \epsilon$ c) $C > 1 - C^*$

Fig. 7. Growth of metallic regions (shaded) with increase of metallic volume fraction C. Below the percolation threshold C^* there are isolated metallic regions and no conductance, a.). Above the percolation threshold a metallic path crosses the material and conduction occurs, b.). Above $1 - C^*$ the insulating regions are disjoint, c.).

exceeds C^*, a finite fraction of the material (equal to the percolation probability) is connected into a conduction path leading across the sample, Fig. 7b, and σ becomes an increasing function of C. Finally, when C exceeds $1 - C^*$ the insulating regions are disjoint and isolated as in Fig. 7c. Thus a metal-nonmetal transition occurs at the percolation threshold C^*. The dependence of the macroscopic conductivity σ on C^* is sketched in Fig. 8.

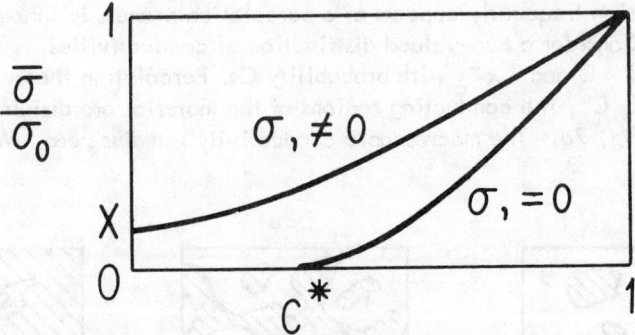

Fig. 8. Ratio of the macroscopic conductivity $\bar{\sigma}$ to the higher σ_0 of the two local conductivities σ_0 and σ_1 in a random binary medium. C is the volume fraction having local conductivity σ_0. There is a percolation threshold C* below which $\bar{\sigma}$ is zero when σ_1 vanishes, but there is a finite C = 0 intercept at X = σ_1/σ_0 when σ_1 is finite.

IV. SOME THEORETICAL RESULTS

We are interested in problems which, within the limits we have imposed, are still quite general. The local conductivity σ can have a general distribution. We are also interested in a wide range of response functions including not only the conductivity but also the magnetoconductivity tensor $\bar{\sigma}$ (H), the thermal conductivity $\varkappa$, the thermoelectric power S, the real and imaginary parts of the dielectric function, $\epsilon_1(\omega)$ and $\epsilon_2(\omega)$, respectively. What are the tools available to us?

The first set of tools comes from percolation theory, scaling theory and renormalization group analysis, the availability of which is a consequence of the analogy of percolation and conduction to critical phenomena. A second set is provided by numerical simulations of the transport properties of model systems, and a third set by effective medium theory (EMT) and related methods of analysis. There are generalizations and improvements of the effective medium theory which we shall not review here. Both the EMT and its generalizations are discussed at length elsewhere in this volume.

Turning first to percolation theory, Fortuin and Kasteleyn[21] pointed out that the percolation problem could be related to the statistical mechanics of the Potts model, a generalization of the Ising model. Thus the onset of percolation at the percolation threshold is a phase transition and the percolation threshold a critical point. One infers immediately that the percolation probability must show scaling behavior

$$P(p) = (p - p^*)^s \qquad (31)$$

with, as we shall see, $s \lesssim 1$. The macroscopic conductivity $\bar{\sigma}$ also shows scaling behavior.[22,2,23,6] Consider a binary medium in which the local conductivity σ has the value σ_0 with probability C and σ_1 with probability $1 - C$. Suppose that $\sigma_1 \ll \sigma_0$, i.e.

$$X = \sigma_1/\sigma_0 \ll 1, \qquad (32)$$

so that we are near the percolation problem in which $X = 0$. It is found that

$$\begin{aligned}(C)\bar{\sigma}(C) &\propto \sigma_0 (C - C^*)^\beta & C > C^* \\ (C)\bar{\sigma}(C) &\propto \sigma_1 (C^* - C)^{-\alpha} & C < C^* \\ (C^*)\bar{\sigma}(C^*) &\propto \sigma_0 X^\delta \\ \delta &= \frac{\beta}{\alpha + \beta}\end{aligned} \qquad (33)$$

as depicted in Fig. 9. The exponents α, β, δ are termed critical exponents.

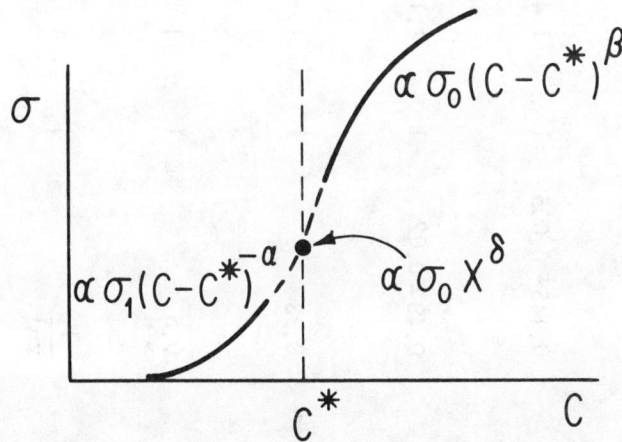

Fig. 9. σ versus C near the percolation threshold C^* for a random binary medium having the values σ_0 with probability C and σ_1 with probability $1 - C$ for the local conductivity, $\sigma_1 \ll \sigma_0$.

The relation in (33) connecting them is termed a scaling relation. Table I gives percolation thresholds, critical exponents, and the scaling relation for several known cases.[10] Apart from certain cases in which the exponents are obtained exactly, the entries in Table I derive from numerical simulations.

TABLE I. CRITICAL EXPONENTS OF PERCOLATION CONDUCTIVITY

		C*	α	β	$\frac{\beta}{\alpha+\beta}$	δ
Three-Dimensional Bond Percolation	(a)	0.25	1	1.6 ± 0.1	0.615 ± 0.15	0.67 ± 0.08
Three-Dimensional Continuous Percolation	(b)	0.145 ± 0.005	1	1.4 ± 0.05	0.585 ± 0.01	0.65 ± 0.05
Two-Dimensional Bond Percolation	(c)	0.48 ± 0.02	1	1.1 ± 0.1	0.52 ± .02	0.51 ± 0.01
Three-Dimensional EMT		0.33	1	1	0.5	0.5
Two-Dimensional EMT		0.5	1	1	0.5	0.5
Bethe Lattice of Co-ordination Number z	(d)	$\frac{1}{z-1}$	1	1	0.66	≳ 0.6

(a) Refs. 22, 2, 6, and 10.
(b) Refs. 2 and 10.
(c) Refs. 24 and 10.
(d) Refs. 25 and 23.

Stinchcombe and Watson[26] in 1976 obtained $\beta = 1.13 \pm 0.09$ for the two-dimensional bond-percolation problem by the use of real space renormalization group techniques, in agreement with the numerical simulation.

The numerical simulations of conduction in continuous random media proceed as follows.[2,6,8,9,10,11] The problem is first stated in continuous form

$$\vec{j} = \sigma \vec{E} \qquad (34a)$$

$$\vec{\nabla} \cdot \vec{j} = 0 \qquad (34b)$$

$$\vec{E} = -\vec{\nabla} V \qquad (34c)$$

Then space is discretized into a set of points i on a lattice with voltages V_i. The current is channeled into nearest neighbor links between lattice points, I_{ij} flowing from i to j. Eq. (34a) becomes Ohm's law

$$V_j - V_i = r_{ij} I_{ij} \qquad (34a')$$

and the equation of continuity Kirchhoff's law

$$\sum_j I_{ij} = 0 \qquad (34b')$$

Here r_{ij} is the resistance of the link between $i + j$, and the summation on j is over all nearest neighbors of i. Fig. 10 shows discretization into a two dimensional square lattice. In an actual lattice problem, the solution of (34a') and (34b') would proceed numerically after random assignment of the r_{ij}. If the simulation is to represent a continuous random medium, however, it is necessary to introduce correlation among the values of neighboring resistances and to pass to the limit in which the range of the correlation is large compared to the lattice separation. Similar algorithms can be developed for the discretization of the other transport problems of interest to us, e.g. $\vec{\sigma}$ (H) and S.

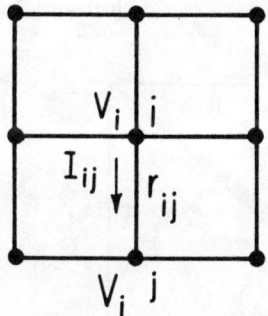

Fig. 10. Discretization of current I, voltage V, and resistance r in a continuous random conducting medium.

The numerical simulations are tedious and can be costly. It is of course far more convenient to use approximate analytic or nearly analytic theories instead. These are of limited accuracy, however, and the numerical simulations provide us with the conditions under which such approximate theories can be used. The most widespread of these is the effective medium theory (EMT).[27] One imagines a spherical region of radius b within which the current density, electric field, and conductivity take on their local values $\vec{j}'$, $\vec{E}'$, and σ'. The probability distribution of σ' being $P(\sigma')$. Outside that region the actual material is replaced by a homogeneous effective medium with conductivity σ equal to the actual macroscopic conductivity, and current density $\vec{j}$ and field $\vec{E}$ asymptotically far away. One solves the electrostatic problem thus posed for $\vec{j}'$ and $\vec{E}'$ and then imposes the requirement that on average the current and field obey the macroscopic constitutive equation everywhere, in particular,

$$\langle \vec{j}' \rangle = \sigma \langle \vec{E}' \rangle \tag{35}$$

Eq. (35) imposes a self consistency requirement on the macroscopic conductivity σ, the so-called effective medium condition

$$\int \frac{\sigma' - \sigma}{\sigma' + 2\sigma} P(\sigma') d\sigma' = 0, \tag{36}$$

from which σ can be directly determined.

An effective medium theory can be done for each of the linear response functions, as can a numerical simulation.[2,8,9,11] As first obtained by Kirkpatrick[22] for σ, the general result for binary media is that provided the ratio of the smaller to larger value of the local transport coefficient is greater than about 1/30, the EMT is accurate for all C. Otherwise it is accurate only for $C \gtrsim 0.4$. A comparison of the results of numerical simulations with the EMT is shown in Fig. 11.

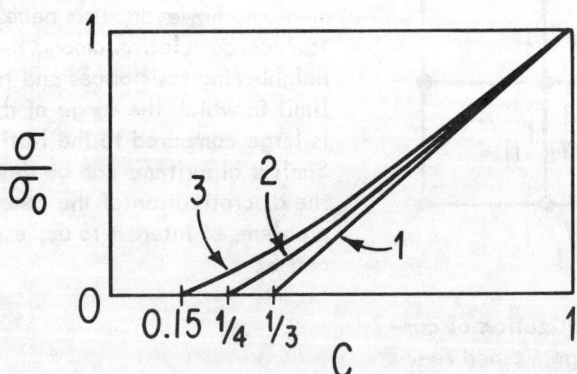

Fig. 11. Conductivity of a random binary medium in which the probability of a local conductivity σ_0 is C and of vanishing conductivity 1 – C: 1. EMT; 2. Simple cubic resistor lattice (numerical simulation), 3. Continuous random material (numerical simulation).

V. SOME APPLICATIONS TO METAL-NONMETAL TRANSITIONS

The conductivity of expanded liquid mercury drops by a factor of about 30 as the mass density is decreased from 9.3 to 8.0 g/cc. into the nonmetallic range of conductivities. Other electronic properties indicate clearly that the conduction mechanism is not a conventional one. The data on σ, the Hall constant R, the Hall mobility μ, and the Knight shift K are quantitatively consistent with a model[1,7] in which the liquid is microscopically inhomogeneous because of density fluctuations in the range 8.0-9.3 g/cc. At low densities there is an energy gap in the local density of states around the Fermi level; at higher densities the material is locally metallic. This picture has not yet been shown to be unique, but it is internally consistent. In particular, the conditions for a local electronic structure, for local electronic transport, and for the neglect of tunneling are all well met. One striking result of the analysis is that the value of b is 15Å.

In liquid Te, there is a five-fold drop of the conductivity in a range of temperatures around 1000K. Once again other transport data indicate that the transport mechanism is not conventional. The data for σ, R, μ, K, the nuclear relaxation rate T_1^{-1} and X-ray and neutron scattering intensities are all quantitatively consistent with a model[5] in which the liquid is microscopically inhomogeneous because of bonding fluctuations. At lower temperatures the Te atoms are two-fold coordinated and the material is a semiconductor. At higher temperatures a nonbonding electron is promoted, the Te atoms become three-fold coordinated, and the material is metallic. The metal-nonmetal transition is gradual, with mixed bonding configurations occurring over a range of temperatures. Our analysis of the data suggests that the short correlation length for the coordination number is at least twice the interatomic separation.

In Li and Na liquid ammonia solutions, the conductivity changes by more than three orders of magnitude over the concentration range 1 to 10 mole percent metal (MPM) at about 10K above the consolute temperature. The data on σ, R, μ, K, S, ϵ_1, ϵ_2, magnetic susceptibility, nuclear resonance, and velocity of sound are all consistent with a model[3,4] of the solutions in which there is microscopic inhomogeneity associated with concentration fluctuations. Variation of the concentration distribution leads to a gradual metal-nonmetal transition. Our analysis suggests that a binary concentration fluctuation model is sufficient. The nonmetallic regions have a local value of concentration equal to 2.3 MPM, the metallic regions 9 MPM, and C varies linearly with the mean concentration from zero at 2.3 MPM to unity at 9 MPM. The bimodality in the concentration distribution may not be needed, and the large concentration fluctuations may well be a critical phenomenon.[17] However, the values of b which emerge from our analysis, particularly the value of 30Å for the Na solution, are so large that more appears to be going on than simple critical fluctuations.

Granular metals, prepared by coevaporation of a metal and an oxide as discussed by Abeles and by Sheng at this conference, show a metal-nonmetal transition of the percolative type with a remarkably large value of C*, 40-50% as opposed to the value of 15% expected for continuum percolation in three dimensions.[28] The films show a grainy structure, consisting of metal particles surrounded by oxide, and the grain size is small compared to film thickness so that conduction is three dimensional. The conductivity exponent is also large, 1.9 as opposed to the universal value of 1.4 expected, Table I. It has been suggested that the large percolation threshold arises from a tendency of the metallic grains to become coated with oxide during their formation. This suggestion is not well supported by the morphology of the films as revealed by electron microscopy. One would have expected internal boundaries of oxide within large metallic grains, which is not observed. Instead Jortner, Webman and myself[29] have proposed a model in which the segregation of the metal into grains is incomplete so that some metal remains randomly dispersed in the oxide matrix. We are able to reproduce the conductivity data quantitatively with this model, obtaining the large values of C* and explaining the large exponent.

These granular metals offer interesting examples of composite materials. There are many other examples with properties of great potential interest, scientifically and technically. An understanding of these properties can be achieved with the tools reviewed herein and discussed in detail during this conference. We shall allude here to one further example treated by Davidson and Tinkham[30] and discussed at this conference by Tinkham although a superconducting to normal and not a metal-nonmetal transition is involved. Wires of Nb, of Sn, and of Cu are drawn down together and heat treated so as to produce an inhomogeneous, anisotropic mixture of superconducting and normal metal. The resistance of the superconductor vanishes below its transition temperature. Thus the resistance presents a percolation problem. When the volume fraction of superconductor falls below the percolation threshold, the resistance becomes finite, or, conversely, there is an onset of superconductivity at the percolation threshold.

REFERENCES

1. M. H. Cohen and J. Jortner, Phys. Rev. A $\underline{10}$, 978 (1974).
2. I. Webman, J. Jortner, and M. H. Cohen, Phys. Rev. B$\underline{11}$, 2885 (1975).
3. M. H. Cohen and J. Jortner, J. Phys. Chem. $\underline{79}$, 2900 (1975).
4. J. Jortner and M. H. Cohen, Phys. Rev. B$\underline{13}$, 1548 (1976).
5. M. H. Cohen and J. Jortner, Phys. Rev. B$\underline{13}$, 5255 (1976).
6. I. Webman, J. Jortner, and M. H. Cohen, Phys. Rev. B$\underline{14}$, 4737 (1976).
7. M. H. Cohen and J. Jortner, Phys. Rev. B$\underline{15}$, 1227 (1976).
8. I. Webman, J. Jortner, and M. H. Cohen, Phys. Rev. B$\underline{15}$, 1936 (1976).
9. I. Webman, J. Jortner, and M. H. Cohen, Phys. Rev. B$\underline{15}$, 5712 (1977).
10. I. Webman, J. Jortner, and M. H. Cohen, Phys. Rev. B$\underline{16}$, 2593 (1977).
11. I. Webman, J. Jornter, and M. H. Cohen, Phys. Rev. B$\underline{16}$, 2959 (1977).

12. S. R. Broadbent and J. M. Hammersley, Proc. Camb. Phil. Soc. 53, 629 (1957).
13. H. L. Frisch and J. M. Hammersley, J. Soc. Ind. Appl. Math. 11, 894 (1963).
14. V.K.S. Shante and S. Kirkpatrick, Adv. Phys. 20, 325 (1971).
15. J. W. Essam, in "Phase Transitions and Critical Phenomena", eds. C. Domb and M. S. Green, Academic Press, London, 1972 Vol. 2 p. 197.
16. R. Zallen, in Proc. 13th IUPAP Conf. on Statistical Physics, Haifa, 1977.
17. P. Damay and P. Schettler, J. Phys. Chem. 79, 2930 (1975).
18. H. E. Stanley, "Introduction to Phase Transitions and Critical Phenomena", Oxford University Press, 1971.
19. N. F. Mott, "Metal-Insulator Transitions", Taylor and Francis, London, 1974.
20. R. Zallen and H. Scher, Phys. Rev. B4, 4471 (1971).
21. C. M. Fortuin and P. W. Kasteleyn, Physica (Utrecht) 57, 536 (1972); P. W. Kasteleyn and C. M. Fortuin, J. Phys. Soc. Jpn. Suppl. 26, 11 (1969).
22. S. Kirkpatrick, Phys. Rev. Lett. 27, 1722 (1971); Rev. Mod. Phys. 45, 574 (1973).
23. J. P. Straley, J. Phys. C 9, 783 (1976); Phys. Rev. B15, 5738 (1977); and present volume.
24. S. Kirkpatrick, Phys. Rev. Lett. 36, 69 (1976).
25. R. B. Stinchcombe, J. Phys. C 6, L1 (1973); J. Phys. C 7, 179 (1974).
26. R. B. Stinchcombe and B. P. Watson, J. Phys. C 9, 3221 (1976).
27. D.A.G. Bruggeman, Ann. Phys. (Leipz.) 24, 636 (1935); R. J. Landauer, J. Appl. Phys. 23, 779 (1952); V. I. Odehlevskii, J. Tech. Phys. (USSR) 21, 678 (1951); H. J. Juretschke, R. Landauer, and J. A. Swanson, J. Appl. Phys. 27, 838 (1956); Morrel H. Cohen and J. Jortner, Phys. Rev. Lett. 30, 696 (1973).
28. B. Abeles, H. L. Pinch, and J. I. Gittleman, Phys. Rev. Lett. 35, 286 (1975).
29. M. H. Cohen, J. Jortner, and I. Webman, unpublished.
30. A. Davidson and M. Tinkham, Phys. Rev. B13, 3261 (1976).

DISCUSSION

S. NAM (University of Dayton): You have emphasized the low q features of the structure factor. What values should an experimentalist look for?

COHEN: You should look at the inverse of $15\overset{\circ}{A}$, i.e. values of scattering wave vector around tenths of inverse $\overset{\circ}{A}$.

G. PIKE (Sandia Labs): There are two ways of eliminating the effect of lattice spacing and producing a continuous medium. One way is to consider correlations as you have done and the other is to consider long range interactions. The two methods apparently do not give identical answers. The value of 15% for the percolation threshold comes from the correlation length work which you have done. In the correlated bond problem which Kirkpatrick introduced a value near 10% is obtained. If long range interactions are considered directly, such as I have done, or if the regular lattice problem is extended to large interaction distances, a percolation threshold near 30% is found. In the two dimensional cases the differences are even larger. There may not be something which can be uniquely defined as a continuous medium in the sense being used in this problem.

COHEN: This point has been examined critically by Zallen, who comes to a conclusion opposite from yours. The matter is not settled, but the preponderance of evidence is beginning to accumulate on the side of the 15% number.

J. P. STRALEY (University of Kentucky): Where exactly in the data you have shown does the metal-insulator transition take place?

COHEN: You can put the metal-nonmetal transition anywhere you like. It is essentially a continuous transition and to identify some specific value of the conductivity with the transition one must invoke a specific theoretical interpretation. The particular interpretation which we have used puts the percolation transition near 3.5 mole percent metal.

STRALEY: What is the ratio of conductivities?

COHEN: It is about three orders of magnitude. It is smaller than that for $Li-NH_3$ and larger for $Na-NH_3$. One is always the prisoner of the experimentalist in these matters: one interprets the data one has and, as with theories, experimental results change with time. The latest data is not always the correct data. The data of 1926 on the conductivity of metal-ammonia solutions is about as accurate as anything that has been done. The Hall effect data has recently been revised considerably. We need more thermoelectric power data. As a result, if the same theory is used to fit new data, the fit may not be as satisfactory. I don't think the results I've talked about will change grossly.

N. W. ASHCROFT (Cornell University): I'm reminded of the comment that was made along these lines at the Liquid Metals Conference. Someone said exactly what you have said about some experimental measurements having a habit of changing with time and a French participant said "Ah, yes, put then there are those measurements which, like good wine, improve with age."

B. ABELES (Exxon Research and Development): In the cermets the metal-nonmetal transition takes place at roughly 60 volume percent metal. The explanation is that there is a strong correlation between metal and insulator: the insulator tends to coat the metal grains.

COHEN: We have assumed that outside the short correlation length b, there was no correlation. We have tried to understand quantitatively the values for both the percolation threshold and the critical exponent. (The exponent is strange, as well.) If the cermets are assumed to consist of coated grains plus a fine, random, distribution of metal, both the threshold and the exponent can be reproduced quite easily. Of course, in doing this model, we have two parameters to fit and we have introduced two new parameters into the problem.

SCATTERING THEORY AND EFFECTIVE MEDIUM APPROXIMATIONS TO HETEROGENEOUS MATERIALS

J. E. Gubernatis
Los Alamos Scientific Laboratory, Los Alamos, NM 87545

ABSTRACT

The formal analogy existing between problems studied in the microscopic theory of disordered alloys and problems concerned with the effective (macroscopic) behavior of heterogeneous materials is discussed. Attention is focused on 1) analogous approximations (effective medium approximations) developed for the microscopic problems by scattering theory concepts and techniques, but for the macroscopic problems principally by intuitive means, 2) the link, provided by scattering theory, of the intuitively developed approximations to a well-defined perturbative analysis, 3) the possible presence of conditionally convergent integrals in effective medium approximations.

INTRODUCTION

Many physical phenomena are associated with inhomogeneous medium. For example, a solid state physicist may study the electrical properties of disordered alloys; a geologist, the mechanical properties of polycrystalline rocks. Other scientists may ponder the twinkling of stars or water-seepage through concrete.

Various theoretical approaches to these diverse phenomena generally exhibit two common features. One feature is a statement of physics, usually in the context of a model. This statement may be the equation of motion for a solid continuum, the time independent Schrödinger equation, Helmholtz's equation, etc. These various equations regulate the dynamical variables and have as parameters stochastic variables associated with the inhomogeneous medium. The second common feature is an average of the dynamical variables or their products over the distribution of the stochastic parameters. Even when the statistical information is complete, only an approximation to the averaging, which for some problems is called an effective medium approximation, is generally possible. Most often effective medium approximations are developed by intuitive means.

In solid state physics in the study of disordered alloys, theorists stated the physics of their problem in terms of an integral equation and analyzed this equation by techniques developed in the quantum mechanical theory of scattering. This integral equation is equivalent to a perturbation series (an infinite series), and various effective medium approximations were developed usually by approximating the average of each term in the series and then summing an infinite series of terms. This approach, scattering theory with infinite order perturbation summation, developed perturbatively several effective medium approximations. Some of these approximations have direct analogs to approximations developed intuitively for other phenomena.

I will illustrate the application of the scattering theory approach to the computation of the effective dielectric constant of a polycrystal. I will state the problem in the form of an integral equation, recover several well-known intuitive approximations, and indicate, but not demonstrate, the connection of the approximations to perturbation theory. This example and its discussion hopefully clarify the meaning of several commonly used effective medium approximations.

For this same problem, I will indicate the existence of conditionally convergent terms that may appear in an improperly constituted perturbation series. These terms can cause the effective property or approximations to the effective property of the material to depend on the shape of the material. Batchelor[1-3] first noticed the possibility of such terms in an analogous problem and suggested a procedure for their removal. His procedure is discussed in light of standard practices in scattering theory.

STATEMENT OF THE PROBLEM

The example under consideration is a polycrystalline material that has a dielectric constant tensor $\underline{\epsilon}(\underline{r})$ which in general changes from grain to grain because of changes in material type, orientation of the grain, or grain size and shape. However, when measured, the material as a whole behaves effectively as a homogeneous medium with a dielectric constant $\underline{\epsilon}^*$. The problem is to calculate $\underline{\epsilon}^*$ from statistical information about $\underline{\epsilon}(\underline{r})$.

For a static problem the physics follows from

$$\underline{\nabla} \cdot \underline{D} = 0 \qquad (1)$$

where $\underline{D}$ is the electric displacement field. Modelling of the medium begins with the constitutive relation

$$\underline{D}(\underline{r}) = \underline{\epsilon}(\underline{r}) \, \underline{E}(\underline{r}) \qquad (2)$$

where $\underline{E}$ is the electric field, and the effective dielectric constant is defined by

$$\langle \underline{D} \rangle = \underline{\epsilon}^* \langle \underline{E} \rangle \qquad (3)$$

where the angular brackets denote ensemble averaging.

The problem specified by (1)-(3) is analogous to many other problems. As $\underline{E}$ is a gradient of a potential, replacing $\underline{E}$ by the gradient of temperature and $\underline{D}$ by the heat current re-interprets $\underline{\epsilon}$ as a thermal conductivity. Through similar re-interpretations the problems of effective electrical conductivity, permeability, diffusivity, elastic stiffness, etc. are seen as directly analogous. The problem of effective elastic properties, of course, requires increases in tensorial rank.

It is useful to write $\underline{\epsilon}$ as the sum of two parts

$$\underline{\underline{\epsilon}} = \underline{\underline{\epsilon}}^\circ + \delta\underline{\underline{\epsilon}} \tag{4}$$

where $\underline{\underline{\epsilon}}^\circ$ is some arbitrarily, but in general conveniently, chosen spatially invariant (homogeneous) dielectric constant so that all stochastic variations are contained in the perturbation $\delta\underline{\underline{\epsilon}}$. One can now show the equivalence of (1)-(4) to the integral equation

$$\underline{E} = \underline{E}^\circ + \int d\underline{r}' \, [\delta\underline{\underline{\epsilon}}(\underline{r}')\underline{E}(\underline{r}') \cdot \nabla'] \underline{g}(\underline{r},\underline{r}') \tag{5}$$

where $\underline{g}$ is the Green's function satisfying

$$\nabla \cdot \underline{\underline{\epsilon}}^\circ \underline{g} = \delta(\underline{r} - \underline{r}')$$

and E° satisfies the homogeneous equation

$$\nabla \cdot \underline{\underline{\epsilon}}^\circ \underline{E}^\circ = 0$$

It is also useful to write the integral equation in several different forms. First, with a standard indicial notation

$$E_i = E_i^\circ + \int d\underline{r}' \, G_{ij} \delta\epsilon_{jk} E_k \tag{6a}$$

where

$$G_{ij} = G_{ji} = g_{i,j} = \phi_{,ij}$$

since g_i is expressible as the gradient of a potential ϕ. Next, in an operator notation

$$E = E^\circ + G\delta\epsilon E \tag{6b}$$

where G is an integral operator

$$Gf \to \int d\underline{r}' \, G_{ij}(\underline{r},\underline{r}') f_j(\underline{r}')$$

Equation (6) is a statement of the physics that is identical to (1)-(3).

The integral equation has a formal solution obtained by iteration and represented by the infinite series

$$E = E^\circ + G\delta\epsilon E^\circ + G\delta\epsilon G\delta\epsilon E^\circ + \cdots$$

Different terms contain the perturbation $\delta\epsilon$ to different "powers". With the definition of T, the T-matrix,

$$T = \delta\epsilon + \delta\epsilon \, G \, \delta\epsilon + \delta\epsilon \, G \, \delta\epsilon \, G \, \delta\epsilon + \cdots$$

the following equivalent to (6) is obtained

$$E = E^\circ + GTE^\circ \tag{7}$$

Now T represents the perturbation series, and a formal summation of

T is obvious

$$T = \delta\epsilon (I - G\delta\epsilon)^{-1}$$

Since the explicit character of $\delta\epsilon$ is yet unspecified, (6) and (7) are general. The specification of $\delta\epsilon$ for a polycrystal assumes the following model:

$$\delta\epsilon(\underline{r}) = \sum_\alpha \delta\epsilon^\alpha \Theta^\alpha(\underline{r}) \qquad (8)$$

where $\Theta^\alpha(\underline{r}) = 1$, $\underline{r}$ in grain α
$\phantom{where \Theta^\alpha(\underline{r}) =}\, 0$, otherwise.

The term "grain" is used in the broadest possible sense. For example, the region α could be a pore.

To correspond to the piecewise behavior of $\delta\epsilon$, it is convenient to define a t-matrix associated with grain α

$$t^\alpha = \delta\epsilon^\alpha \Theta^\alpha (I - G\delta\epsilon^\alpha \Theta^\alpha)^{-1} \qquad (9)$$

With this definition T is expressible as

$$T = \sum_\alpha t^\alpha + \sum_\alpha \sum_{\beta \neq \alpha} t^\alpha G t^\beta + \sum_\alpha \sum_{\beta \neq \alpha} \sum_{\gamma \neq \beta} t^\alpha G t^\beta G t^\gamma + \cdots \qquad (10)$$

Although (10) is equivalent to (7), it shifts the focus of the perturbation series from $\delta\epsilon$ to t^α. The significance is that a truncation of (7) after the first term produces a perturbation series of first order in $\delta\epsilon^\alpha$, a truncation of (10) after the first term produces a perturbation of infinite order in $\delta\epsilon^\alpha$.

t^α has a physical meaning: If the deviation from homogeneity is confined solely to one region α, then

$$E = E° + G\delta\epsilon^\alpha \Theta^\alpha E$$

which when iterated and summed becomes

$$E = E° + G t^\alpha E°$$

Thus t^α is the T-matrix which solves the single inhomogeneity problem. Consequently, the first term in (10) represents contributions from regions α individually embedded in a homogeneous medium $\epsilon°$. The remaining terms, all involving at least two regions, represent the interaction between regions.

THE AVERAGING

To compute ϵ^*, $\langle E \rangle$ and $\langle D \rangle$ are needed. $\langle E \rangle$ is determined directly from (7)

$$\langle E \rangle = E° + \langle GT \rangle E° \qquad (11)$$

but the determination of $\langle D \rangle$ involves several steps. First, since $\epsilon = \epsilon^\circ + \delta\epsilon$,

$$D = \epsilon^\circ E + \delta\epsilon E$$

Next from a comparison of (6) and (7),

$$\delta\epsilon E = TE^\circ$$

Finally,

$$\langle D \rangle = \epsilon^\circ \langle E \rangle + \langle T \rangle E^\circ \tag{12}$$

Thus, (11) and (12) with (3) yield

$$\epsilon^* = \epsilon^\circ + \langle T \rangle (I + \langle GT \rangle)^{-1} \tag{13}$$

This is an exact equation, independent of the assumed polycrystalline model.

As with most exact equations, exact evaluation is usually impossible. Approximations are needed. In the present case, one sees that approximations to T are especially important, and one possible approximation is to truncate (10) after the first term.

$$T \simeq \sum_\alpha t^\alpha \tag{14}$$

For the standard problem of isotropic $\delta\epsilon^\alpha$ and spherical grains, the above, when used in (13), yields after several algebraic manipulations

$$\frac{\epsilon^* - \epsilon^\circ}{\epsilon^* + 2\epsilon^\circ} = \sum_j v_j \frac{\epsilon^j - \epsilon^\circ}{\epsilon^j + 2\epsilon^\circ} \tag{15}$$

where v_j is the volume fraction of material type j that has a dielectric constant ϵ^j. Often, a polarizability

$$\frac{4\pi\alpha^j}{3} = \frac{\epsilon^j - \epsilon^\circ}{\epsilon^j + 2\epsilon^\circ}$$

is associated with each grain, and (15) is written as

$$\frac{\epsilon^* - \epsilon^\circ}{\epsilon^* + 2\epsilon^\circ} = \frac{4\pi}{3} \sum_j v_j \alpha^j \tag{16}$$

This equation is now of the form of the famous Clausius-Mossotti equation.

It is important to note that (14) is used in (13), not in

$$\epsilon^* = \epsilon^\circ + \langle T \rangle \tag{17}$$

Since t^α solves the problem of a single grain α embedded in $\epsilon°$, (17) is a simple sum of the average contribution of each grain. From elementary electrostatics, when an isotropic dielectric sphere is placed in a uniform electric field $E°$, the sphere is polarized; t^α is connected with this polarization. Equation (14) sums individual "dipoles" embedded in $\epsilon°$ and does not account for the fact that any given dipole sees a medium in which other "dipoles" are present. The factor $(I + \langle GT \rangle)^{-1}$ in (13) is the Lorentz correction which accounts for the presence and the polarization of other grains by replacing these grains by a uniformly polarized medium. The important point is that in (16) the grain α is embedded not in $\epsilon°$, as in (14), but in a uniformly "polarized" medium. However, interactions between grains are still neglected.

The approximation (14) has been used in many different contexts usually with $\epsilon° = \langle \epsilon \rangle$, for example, in electrical conductivity problems by Maxwell[4], (frequency dependent) dielectric problems by Maxwell-Garnett[5,6], thermal conductivity problem by deVries[7], and in solid state physics by Elliott and Taylor[8]. For elastic problems, Kröner's[9] result bears some relation to (17). In solid state physics, (14) is sometimes called the Average T-matrix Approximation (ATA).

A commonly used approximation is a self-consistent effective medium approximation. "Self-consistency" is a term used in different contexts, often incompatibly. In the present context, the self-consistency means the following: T in (13) depends on $\epsilon°$ through $\delta\epsilon$ and G. If $\epsilon°$ were chosen so that $\langle T \rangle = 0$, then $\epsilon^* = \epsilon°$.

T, let alone $\langle T \rangle$, is not a quantity one expects to evaluate exactly. Self-consistent solutions, however, can be sought for approximations to T. If, for example, one takes

$$T \simeq \sum_\alpha t^\alpha \tag{18a}$$

and requires

$$\langle T \rangle = 0 \tag{18b}$$

then for the standard problem of isotropic $\delta\epsilon^\alpha$ and spherical grains, he finds the following well-known self-consistent effective medium approximation

$$\sum_j v_j \frac{\epsilon^* - \epsilon^j}{\epsilon^* + 2\epsilon^j} = 0 \tag{19}$$

The self-consistency condition is significant. Both the ATA and self-consistent approximation are based on a perturbation series represented by

$$T \simeq \sum_\alpha t^\alpha$$

i.e., the embedding of single grains in a homogeneous medium. Equation (16), however, does not equal (19). The difference is the ATA stops with the above and thereby ignores interactions between grains, while the self-consistent approximation through the additional condition (18b) incorporates interacting grains in an average manner.

The mathematical details of how (18b) incorporates some of the interactions is fully documented elsewhere.[10,11] Physically, one considers a single grain α in a uniformly polarized medium that represents the presence of the other grains. This polarized medium polarizes α, but the polarization of α changes the polarization of the uniform medium, the change in this polarization changes the polarization of α, etc. until the polarization of α and the uniform medium representing the other grains are consistent on the average.

Self-consistent approximations, as defined in the present context, have been used by various investigators, for example, Bruggeman[12] and Landauer[13] for electrical media, Hershey[14], Hill[15], and Budiansky[16] for elastic media. In solid state physics this approximation was derived by Taylor[17] and by Soven[18] and is called the (single site) Coherent Potential Approximation (CPA).

REMARKS

From (10) corrections to ϵ^* beyond the ATA are easily seen to be second order in t. Corrections to ϵ^* beyond the CPA are not as obvious, but are known to be fourth order in t, which implies that the ATA is an approximation to the CPA. The important point is that analysis natural to the scattering theory with infinite order perturbation summation approach can connect condition (18) to a well-defined perturbation analysis and identify the next order corrections: The CPA is exact through the third order in t. Since the self-consistency condition is a formal expression of intuitive and physical conditions stated for a variety of related problems, the ad hoc flavor of many self-consistent effective medium approximations is placed on clearly defined theoretical grounds by their relation to an explicit perturbation series that has immediately identifiable corrections.

On physical grounds the approximation,

$$T \simeq \sum_\alpha t^\alpha$$

because it neglects interactions between grains and clustering effects is expected to work for well-separated grains of different species embedded in $\epsilon°$. For two-phase systems, this means small concentrations of one species hosted in the other. Because the CPA is also based on this same approximation to T, it too is often regarded as a small concentration approximation. This is not completely correct since comparisons with experiment suggest otherwise. Apparently, for many problems the interactions between grains and such clustering effects as "touching" grains are on the average not as important as one would a priori suppose.

For a two-phase medium the CPA is exact for the low concentration limit for either species. One interpretation of the CPA is that of an interpolation formula between these limits, an interpolation performed according to a specific approximation. The interpretation, of course, is not always useful for problems involving porous materials.

Another important point about the CPA is that it is the best possible approximation using statistical information about single grains, e.g. volume fraction, texture, etc. It is the best possible in the sense that it evaluates every perturbation term that can be evaluated with only single grain distribution functions. Corrections to the CPA require knowledge of two-grain distribution functions. To be precise, the CPA is the best single site approximation in the sense that all single-site diagrams are taken into account.

An interesting, new effective medium approximation, called the self-consistent cumulant approximation (SCA), was recently proposed by Hori and Yonezawa.[19-25] When they compared the SCA to the results of a computer experiment on a network of resistors randomly placed on Bravais lattice, they found that near the percolation limit the SCA matched the computed results better than the CPA given in (19). Since clustering phenomena becomes important near the percolation threshold and since the CPA ignores clustering, it would appear that the SCA includes some clustering effects. Indeed, Hori and Yonezawa[22] identify these contributions.

The SCA, however, is based on a special model of a heterogeneous material, called the perfectly disordered material. In this model the value of the dielectric constant at $\underline{r}$ is statistically independent of the value at $\underline{r}'$. (In a polycrystal, the values of $\delta\epsilon$ at different points within each grain are strongly correlated; hence, a polycrystal does not fit the above definition of a perfectly disordered material.) Additionally, the SCA also based on a infinite summation of a perturbation series that includes various terms in the series more than once. In the jargon of scattering theory, the propagator is not renormalized in a self-consistent fashion so different diagrams are multiply counted. The physical and mathematical basis of the SCA needs further study.

There is more than one way to relate effective medium approximations to perturbation theory. Elliott et al[10] and Yonezawa and Morigaki[11] review these ways for solid state physics problems. In a series of papers[19-25] and with a specific approach, Hori and Yonezawa do an extensive analysis of electrical, thermal, and magnetic problems. Gubernatis and Krumhansl[26,27] discuss the analysis as applied to elastic problems. Other references[28-40] treat a variety of problems both from the intuitive and more formal viewpoints. In some cases comparisons with experiment are given. This list is not close to being definitive.

CONDITIONALLY CONVERGENT INTEGRALS

Several years ago, Batchelor[1-3] noted that if one attempts to calculate the effective properties of a heterogeneous material by looking for a solution in terms of an expansion of the volume fraction v of particles in a "dilute suspension", integrals that are only conditionally convergent can arise. If such integrals are evaluated for a finite volume which is allowed to go to infinity without change of shape, then the limit depends on the shape of the volume chosen.[41,42] Thus, the appearance of such integrals seemingly makes the effective property dependent on the shape of the sample. In the effective dielectric problem discussed above, conditionally convergent integrals with integrands $O(r^{-3})$ arise from the dipole character of the G_{ij} in (6a),

$$G_{ij} = \frac{1}{4\pi\epsilon^\circ} \frac{\partial}{\partial x_i} \frac{\partial}{\partial x_j} \frac{1}{|\underline{r} - \underline{r}'|} \qquad (20)$$

(the ith component of an electric field at $\underline{r}$ because of the jth component of dipole placed at $\underline{r}'$ in a medium with dielectric constant ϵ°). To be properly convergent, the integrand at large r must vary as $r^{-(3+\eta)}$ where $\eta > 0$.

Batchelor suggested a procedure[1-3,41-3] to circumvent the apparent difficulties to $O(v^2)$. (If v is small, terms $O(v^2)$ exhibit the first order effects of interaction between two particles.) Qualitatively, his procedure is as follows: Subtract from the given conditionally convergent integral another conditionally convergent integral whose integrand has the same asymptopic behavior as the integrand of the integral in question and whose ensemble average is zero. Then the ensemble average of the given integral is unchanged, but the troublesome asymptopic behavior of the integrand is removed. The result is now shape independent.

The source of difficulty is the long-ranged nature of the dipole field. Traditionally, in scattering theory (and many-body theory), long-ranged interactions are replaced by shorter-ranged screened interactions obtained by a partial, but infinite, summation of perturbation series. Possibly, such a renormalization is related to Batchelor's intuitive approach and would be a systematic way to generalize and extend his procedure.

A question of more immediate interest is whether the effective constant defined by (13) is well-defined, i.e. shape independent. I believe it is.

Briefly, relying heavily on results of Yonezawa and coworkers [11,22], I reason in the following manner: First, write* $\langle GT \rangle = G\langle T \rangle$, then from (13) define $\epsilon^* \equiv \epsilon^\circ + \Sigma$ where

$$\Sigma = \langle T \rangle (I + G\langle T \rangle)^{-1} \qquad (21)$$

is analogous to the self-energy that appears in the solid state physics of disordered alloys. One has that

---

*Although T contains all the statistical information, I prefer, because of G being an integral operator, to write the average of GT as $\langle GT \rangle$ to indicate that the integration GT is averaged and not that the averaged T is integrated. Convention, along with Yonezawa and coworkers, uses $G\langle T \rangle$.

$$\Sigma = \Sigma^{(1)} + \Sigma^{(2)} + \Sigma^{(3)} + \cdots \quad (22)$$

where

$$\Sigma^{(1)} = \langle \delta\epsilon \rangle \quad (23a)$$

$$\Sigma^{(2)} = \langle \delta\epsilon G \delta\epsilon \rangle - \langle \delta\epsilon \rangle G \langle \delta\epsilon \rangle \quad (23b)$$

$$\Sigma^{(3)} = \langle \delta\epsilon G \delta\epsilon G \delta\epsilon \rangle - \langle \delta\epsilon G \delta\epsilon \rangle G \langle \delta\epsilon \rangle$$
$$- \langle \delta\epsilon \rangle G \langle \delta\epsilon G \delta\epsilon \rangle + \langle \delta\epsilon \rangle G \langle \delta\epsilon \rangle G \langle \delta\epsilon \rangle \quad (23c)$$
$$\vdots$$

or in terms of cumulants

$$\Sigma^{(1)} = \langle \delta\epsilon \rangle_c \quad (24a)$$

$$\Sigma^{(2)} = \langle \delta\epsilon G \delta\epsilon \rangle_c \quad (24b)$$

$$\Sigma^{(3)} = \langle \delta\epsilon G \delta\epsilon G \delta\epsilon \rangle_c - \langle \delta\epsilon G \langle \delta\epsilon \rangle_c G \delta\epsilon \rangle_c \quad (24c)$$
$$\vdots$$

where $\langle \cdots \rangle_c$ denotes the cumulant. Equation (23) is composed of both reducible and irreducible terms where the "irreducible" has the usual meaning: A diagram that cannot be divided into two separate parts just by cutting the propagator (the Green's function) once. (Yonezawa called these proper diagrams.) On the other hand, (24) is composed of only irreducible terms.

Only the reducible diagrams, however, can be conditionally convergent. The integrands of reducible terms are the product of G times a function which may tend to a constant as $r \to \infty$ leaving an integrand $O(r^{-3})$ and the integral conditionally convergent. This remark is exemplified by (23b): If the $\delta\epsilon$ refer to different grains, $\langle \delta\epsilon G \delta\epsilon \rangle$ is reducible, and the principal contribution to the term is the integration

$$\iint d\underline{r} d\underline{r}' \, G(\underline{r},\underline{r}') \, P_2(\underline{r},\underline{r}')$$

where $P_2(\underline{r},\underline{r}')$ is the probability of finding a grain centered at $\underline{r}$ if one is centered at $\underline{r}'$. As $|\underline{r} - \underline{r}'| \to \infty$, $P_2 \to$ constant, i.e. the grains become statistically independent, and the integrand is $O(r^{-3})$. If the $\delta\epsilon$ refer to the same grain, $\langle \delta\epsilon G \delta\epsilon \rangle$ is irreducible, and the finite size of the grain eliminates any question of conditional convergence. The second part of (23b), $\langle \delta\epsilon \rangle G \langle \delta\epsilon \rangle$, is also conditionally convergent; the $\delta\epsilon$'s are averaged independently and the integration involves simply the bare G with $\delta\epsilon$.

The two terms in (23b) combine to give (24b) which is irreducible and absolutely convergent. Equation (24b) is proportional to

$$\iint d\underline{r} d\underline{r}' \, G(\underline{r},\underline{r}')[P_2(\underline{r},\underline{r}') - P_1(\underline{r})P_1(\underline{r}')]$$

where $P_1(\underline{r}) = v$ is the one-point correlation function. As $|\underline{r} - \underline{r}'| \to \infty$

$$P_2(\underline{r},\underline{r}') - P_1(\underline{r})P_1(\underline{r}') \to 0$$

and the integral is well-behaved since for any physically reasonable specification of P_2, it now goes to zero at least as fast as $r^{-(3+\eta)}$ where $\eta > 0$. A simple and, in the present context, an important property of cumulants is illustrated: Contributions from statistically independent stochastic variables are not present.

Irreducibility with the non-local character of the Green's function imply another mechanism to produce properly convergent integrals. A general character of irreducibility is convoluted multiple integrals. In individual terms or portions of such terms involving multiple integrations, a given integrand because of the non-locality of G depends on at least one other integration also involving a G. As a result, the entire integration falls off faster than r^{-3}. This is illustrated in the second term in (24c) which is proportional to

$$\iint d\underline{r}d\underline{r}' [P_2(\underline{r},\underline{r}') - P_1(\underline{r})P_1(\underline{r}')] \int d\underline{r}'' G(\underline{r},\underline{r}'')G(\underline{r}'',\underline{r}')$$

In light of the above remarks, I believe an examination of the $\Sigma^{(i)}$ to all orders indicates that ϵ^* is well-defined, i.e. shape independent. The key is the ability of sum $(I + G\langle T \rangle)^{-1}$ to subtract out the conditionally convergent terms in $\langle T \rangle$ on a one-to-one basis. The expansion of Σ in cumulants is a convenient, but not necessary, vehicle to discuss this. If (17) was the basis for defining ϵ^*, which essentially is Batchelor's basis, ϵ^* would be inherently shape dependent and hence ill-defined.

One should note that implicit in the above discussion is the need for caution in approximating (13). The CPA, (18), is directly expressible as a sum of corrected cumulants and thus is expected to be a shape independent approximation.

Admittedly, I have not presented a proof; a proof has been developed by Baker.[44] My intention was to be heuristic. The question of improperly convergent terms in the perturbation series has gone unnoticed for nearly eighty years and the existence of these difficulties is still unknown to many interested in the effective behavior of heterogeneous materials. I myself just heard of the difficulties several weeks ago. Hopefully, the problems now have the attention and interest of a wider community.

ACKNOWLEDGEMENTS

This work was performed under the auspices of USERDA. The first part of this paper was adapted from a presentation made at the Second International Symposium on Continuum Models of Discrete Systems (Mont Gabriel, Quebec, June 26 - July 2, 1977). This presentation benefited from helpful comments by J. A. Krumhansl and R. Silver. It was at the Symposium I first heard of the apparent problems of conditionally convergent integrals. (The interested are referred to papers by D. J. Jeffery and J. R. Willis in the proceedings of that meeting.) In this paper the section on conditionally convergent integrals benefited from a helpful discussion with George A. Baker, Jr.

REFERENCES

1. G. K. Batchelor, J. Fluid Mech., **52**, 245 (1972).
2. G. K. Batchelor and J. T. Green, **56**, 401 (1972).
3. G. K. Batchelor, Ann. Rev. Fluid Mech., **6**, 227 (1974).
4. J. C. Maxwell, Treatise on Electricity and Magnetism, (Oxford Press, Oxford), p. 365.
5. J. C. Maxwell-Garnett, Phil. Trans. R. Soc., **203**, 385 (1904).
6. J. C. Maxwell-Garnett, Phil. Trans. R. Soc., **205**, 237 (1904).
7. D. A. de Vries, Int. Inst. of Refrigeration, (Annexe 1952) **32**, 115 (1952).
8. R. J. Elliott and D. W. Taylor, Proc. R. Soc. Lond. A, **296**, 161 (1967).
9. E. Kröner, J. Mech. Phys. Solids, **15**, 319 (1967).
10. R. J. Elliott, J. A. Krumhansl, and P. L. Leath, Rev. Mod. Phys., **46**, 465 (1974).
11. F. Yonezawa and K. Morigaki, Prog. Theor. Phys. Suppl., **53**, 1 (1973).
12. D. A. G. Bruggeman, Annln. Phys., **24**, 636 (1952).
13. R. A. Landauer, J. Appl. Phys., **23**, 779 (1952).
14. A. V. Hershey, J. Appl. Mech., **21**, 236 (1954).
15. R. J. Hill, J. Mech. Phys. Solids, **13**, 89 (1965).
16. B. Budiansky, J. Mech. Phys. Solids, **13**, 213 (1965).
17. D. W. Taylor, Phys. Rev., **156**, 1017 (1967).
18. P. Soven, Phys. Rev., **156**, 809 (1967).
19. M. Hori, J. Math. Phys., **14**, 514 (1973).
20. M. Hori, J. Math. Phys., **14**, 1942 (1973).
21. M. Hori and F. Yonezawa, J. Math. Phys., **15**, 2177 (1974).
22. M. Hori and F. Yonezawa, J. Math. Phys., **16**, 352 (1975).
23. M. Hori, J. Math. Phys., **16**, 1777 (1975).
24. M. Hori, J. Math. Phys., **18**, 487 (1977).
25. M. Hori and F. Yonezawa, J. Phys. C, **10**, 229 (1977).
26. E. Domany, J. E. Gubernatis, and J. A. Krumhansl, J. Geo. Res., **80**, 4851 (1975).
27. J. E. Gubernatis and J. A. Krumhansl, J. Appl. Phys., **46**, 1875 (1975).
28. W. F. Brown, Jr., J. Chem. Phys., **23**, 1514 (1955).
29. D. Stroud, Phys. Rev. B, **12**, 3368 (1975).
30. R. E. Smith, G. B. Spence, J. E. Gubernatis, and J. A. Krumhansl, Carbon, **14**, 185 (1976).
31. D. M. Wood and N. W. Ashcroft, Phil. Mag., **35**, 269 (1977).
32. P. H. Dederichs and R. Zeller, Z. Phys., 259, 103 (1973).
33. J. Korringa, J. Math. Phys., **14**, 509 (1975).
34. S. Kirkpatrick, Rev. Mod. Phys., **45**, 574 (1973).
35. W. L. Bragg and A. B. Pippard, Acta. Cryst., **6**, 865 (1953).
36. J. Bernasconi, Phys. Rev. B, **9**, 4575 (1974).
37. J. Bernasconi and H. J. Wiesmann, Phys. Rev. B, **13**, 1131 (1976).

38. M. Beran, J. Appl. Phys., 39, 5712 (1968).
39. R. S. Smith, J. Appl. Phys., 27, 824 (1956).
40. Y. Yuge, J. Stat. Phys., 16, 339 (1977).
41. D. J. Jeffery, Proc. R. Soc. London A, 335, 355 (1973).
42. D. J. Jeffery, Proc. R. Soc. London A, 338, 503 (1974).
43. J. R. Willis and J. R. Action, Q. J. Mech. Appl. Math., 29, 163 (1976).
44. G. A. Baker, Jr., unpublished.

DISCUSSION

M. H. COHEN (Univ. of Chicago): I believe your improper integrals correspond to surface depolarization effects. These effects are actually included in the definition of the macroscopic fields, so when the calculations are done systematically one can subtract them out at the beginning. If you do not do this subtraction, then they absolutely must come out at the end!

GUBERNATIS: I recently heard about this problem about six weeks ago at a continuum mechanics conference. The approach is one used by Jeffery and Willis, who were examining the integrations along one surface (which goes to infinity) and were making statements about what has to happen on the surface. The problem is that not many non-physicists have ever heard of the Clausius-Mossotti equation, so that to couch the problem in the kind of language you were using takes a little time for non-physicists to appreciate.

COHEN: Although most physicists have heard of the Clausius Mossotti relation, not many really understand its subtleties.

A. B. PIPPARD (Cavendish Lab.): This problem is illustrated by the well known paradox of screening in metals. An electron is always surrounded by a screening cloud which exactly neutralizes it; the electron, therefore, can never carry any current! Here again the answer concerns the convergence of the dipole screening field. In order to get any current the boundary conditions must be fixed properly. The great strength of the Effective Medium Approximation is that it avoids this difficulty; the great weakness of the approximation is that one cannot extend it very far.

D. McKENZIE (Sydney Univ.): I would like to stand up for Lord Rayleigh, who did not have this problem because he was clever enough to use long needles for which the depolarization field does not exist.

D. BERGMAN (Tel-Aviv Univ.): Has anyone ever tried to apply a self-consistent approach, such as the Coherent Potential Approximation or Effective Medium Approximation, to an anisotropic system to see if the percolation threshold is more accurately predicted?

GUBERNATIS: I have applied the EMA to anisotropic elastic materials, although there is no direct connection to the percolative processes which you mentioned. In a porous medium, however, the bulk modulus goes to zero when the porosity becomes great enough. The EMA predicts such a transition although at a concentration for which the material still retains some strength. Thus there is some analogy to percolation in the elastic constants, although the connection to percolation theory is not clear. The medium I considered consisted of polycrystalline carbon with about eight percent epoxy (used as a binder) and about thirty percent voids. The anisotropy was in the elastic properties of the carbon grains.

COHEN: This material does not seem to correspond to the assumptions of randomness required by the Effective Medium Approximation. In a three phase medium where one constituent coats the grains of a second constituent, the effect is to increase the percolation fraction. The system appears to resemble the cermets.

GUBERNATIS: Although I don't want to get into the details of polycrystalline carbon, the medium is actually more complicated than you have described.

THE GEOMETRY OF THE PERCOLATION THRESHOLD

Scott Kirkpatrick
IBM Research Center, Yorktown Heights, N.Y. 10598

ABSTRACT

Computer-generated pictures are presented of the connected component ("infinite cluster") found at concentrations just above the threshold for 2D. site percolation in large (400x400 site) lattices. For each case, we also show the "backbone" of the cluster, the smaller set of sites through which a current may flow. The simulations are contrasted with the model of conduction just above threshold due to Skal and Shklovskii and to de Gennes. That model is found to be inconsistent with the observed critical behavior of the conductivity in 2D and 3D models, but may apply to percolation in 4D and above. We show that a proper treatment of inhomogeneity on scales smaller than the coherence length is necessary to account for the observed conductivity and backbone volume just above threshold, and introduce a self-similar model which accounts reasonably well for these properties.

INTRODUCTION

A percolation threshold,[1-3] in which a mixture of conducting and insulating material abruptly ceases to conduct at some sharply defined and reproducible composition, can be described mathematically as a form of second order phase transition,[4] with associated critical phenomena. This introduces complications not envisioned by the effective medium approaches which this conference has focussed upon.[5] An effective medium description of a composite material assumes the material to be homogeneous over distances greater than some grain size, ℓ_0, and describes the system by some locally uniform conductance, $\sigma_0(r)$, on all scales less than ℓ_0.

Near a percolation threshold, the appropriate grain size is ξ, the coherence length, defined by

$$G(r - r') - P(p)^2 \propto \exp(-|r - r'|/\xi) \quad , \tag{1}$$

where $G(r - r')$ is the probability that the points r and r' are both conducting and connected to each other by conducting material, p is the volume fraction of conducting material, and P(p) is the percolation fraction. Properties averaged over regions of dimensions $\gg \xi$ should be spatially uniform. However, the conduction process is quite non-uniform

1a. The largest connected component, or "infinite cluster", in a 400x400 site 2D square lattice with .66 of the sites present is shown as white. Black denotes non-conducting sites, whether absent or conducting but isolated. For this sample, $P(p) = .635$. Sites are deemed connected when they are adjacent in the horizontal or vertical directions, but not diagonally.

over distances $\lesssim \xi$, as will be discussed in detail below.

Figs. 1a - 4a give examples of the evolution of the infinite clusters as p approaches its threshold value, p_c, from above. The figures show site percolation in 2D square lattices of 400x400 sites, with white representing the material available for conduction. The percolation threshold for this model occurs at $p = .591$. The concentrations shown approach p_c in steps of 0.02. To eliminate dependence on boundary conditions the opposite

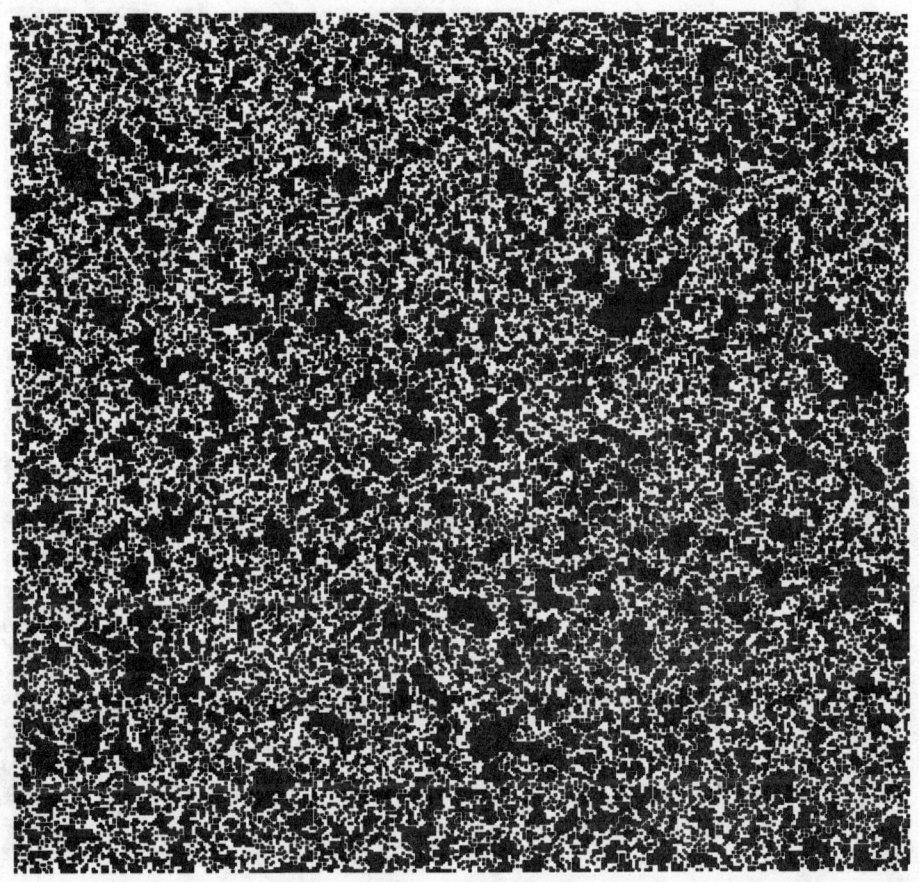

1b. The largest biconnected component of the cluster shown in Fig. 1a. For this sample, $B(p) = .506$.

edges of each sample are considered to be connected, and the computer program simply finds and plots the largest connected cluster in the sample, without checking whether that cluster contains conducting paths which cross the sample. (In fact, this has always been the case for $p \gtrsim 0.60$ in samples 250x250 or larger.)

A current flowing through the sites remaining in Figs. 1a - 4a would be quite spatially non-uniform, and no current at all will flow through many of the sites. Skal and Shklovskii[6] and de Gennes[7] have recently emphasized

2a. Largest connected component in a 2D square lattice, as in Fig. 1a but with p = .64, P(p) = .601.

that current flow will be confined to the "backbone" of the infinite cluster, while "tag ends", which connect to the backbone at only one point, will be subject to no potential drop and carry no current.

A precise definition of the backbone of the infinite cluster (this is left somewhat unclear in Refs. 6 and 7) is that it is the largest biconnected component of the dilute lattice. A set of lattice sites is said to be biconnected if every pair of sites can be linked by at least two distinct conducting paths. We shall denote the fraction of sites in the backbone as B(p), by analogy with P(p). An efficient algorithm for identifying the biconnected components of a cluster has been given by Hopcroft and Tarjan.[8] This was

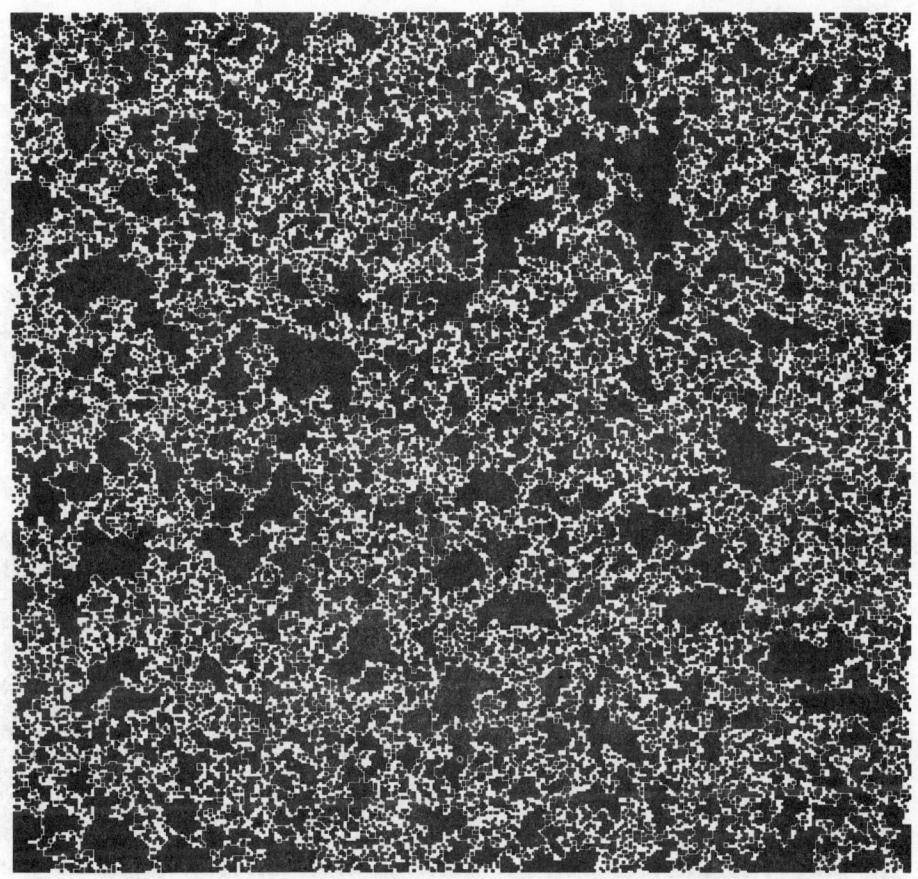

2b. Largest biconnected component in Fig. 2a. $B(p) = .436$.

used to isolate the backbone of each cluster shown in Figs. 1a - 4a, with the results shown in Figs. 1b - 4b. Although each of Figs. 1b - 4b was obtained from the corresponding figure of the set 1a - 4a, there is no relation between the four examples other than the steadily decreasing value of p. In each case a different seed was supplied to the random number generator to determine the arrangement of sites present and absent.

Fig. 1a, which shows the connected component for $p = .66$, appears fairly uniform to the eye. Only 2.5 per cent of the sites are present but isolated from the infinite cluster. The most conspicuous evidence of these missing sites is a low density of dark regions which appear to be convex and ≲ 20 sites in linear dimension. However, closer inspection shows that many of

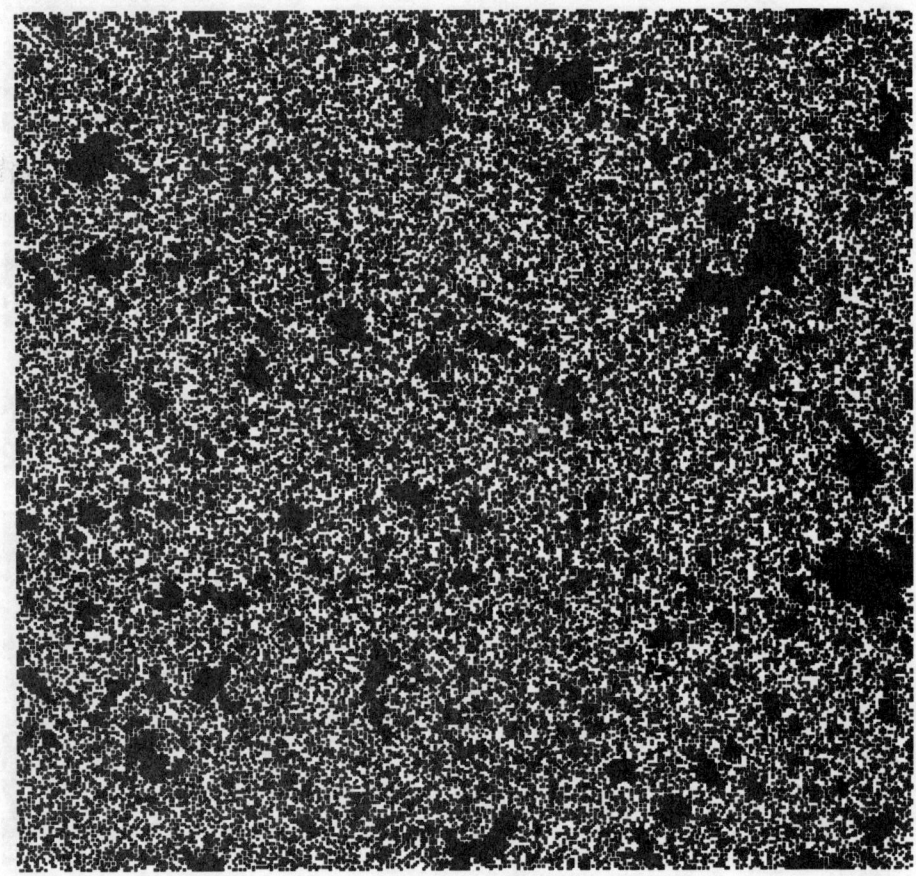

3a. Largest connected component in a 2D square lattice with p = .62, P(p) = .549.

the dark regions are not convex but are tenuously linked to their neighbors. Examination of the backbone (Fig. 1b) confirms this. Current-free regions extending up to ≲ 80 sites are found, and these have quite contorted shapes.

Decreasing p to .64 causes only a slight apparent coarsening in Fig. 2b. Now about 4 per cent of the sites are present but isolated. Some of the larger regions of isolated material have coalesced and the dark regions have more intricate shapes, but the picture appears to be homogeneous on scales of ≳ 40 sites. The backbone (Fig. 2b), however, shows marked coarsening when compared with Fig. 1b. Nonconducting regions occur with ≲ 100

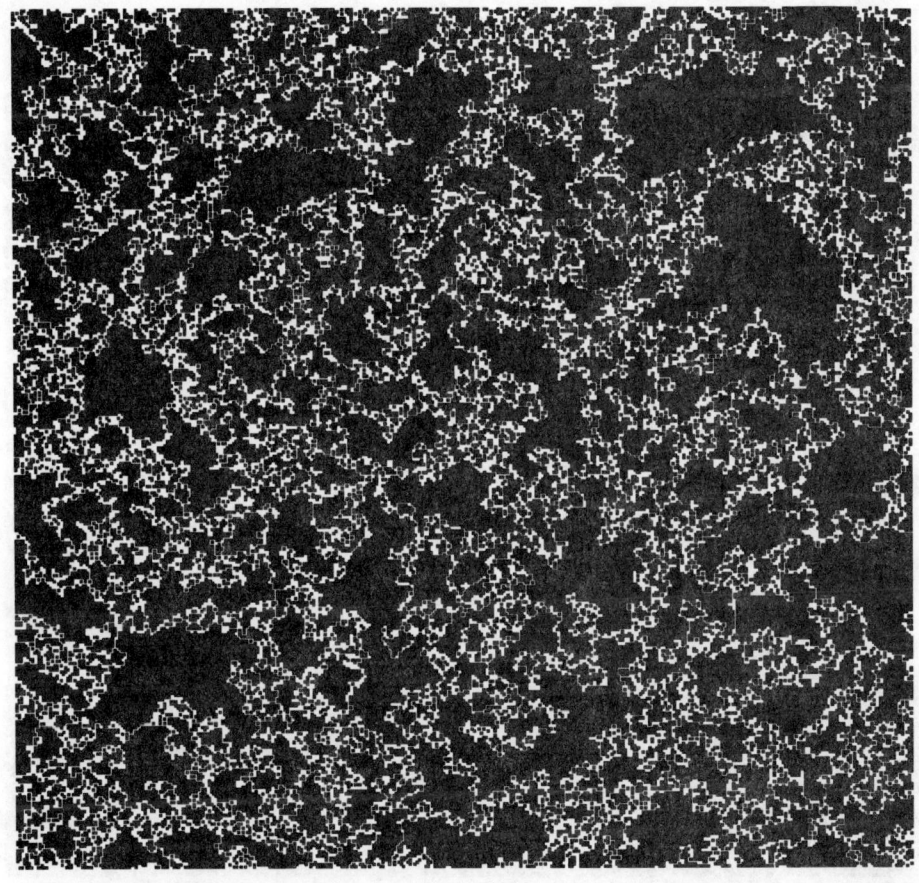

3b. Largest biconnected component in the cluster of Fig. 3a. B(p) = .338.

sites linear dimensions, generally long and stringy in shape. The conducting material in between is sparser than that in Fig. 1b.

In Fig. 3a, for p = .62, the infinite cluster looks even more ragged, with a few isolated regions of ~ 100 sites across, more regions of roughly 40 sites width, and a still larger number of smaller isolated areas. The current carrying paths which make up the backbone in Fig. 3b are still more tenuous, and are now mostly 1-3 sites thick. The largest non-conducting regions in this figure are roughly one-third the sample dimensions, so one would have to average over several regions of at least the present sample

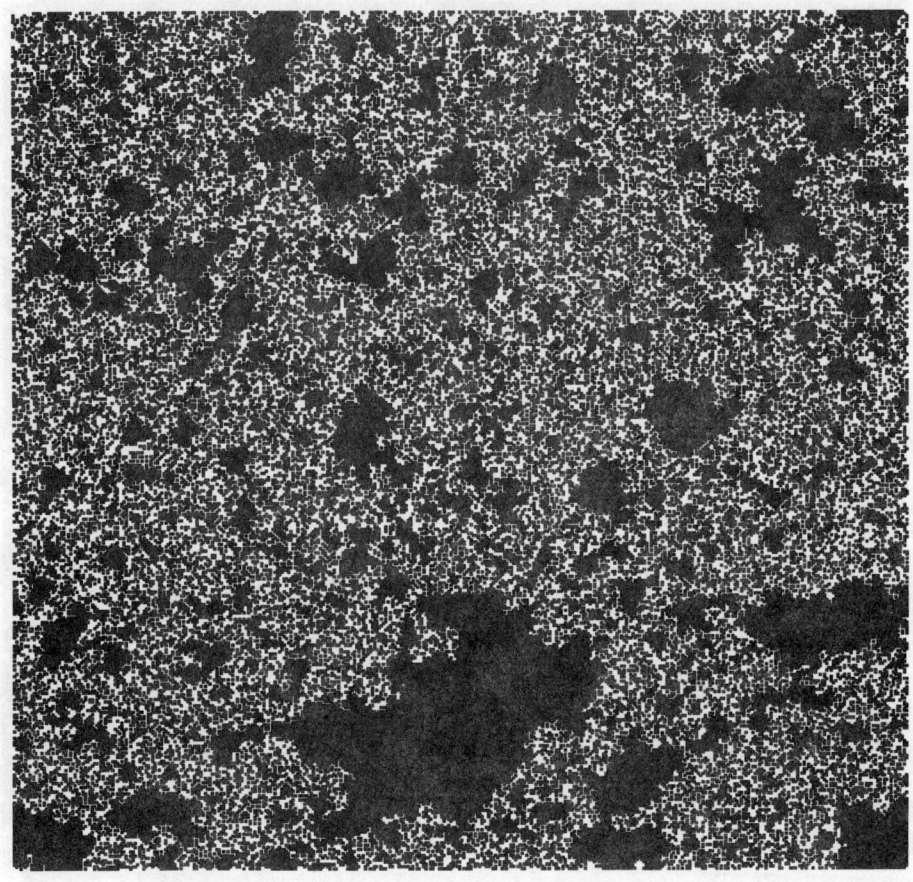

4a. Largest connected component in a 2D square lattice with p = .60, P(p) = .461.

size (400 sites) to get an accurate calculation of average transport properties of the infinite system this close to the percolation threshold.

The close approach of the percolation threshold is especially evident in both Figs. 4a and 4b, which show p = .60. In Fig. 4a, nearly 14 per cent of the sites are present but isolated from the rest, and dark regions are seen which are comparable to the sample dimensions. Fig. 4b shows one non-conducting inclusion which completely crosses the sample. (Because of the periodic boundary conditions, the sample remains conducting in both the vertical and horizontal directions.) Note that the individual strands of the

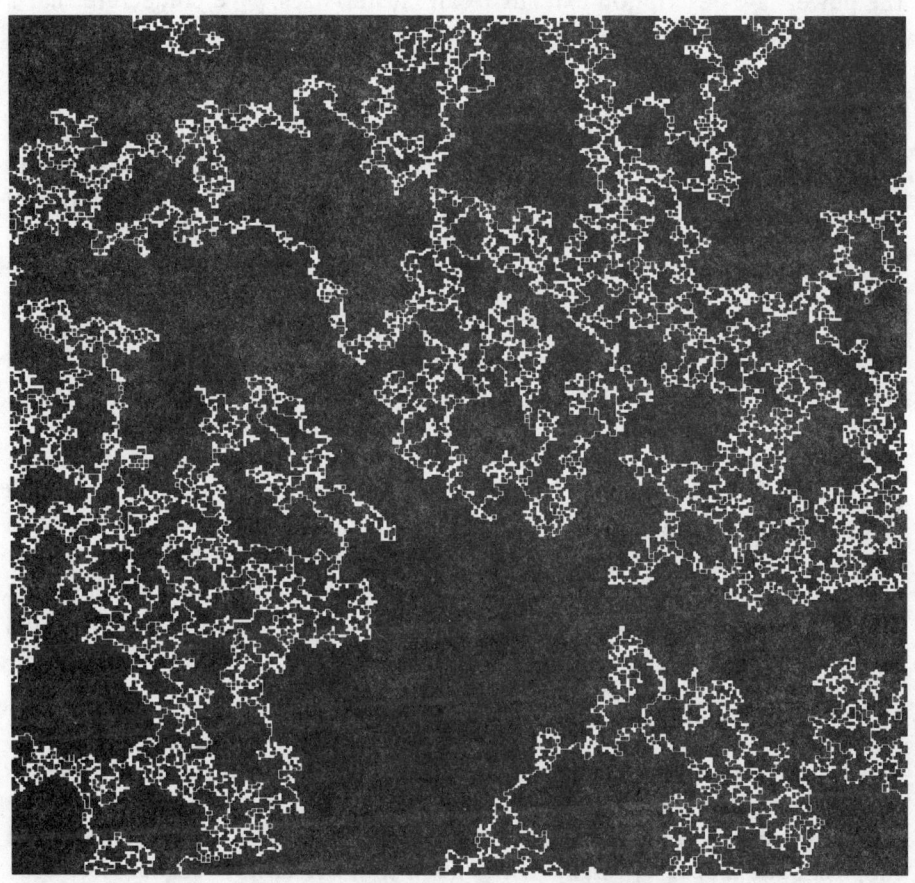

4b. Backbone of the cluster shown in Fig. 4a. B(p) = .203.

backbone are still 1-3 sites thick, and that they enclose non-conducting regions of all sizes, from a few sites across to 200-400 sites in extent.

It is this last feature which most resembles the self-similarity, or scale invariance found in the thermal fluctuations which accompany a conventional second-order phase transition. Loosely speaking, small sections of Fig. 4b are similar to the whole, in the sense that it would be difficult to tell which was which. There may be a more precise way of establishing a scale invariance for the backbone. Leath[9], in studying computer generated connected clusters below p_c finds that their probability distribution is algebraic in the cluster size up to a size which corresponds to the coherence length, ξ, then decreases exponentially for larger sizes. At p_c, of course, the power law description extends to all cluster sizes. We conjecture that a similar probability law describes the non-conducting inclusions seen in Figs. 1b - 4b.

Finally, we note that the great disparity between the volume of material in the "tag ends" and the volume which takes part in conduction, as seen in Figs. 1-4, will have observable consequences in several sorts of physical phenomena. For example, even when the conductivity has fallen to a low value, and the backbone volume has become small, the tag ends provide extremely good capacitative coupling between parts of the infinite cluster which are widely separated along the backbone. This may be the principal contribution to the anomalous singularity in the dielectric constant which Bergman[10] predicts must occur at the percolation threshold. In a dilute magnetic system (typically an alloy of magnetic and non-magnetic atoms in a regular insulating matrix), one would expect the backbone regions to be well ordered at low temperatures, while the tag ends would behave like finite clusters which feel the exchange field of the backbone only at one external site. This sort of analysis has in fact been used by Birgeneau et. al.[11] to interpret the rather large temperature-dependent diffuse scattering seen in some ideal dilute magnetic alloys below their apparent ordering temeratures.

NODES AND LINKS

In order to predict transport properties of systems undergoing a percolation threshold it is desirable to develop models of the geometry of the infinite cluster sufficiently simple to permit calculation. The earliest and simplest such model[12] was the assumption that the infinite cluster should be thought of as a 1D "channel" of cross section $P(p)$, with a local conductivity, σ_o, inside the channels. This picture predicts $\sigma(p) = P(p)\sigma_o$, a result which overlooks[13] the uselessness of the "tag ends" which are counted in $P(p)$, disregards the space charge effects caused by pushing the current aroung obstacles (and as a result fails to agree with effective medium theory as $p \to 1$),[14] and, at any rate, was quickly disproved by experimental evidence.[13-16] $P(p) \sim (p - p_c)^\beta$ is found to vanish at p_c with infinite slope, i.e. $0 \leq \beta \leq 1$, while $\sigma(p) \sim (p - p_c)^t$ is found to have $t > 1$, which implies vanishing slope at p_c.

De Gennes[7] and Skal and Shklovskii[6] have proposed picturing the current paths near threshold as a homogeneous network of nodes connected by effective links. A node is any site from which three or more distinct paths lead off to the boundaries of the sample. It is presumed that nodes which are close together can be lumped, leaving the typical node-node separation to be $\xi \sim |p - p_c|^{-\nu}$. The path length along the sites linking nodes may exceed ξ. If we define this length to be $L(p)$, another exponent, ζ, is introduced[16] to describe the divergence of L: $L(p) \sim (p - p_c)^{-\zeta}$. Since

$$L(p) \geq \xi(p) , \qquad (2)$$

we require

$$\zeta \geq \nu . \qquad (3)$$

De Gennes[7] has suggested that $\zeta = 1$, regardless of dimension. Skal and Shklovskii[6] have also considered the possibility that the effective cross section of the links, $A(p) \sim (p - p_c)^{\zeta'}$, could modify critical behavior near threshold. Since $A(p)$ must be finite, we require that

$$\zeta' \geq 0 . \qquad (4)$$

The conductance is now estimated by neglecting all features on a scale $<< \xi$, on the assumption that they will not alter critical properties. The result is that $\sigma(p)$ is given by the conductivity per link times the appropriate length scale:

$$\sigma(p) \sim (A(p)/L(p)) \xi^{-(d-2)} \qquad (5a)$$
$$\sim (p - p_c)^{(d-2)\nu + \zeta + \zeta'} . \qquad (5b)$$

We observe that the physical constraints (2) and (3) require the conductivity exponent, t, to satisfy

$$t \geq (d - 1)\nu , \qquad (6)$$

within this picture.

The exponent ν can be estimated quite accurately from computer simulations of $P(p)$, which have been done on lattices of up to 10^7 sites. The results are[17] $\nu = 1.2$-1.3 (2D), .8-.9 (3D), and $.67 \pm .05$ (4D). Currently accepted values of t are:[15,16] $t = 1.1$ (2D), 1.6-1.7 (3D), and 2.2 (4D)[18]. Consequently, the inequality (6) is not satisfied in 2D and is barely satisfied in 3D, where it requires $\zeta \approx \nu$ and $\zeta' \approx 0$. In 4D it is satisfied, and we obtain $\zeta + \zeta' \approx 0.9$.

The proposed divergences described by ζ and ζ' can also be checked by direct inspection of the backbone pictures Figs. 1b-4b, at least for 2D. We have already noted that the cross section of a link remains constant at one or two sites over a range of concentrations, suggesting $\zeta' \approx 0$. The links all seem to be fairly direct, not involuted, hence $\zeta \approx \nu$. The Russian workers have reached similar conclusions from computer studies in 2D and 3D,[6] and propose as a result that (6) should be an equality.[19] Dasgupta et. al.[20] have observed that (3) fails in 2D, and comment as well that the

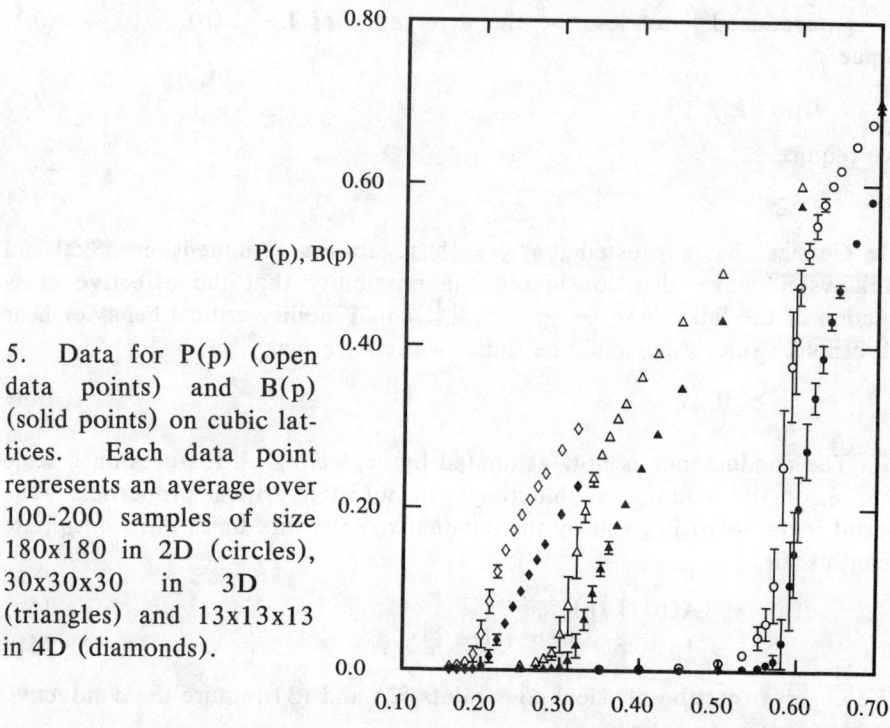

5. Data for P(p) (open data points) and B(p) (solid points) on cubic lattices. Each data point represents an average over 100-200 samples of size 180x180 in 2D (circles), 30x30x30 in 3D (triangles) and 13x13x13 in 4D (diamonds).

high-dimensionality result, $\zeta = 1$, cannot describe the 3D exponents.

We have obtained quantitative data on B(p), the fraction of sites in the backbone, for site percolation in 2D, 3D and 4D. This provides a further test of the node-and-link model and other simple pictures of the conduction paths near the percolation threshold. The data for P(p) and B(p) for reasonably large sample sizes are shown in Fig. 5 for each of the three dimensionalities studied. Similar results are observed in each case: B(p) is significantly less than P(p) for concentrations 10-20 per cent above p_c, and the ratio B(p)/P(p) appears to tend to zero at threshold, indicating that the current must flow through a vanishingly small portion of the actual infinite cluster. Consequently we must define an additional critical exponent to describe the behavior of B(p) near threshold: $B(p) \sim (p - p_c)^{-\beta'}$.

Because of the coarseness of the backbone near threshold, as seen in Figs. 3b and 4b, it is difficult to obtain β' from our data by simply fitting the results in Fig. 5 to a power law form. Besides the scatter in the points, the principal obstacle is the effect of finite sample size, as is evident for the 2D case in Fig. 6. However, one can use this size dependence to extract β' by making an additional scaling assumption. If we assume that close to

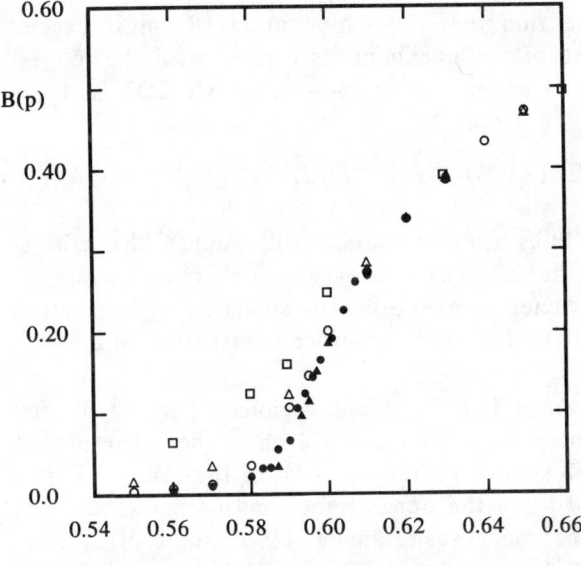

6. Data for B(p) for 2D square lattices of four sizes: 48x48 (squares, 200 samples per data point); 96x96 (open triangles, 200 samples per point); 180x180 (open triangles, 100 samples per point); 250x250 (dots, 60 samples per point) and 400x400 (solid triangles, 60 samples per point).

threshold, for a sample of linear dimension N, B(p,N) depends only upon the ratio $N/\xi(p) \sim N |p - p_c|^\nu$, we can write

$$B(p,N) \sim N^{-\beta'/\nu} g[(p - p_c)N^{1/\nu}] \quad , \qquad (7)$$

where g is a (not especially interesting) function which may depend upon the type of lattice studied, and the prefactor ensures that the correct power law behavior is obtained for large N.

This sort of finite-size scaling has been found to hold for P(p,N) and other averages over the connected cluster size distribution.[17] As a result, p_c and ν are known quite accurately for 2, 3, and 4D site percolation. To determine β'/ν requires only data at p_c, since setting $(p - p_c) = 0$ in (7) leaves the prefactor as the sole source of size-dependence in B. (The same shortcut has been used to study the specific heat exponent in models of phase transitions for which the exact transition temperature is known.[21]) The results are: $\beta' = .5$-$.6$ (2D); $.9 \pm .1$ (3D); and $1.1 \pm .1$ (4D).

Nodes and links on the scale of ξ can account for a vanishingly small fraction of this backbone volume. If we estimate this contribution to B(p) as

$$(A(p)L(p)) \xi^{-d} \sim (p - p_c)^{d\nu - \zeta + \zeta'} \quad , \qquad (8)$$

substituting (3) and (4) we find that the exponent in (8) must exceed $(d - 1)\nu$. Thus the volume of the backbone associated with the longest links vanishes at threshold at least as fast as $(p - p_c)^{1.3}$ (in 2D), as $(p - p_c)^{1.7}$ (in 3D) or as $(p - p_c)^{1.8}$ (in 4D).

A SELF-SIMILAR MODEL

To study whether the links shorter than ξ will modify the critical behavior of $\sigma(p)$ as well as that of B(p) we consider a simplified model of the backbone which is constructed to guarantee the similarity of fluctuation phenomena as observed on all scales from the lattice constant up to ξ.

The construction is shown in Fig. 7. Since regions separated by distances $\gg \xi$ are independent, we consider a cell ξ on a side. The largest non-conducting region will fill some fraction, $(1 - f)$, of the cell. In Fig. 7 we have taken $f = 3/4$, and leave the upper right hand corner of the cell empty. Next, we require that on a scale smaller by a factor of 2, non-conducting regions again fill $(1 - f)$ of the volume which was regarded as conducting when viewed on the larger scale. In Fig. 7, three non-conducting regions of size $\xi/4$ result. Continuing the process gives nine non-conducting regions of size $\xi/8$, 27 regions of size $\xi/16$, and so forth, all randomly arranged. The process must terminate when we reach the scale of the lattice constant. The result is a backbone in which the distribution of non-conducting regions is algebraic in the region size at first, then cuts off sharply above ξ.

To calculate the critical behavior of the backbone fraction in this model we shall sum the lengths of the bonds uncovered at each of the decreasing scales from ξ to 1. The initial stage of the process gives $c\xi$, for some constant, c. At the next stage we must sum contributions from $2^d f$ smaller cells, but each contribution is of order $c\xi/2$. Further decreases in scale give a geometric series, $c\xi[1 + 2^{d-1}f + (2^{d-1}f)^2 + (2^{d-1}f)^3 + ...]$. The series terminates when the scale of the lattice constant is reached; the highest power is $\sim \ln_2 \xi$. If $2^{d-1}f < 1$, the series converges to a constant, and the longest bonds dominate. The more interesting case is $f > 2^{1-d}$, as occurs in Fig. 7, where the series diverges and its highest order term is dominant. Normalizing the volume calculated in this way to ξ^{-d}, we obtain

$$B(p) \sim \xi^{1-d} (2^{d-1}f)^{\ln_2 \xi} \sim \xi^{1-d} \xi^{\ln_2(2^{d-1}f)}$$

$$\sim \xi^{\ln_2 f} \sim (p - p_c)^{+\nu |\ln_2 f|}. \tag{9}$$

For the example shown in Fig. 4, our choice of $f = 3/4$ gives $\beta' = .54$, as observed in 2D. This sort of a picture can readily account for the observed values of β', since $2^{1-d} < f < 1$ implies $0 < \beta' < (d - 1)\nu$.

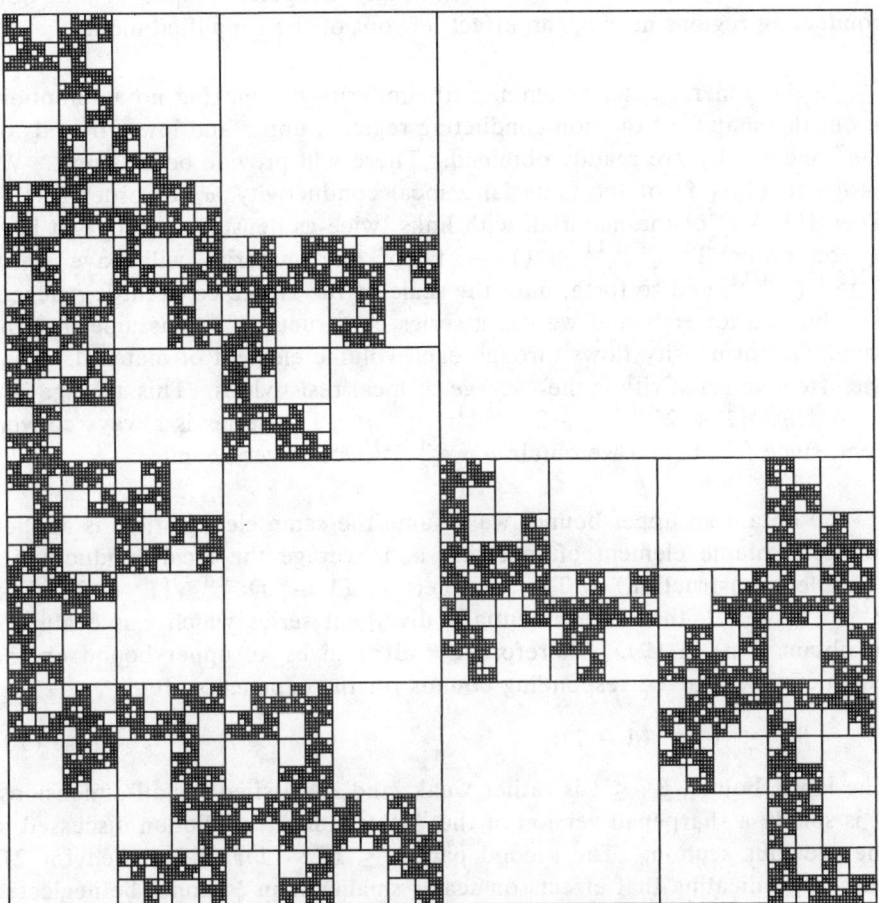

7. An alternative to the homogeneous "nodes and links" picture of the backbone. The construction is described in the text. Current-carrying material is indicated by black lines in this figure, the reverse of the convention used in Figs. 1-4.

Finally, we shall relate the conductivity exponent, t, to f in the self-similar backbone picture. Unfortunately, to determine the conductivity of a backbone such as Fig. 7 requires either numerical calculation or study of the behavior of a finite number of iterations of a lattice renormalization procedure for the resistor network problem.[16,22] Computer solution to obtain the conductivity of cells constructed as in Fig. 7 (with f = .75) indicates that σ_0 decreases roughly as $\xi^{-.66}$ for cells 32x32 and larger. Taking Fig. 7 as a model for the 2D backbone thus predicts t ≈ .9, which is not too far below the observed[15] value, t ≈ 1.1 ± 0.05. The remaining difference may be due to the increasingly irregular shapes of the non-conducting regions near p_c, an effect left out of the simplified model.

In the general case, assuming self-similarity but making no assumptions about the shapes of the non-conducting regions, upper and lower bounds on the conductivity are readily obtained. These will provide bounds on t. We assign to $(1 - f)$ of the material a local conductivity, σ_0 of order $\xi^{-(d-1)}$. Then $f(1 - f)$ of the material, with links twice as densely spaced, will have σ_0 of order $2^{d-1}\xi^{-(d-1)}$, $f^2(1 - f)$ of the material will have $\sigma_0 \sim 2^{2(d-1)}\xi^{-(d-1)}$, and so forth, until the scale of the lattice constant is reached. To obtain a lower bound we use a series construction, and assume that the same current density flows through each volume element of material. Then the effective resistivity is the average of local resistivities. This average is $\sim (1 - f)\xi^{d-1}[1 + 2^{-(d-1)}f + 2^{-2(d-1)}f^2 + ...]$. The series is always convergent, since $f \leq 1$, and we obtain $\sigma \sim \xi^{-(d-1)}$ as a lower bound.

To obtain an upper bound, we assume the same electric field is applied to each volume element of material, and average the local conductivities (parallel construction). The result is $\sim (1 - f)\xi^{-(d-1)}[1 + 2^{d-1}f + 2^{2(d-1)}f^2 + ...]$, the same potentially divergent series which was evaluated to obtain B(p) in (9). Therefore $\xi \propto B(p)$ gives an upper bound on the conductivity. The corresponding bounds for the exponents are

$$\beta' \leq t \leq (d - 1)\nu . \qquad (10)$$

The lower bound, $\beta' \leq t$ is rather weak, and is satisfied for all dimensions. It is simply a sharpened version of the "1D channel" prediction discussed in the previous section. The second part, $t \leq (d - 1)\nu$, is satisfied for 2D and 3D, indicating that effects on scales smaller than ξ cannot be neglected in these physically relevant cases. However, the inequality (6), which goes in the reverse direction, is satisfied from 4-6D. In the latter case, as the tree-like limit of higher dimensionality is approached, the divergent path length between nodes seems to be the dominant effect.

Some readers may recognize Fig. 7 as a "fractal," a term coined by Mandelbrot[23] to describe objects with fractional dimension. Such an object has a measured volume which increases as some non-integral power of its linear dimension if the lowest scale of resolution is kept constant. In our case, if the volume of a cell of backbone ξ sites on a side $\sim \xi^{d'}$, d' is to be identified as the fractal dimension of the backbone, and the description is assumed to be meaningful on scales $\lesssim \xi$. Since B(p) must be normalized to the cell volume, $B(p) \sim \xi^{d'-d}$. By taking the ratio of exponents, β'/ν, we obtain the codimension, $d - d'$, the difference between the fractal dimension of the backbone and the dimension of the space in which it is embedded. Our data for 2D, interpreted in this way, suggest that the backbone seen evolving in Figs. 1b-4b is an object with fractal dimension $d_B = 2 - 0.55/1.3 \approx 1.58$.

A different (larger) fractal dimension has been assigned to the infinite cluster at p_c by Mandelbrot[23] and by Stanley.[24] Both find $d_P \approx 1.77$. However, they argue that $d_P = \gamma/\nu$, where γ is the exponent governing the divergent mean cluster size below p_c. The scaling relation $d\nu = \gamma + 2\beta$ then gives $d - d_P = 2\beta/\nu$, not β/ν as one might have guessed by arguing that the infinite cluster is self-similar on scales $\lesssim \xi$. We have been unable to resolve this discrepancy. One possible resolution is that the infinite cluster, being a combination of backbone and tag ends, may not be self-similar on any range of length scales once $p > p_c$. We would still expect that the inferred codimension of the backbone, β'/ν, should be greater than that of the infinite cluster at p_c, since the backbone contains much less material. The data for β and β' in 2D-4D are indeed compatible with this requirement, $\beta' > 2\beta$.

REFERENCES

1. S. R. Broadbent and J. M. Hammersley, Proc. Camb. Phil. Soc. **53**, 629 (1957); H. L. Frisch and J. M. Hammersley, J. SIAM **11**, 894 (1963).
2. For reviews, see J. W. Essam, in *Phase Transitions and Critical Phenomena, Vol 2*, C. Domb and M. S. Green, eds. (Academic Press, New York 1972) p. 197, and V. K. S. Shante and S. Kirkpatrick, Adv. in Phys. **20**, 325 (1971).
3. Conduction near threshold is reviewed in S. Kirkpatrick, Revs. Mod. Phys. **45**, 574 (1973).
4. P. W. Kasteleyn and C. M. Fortuin, J. Phys. Soc. Japan Suppl **26**, 11 (1969); C. M. Fortuin and P. W. Kasteleyn, Physica **57**, 536 (1972).
5. R. W. Landauer, this conference.
6. A. Skal and B. I. Shklovskii, Fiz. Tekh. Poluprovdn. **8**, 1586 (1974) [Soviet Physics — Semicond. **8**, 1029(1975)].
7. P. G. de Gennes, J. de Phys. Lett. **37**, L1 (1976).
8. J. E. Hopcroft and R. E. Tarjan, Commun. A.C.M. **16**, 372 (1973); R. E. Tarjan, SIAM J. Computing **1**, 146 (1972). The recursive procedure described in these articles was rewritten as a single-stack algorithm to avoid the overhead involved in nested recursion several thousand layers deep.
9. P. L. Leath, Phys. Rev. **B14**, 5046 (1976). Also see D. Stauffer, Z. Phys. **B25**, 391 (1976), and preprints, for comments on alternative forms of the cluster distribution functions.
10. D. J. Bergman, this conference, and Phys. Rev. Lett. (to appear).
11. R. J. Birgeneau, R. A. Cowley, G. Shirane, and H. J. Guggenheim, Phys. Rev. Lett. **34**, 940 (1976).

12. T. P. Eggarter and M. H. Cohen, Phys. Rev. Lett. *25*, 807 (1970), *27*, 129 (1971); T. P. Eggarter, Phys. Rev. *A5*, 2496 (1972).
13. B. J. Last and D. J. Thouless, Phys. Rev. Lett. *27*, 1719 (1971).
14. S. Kirkpatrick, Phys. Rev. Lett. *27*, 1722 (1971).
15. J. P. Straley, Phys. Rev. *B15*, 5733 (1976), and this conference.
16. A. B. Harris and S. Kirkpatrick, Phys. Rev. *B16*, 542 (1977).
17. A. Sur, J. L. Lebowitz, J. Marro, M. L. Kalos, and S. Kirkpatrick, J. Stat. Phys. *15*, 345 (1976) gives the analysis and 3D results. Values of ν in 2D and 4D were obtained by reanalyzing the data of S. Kirkpatrick, Phys. Rev. Lett. *36*, 69 (1976).
18. Obtained from series expansion by A. B. Harris and R. Fisch, Phys. Rev. Lett. *38*, 796 (1977).
19. M. E. Levinshtein, M. S. Shur, and A. L. Efros, Zh. Eksp. Teor. Fiz. *69*, 2203 (1975) [Sov. Phys. JETP *42*, 1120 (1976)*)*.
20. C. Dasgupta, A. B. Harris, and T. C. Lubensky, preprint.
21. E. Domany, K. K. Mon, G. V. Chester and M. E. Fisher, Phys. Rev. *B12*, 5025 (1975).
22. S. Kirkpatrick, Phys. Rev. *B15*, 1533 (1977).
23. B. B. Mandelbrot, *Fractals: Form, Chance and Dimension*, (W. H. Freeman and Co., San Francisco, 1977), and to appear in *Statistical Physics XIII*, (IUPAP 1977, Haifa) ed. by C. G. Kuper.
24. H. E. Stanley, R. J. Birgeneau, P. J. Reynolds, and J. F. Nicoll, J. Phys. *C9*, L553 (1977); H. E. Stanley, preprint.

DISCUSSION

D. LAMBETH (Eastman Kodak): It seems to me that there is no cooperative interaction in the way you have placed the bonds in your model. Perhaps I misunderstood this at the beginning.

KIRKPATRICK: The sites are placed at random but they are not connected unless they have neighbors. It is this feature which introduces the cooperative or the critical behavior.

LAMBETH: In a system where atoms are being deposited onto a substrate, it is expected that the atoms might very co-operatively attach to one another. Would you expect, from the analogy to critical phenomena, to shift the volume fraction at which the percolation threshold occurs without changing the exponents?

KIRKPATRICK: It appears that there is still a percolation threshold with a shifted critical volume fraction as long as there is only gentle attraction between the elements. There are two effects here: a clumping which is a genuine thermal phase transition and the connectivity properties. There have been studies of how quenched randomness affects a phase transition in a magnetic system, but this is a little different, I think, than what you had in mind.

COOPERATIVE PHENOMENA IN RESISTOR NETWORKS AND INHOMOGENEOUS CONDUCTORS*

Joseph P. Straley
University of Kentucky, Lexington, Ky. 40506

ABSTRACT

Near the conduction threshold, the electrical conductivity of a random network exhibits power law behavior. An exponent theory modelled on the scaling theory of critical phenomena will be described, which relates the power laws which describe the various ways to approach the threshold. The particular case of finite sample size will be given special attention. The resistor lattice exponents are also relevant to more general inhomogeneous conductors. The dimensionality dependence of the conduction exponents will be reviewed.

THE CONDUCTION THRESHOLD AS A SINGULAR POINT

As a model of an inhomogeneous conductor we might build a large cubical array of resistors, arranged to connect the nearest neighbors of a simple cubic lattice. We could represent the inhomogeneity by choosing the resistors at random to have conductance b (with probability p) or a (with probability 1-p). If $b \gg a$, this model exhibits an conductivity transition[1] near $p_c = 0.247$. For $p > p_c$, the cube has conducting (i.e. all b) paths between opposite faces. These become less likely as the threshold is approached, and the conductivity is described by a power law: $\sigma \sim b (p - p_c)^t$, with $t \sim 1.7$. For $p < p_c$ every path contains a few a's, so that the conductivity is proportional to a, and appears to diverge (and does, if b is infinite) as the threshold is approached from below:[2] $\sigma \sim a (p_c - p)^{-s}$, where $s \simeq 0.7$ for the case under discussion.

These observations show that the specific conductivity, Σ, regarded as a function of p and the conductivity ratio a/b, has a singular point at $p = p_c$ and $a/b = 0$. The singularity manifests itself as nonanalytic power law behavior as the singular point is approached, and different power laws are seen as the singular point is approached in different ways.[2,3] An extensive body of theory of such singular points has already been developed in connection with the thermodynamic critical points,[4] from which we may borrow profitably.

The principal borrowing is the "scaling" theory itself, which relates the singularities observed as the singular point is approached in various ways. There is an infinity of such approaches possible, but the exponent theory will only require the two exponents s and t to describe them. The theory is available in several different forms; I find the Schofield parametric representation[5] to be the most intuitive. First, we define a new coordinate system consisting of a distance λ, an angle x, and a scale μ, which are related to p, a,

---

*Supported by the National Science Foundation under Grant No. DMR76-11402.

and **b** by

$$p - p_c = \lambda x, \quad (1)$$

$$4\, a/b = \lambda^{s+t}(1-x^2), \quad (2)$$

and

$$2a = \lambda^s(1-x)\mu \quad \text{or} \quad 2\mu = b\lambda^t(1+x). \quad (3)$$

In this coordinate system, λ is the distance to the singular point, and $\lambda = 0$ is the singular point for any choice of **x** and μ. The angle **x** differentiates the different ways to approach the singular point. The scale μ is dimensionally a conductance, and is an estimate of how large Σ will be for the choice of **p**, **a**, and **b**. In terms of these coordinates, the "scaling theory" hypothesis is that

$$\Sigma = \mu F(x), \quad (4)$$

where **F** is a nonsingular function of **x** alone, valid asymptotically close to the conduction threshold. It is readily verified that this hypothesis contains the two power laws described above as special cases ($x = \pm 1$), and interpolates between them for the cases where a/b is finite. Another particular special case is $x = 0$ (i.e. $p = p_c$); Eqs. 1 and 2 predict the behavior $\Sigma \sim a^u b^{1-u}$, with $u = t/(s+t)$.

FINITE LATTICE SIZE

The foregoing statements are only strictly true in the "thermodynamic" limit of an infinite system. The ensemble average conductivity of a finite system cannot show any nonanalytic behavior, although it may otherwise approximate an infinite system quite closely. Therefore the ratio h/L of lattice spacing **h** to sample diameter **L** is another variable on which the average specific conductivity depends, and the singular point only occurs in the limit $h/L \to 0$.

A direct manifestation of finite system size is a sample-to-sample variation in conductivity. Strictly speaking, such fluctuations are always present in measurements on finite systems; but they frequently may be removed by time-averaging, if the system dynamics are such as to cause a given physical sample to pass rapidlty through many internal configurations. In the likely realizations of an inhomogeneous conductor, however, the disorder is fixed in place, and the conductivity fluctuations will inevitably be observed.

An ensemble of finite conductors is thus characterized by a distribution of conductances. We can describe the ensemble by means of a distribution function $P(\Sigma,p,a,b,h/L)$, such that $P\,d\Sigma$ is the probability that a member of an ensemble characterized by the given values of p,a,b, and h/L has conductivity in the range $(\Sigma, \Sigma + d\Sigma)$.

Figure 1 shows how **P** depends on **L**. The system chosen is a two-dimensional square lattice containing 60% unit conductors and 40% non-conductors, with $L = 2^n h$. Periodic boundary conditions were used; that is, Σ is the specific conductivity of an infinite lattice formed by periodically repeating a particular member of the ensemble. For

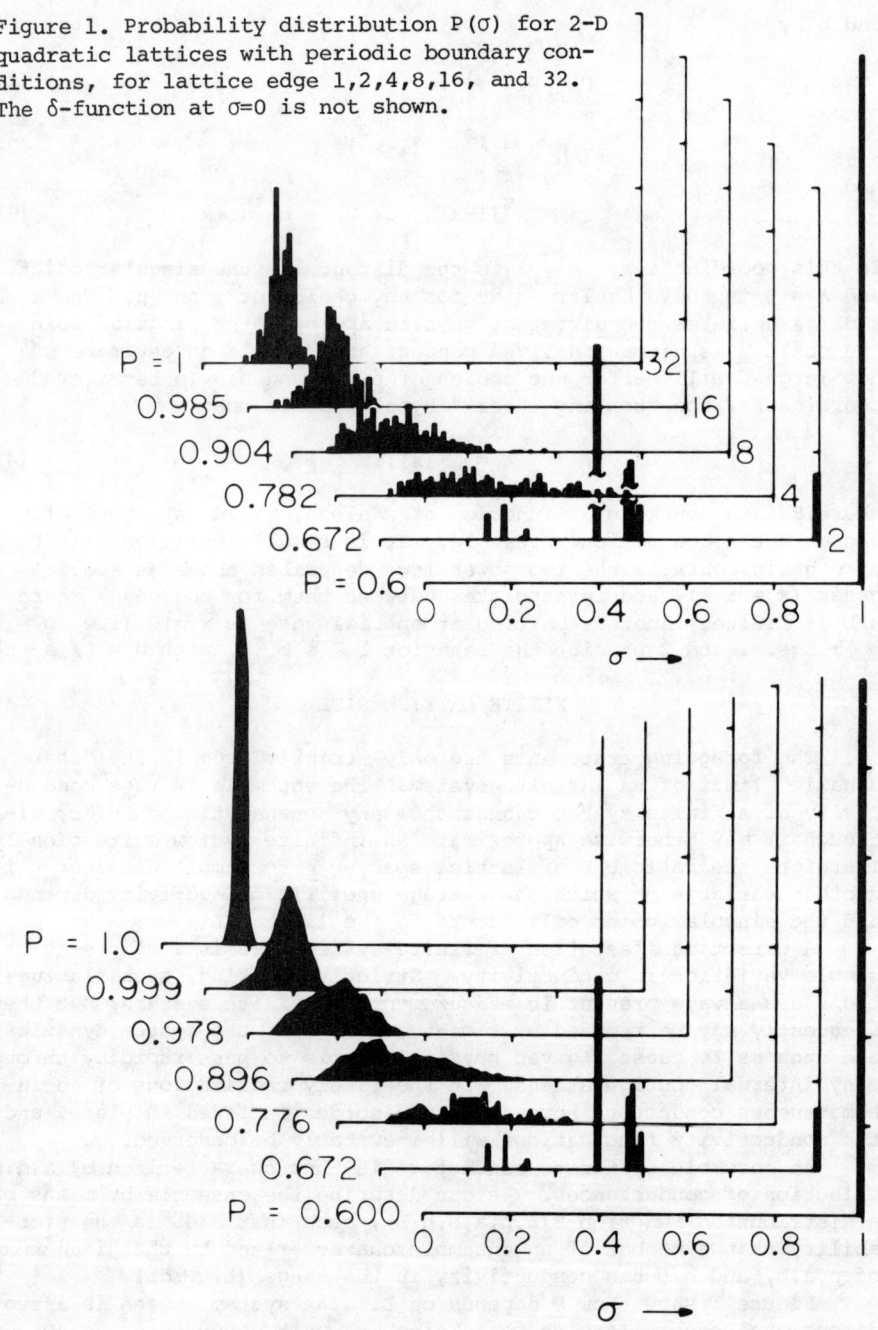

Figure 1. Probability distribution P(σ) for 2-D quadratic lattices with periodic boundary conditions, for lattice edge 1,2,4,8,16, and 32. The δ-function at σ=0 is not shown.

Figure 2. Same as figure 1, but calculated by means of a decimation approximation.

n = 1, $\underline{P}$ is the microscopic distribution of conductances -- it consists of two delta functions at $\sigma = 0$ and $\sigma = 1$, of relative weights 0.4 and 0.6, respectively. In Figure 1 the delta functions at $\sigma = 0$ are not shown, but their weights may be calculated as $1 = \bar{p}$, where $\bar{p}$ is the probability that σ is nonzero, and is given on the figure.

Calculations on large finite lattices are time consuming, making it difficult to get good statistical representation of $\underline{P}$ for large $\underline{L}$. To study this regime we shall introduce a decimation approximation. A lattice of edge 2L is composed of 2^d sublattices of edge L; given the distribution function for the smaller lattice we will approximately calculate the distribution for the larger by assuming that it is a function of the conductivities of the sublattices in the same way that the distribution function for a lattice of edge 2h depends on the $d2^d$ conductors it contains.[23] Thus the sublattice is being represented by a small number of elements (one along each coordinate axis) whose conductance is determined by the distribution P(L), so that the distribution P(2L) may be determined by a relatively simple calculation; the process may then be repeated to calculate P(4L), etc. Figure 2 shows the results of this approximation for the same case as was shown in Figure 1. In comparing these figures the reader should be aware that some of the structure in Figure 1 is sampling error; the figures are in good qualitative agreement.

For the two-dimensional lattice, $p_c = 0.5$, and so a large lattice with p = 0.6 should have a small but finite conductivity. Figures 1 and 2 indicates that this comes about in two steps as the lattice is built up. In the first step, the delta function at zero conductivity is averaged out, given a broad distribution of possible conductances; then in the second step, the distribution narrows and eventually converges to a delta function at the macroscopic (infinite system) Σ. The lattice size required to achieve these two steps depends on $\underline{p}$, and becomes divergently large as p_c is approached. Near p_c the first step is very slow. We may define an averaging length $L(p)$ as being the sample size $\underline{L}$ required for given p such that $\bar{p}$ passes some test, such as $1 - \bar{p} < 10^{-2}$ (or in words, that 99% of all members of the ensemble conduct). It seems likely that L(p) defined this way differs only trivially from the correlation length ξ defined for the percolation problem (e.g. L(p) is a multiple of ξ). Then near p_c, $L(p) \sim \xi \sim |p-p_c|^{-\nu}$.

The second step is roughly independent of $\underline{p}$, especially close to p_c - in the case where $\xi \gg h$, the microscopic texture of the lattice is completely averaged out in the first step, and the second step may be regarded as beginning with structureless building blocks of diameter L(p). The conductivity distribution of the building blocks is quite broad, but its center is near the macroscopic conductivity Σ (the large L limit of the chosen p). As L increases, the distribution rapidly narrows (as Figure 2 suggests). It will be shown in the Appendix that the mean-square width of P decreases by the rule

$$<\delta\sigma^2> \sim \Sigma^2 \, (\xi/L)^d \tag{5}$$

As an example, Figure 3 gives the experimental data of Watson and Leath,[6] with the lines $\Sigma \pm <\delta\sigma^2>^{1/2}$ drawn in. In an attempt

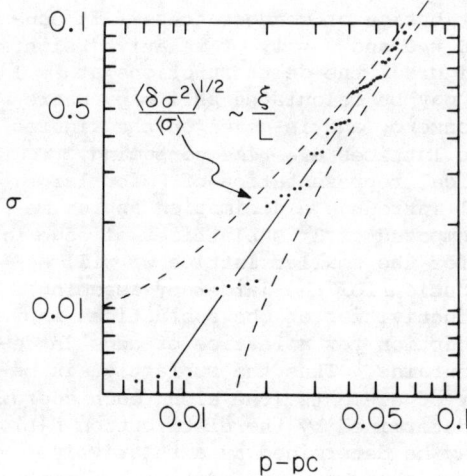

Figure 3. The experimental data of Watson and Leath. The dashed lines are the error estimates.

to avoid introducing an adjustable parameter, Eq. 5 was used as an equality, with $\xi = h((p - p_c)/(1-p_c))^{-\nu}$ (so that $\xi = h$ for $p = 1$); the values chosen for the parameters were $p_c = 0.587$, $\nu = 1.34$, and $L = 137h$.

The agreement of theory and experiment would be worse if the data were reanalyzed to give a smaller $\underline{t}$ exponent (the straight line corresponds to $t = 1.38$), and would be better if a smaller value of ν were chosen. The theory as derived is only valid for $L \gg \xi \gg h$, so that the disagreement for small $p-p_c$ should not be taken too seriously.

INHOMOGENEOUS CONDUCTORS IN GENERAL

Resistor lattices are chiefly for computers to imagine and theorists to think about; there are only a few cases in which they have been constructed as physical entities. We must then ask how the theory is to be translated into the real world of cosputtered gold-silica films[7] or mercury-xenon alloys.[8]

The translation consists of classifying the various aspects and numerical quantities as being "universal", if they are expected to be common for every system, or "nonuniversal", if they depend on microscopic details of the system under discussion. Thus the description afforded by Eqs. 1-3 (or the equivalent homogeneous function representation) is universal, as are the numerical values of the exponents (which depend only upon dimensionality). The values of p_c, a, and b, however, are nonuniversal: they are not determined by this theory; and in absence of other insights into the nature of the conduction threshold, should be treated as adjustable parameters. In particular $\underline{a}$ and $\underline{b}$ should not be identified with the conductivities of the pure phases that have been mixed to form the inhomogeneous conductor: in part the objection here is that these are macroscopic quantities, and may contain cooperative-phenomena corrections from the microscopic values; but it should also be observed that in real inhomogeneous conductors, the inhomogeneity itself may introduce new modes of conduction. Finally it should be noted that the theory presented here only pertains in the near vicinity of the singular point; the predictions should not be taken too seriously if $p-p_c$ or a/b are not small.

Figure 4 displays the prediction of Eqs. 1-3, with $s = 0.7$, $t = 1.7$, $b = 1$, $F = 1$ (constant), and various values of $\underline{a}$. Each curve is composed of three parts: (1) a pure power law $\Sigma \sim \bar{b}(p-p_c)^t$ for

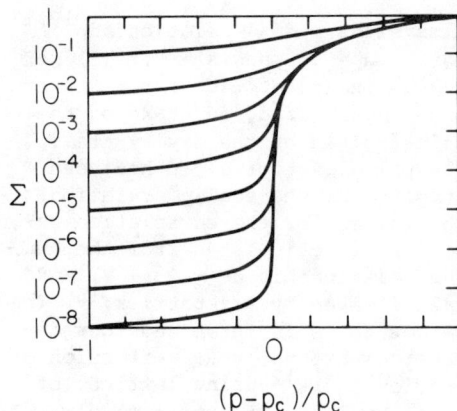

Figure 4. Specific conductivity of a 3-dimensional inhomogeneous conductor, as predicted by eqs. 1 - 3 (with $b = 1$, $a = 10^{-n}$).

$p > p_c$; (2) a pure power law $(p_c-p)^{-s}$ for $p < p_c$; (3) and a transition between these two forms in a narrow region $|p-p_c|^{s+1} < a/b$. In the limit $a \ll b$, the transition region becomes quite small, and the power laws extend over many decades. These predictions (for $a \ll b$) of a narrow transition region and a many-orders-of-magnitude change in Σ over a narrow range of p are fairly firmly rooted in the ideas of the scaling theory, and should appear in every realization of it. There are experiments which find this feature,[9-11] but there are also those which do not.[8] One possible source of the descrepancy can be suggested: the latter case is a thin film (1μm) composed of small granules (~10 nm). Away from the conduction threshold, this is a three dimensional system; but as the threshold is approached, ξ grows until it exceeds the film thickness. Thereafter the conducting network is effectively two dimensional. This cross-over from bulk 3-D behavior to 2-D behavior (with a different p_c) may account for some of the thin film phenomena.

DIMENSIONALITY DEPENDENCE OF THE EXPONENTS

Table I Dimensionality dependence of the conduction exponents

d	s	t	u	ν	
1	1	(0)	0	1	
2	1.1 ± .05[a]	1.1 ± .05[a]	1/2[a]	1.34[b],	1.33 ± .04[c]
3	0.70 ± .02[c]	1.725 ± .01[e]	0.72 ± .02[a]	0.82[b],	0.9 ± .05[c]
4	0.6 ± .1[d]			0.66[f]	
5				0.52[f]	
6 - ∞	0[g]	3[h]		0.5[f]	

[a]Ref.12 [b]Ref.16 [c]Ref.17 [d]this work [e]Ref.18 [f]Ref.15 [g]Ref.19
[h]Ref.20

Table I and Figure 5 present the best available results for the conduction exponents of resistor lattices. Most of these are due to numerically solving large finite lattices, but several theoretical constraints may be imposed: (1) The one dimensional resistor lattice may be solved exactly. The exponent t is not actually defined in this case, since $p_c = 1$; but the value $t = 0$ has been supplied by means of the scaling relationship between s,t, and u.   (2) In two

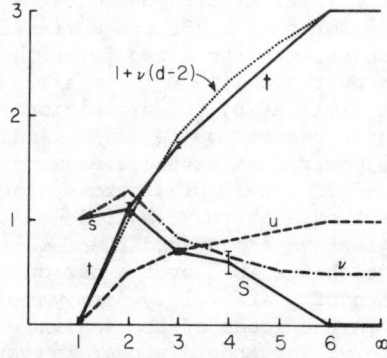

Figure 5. Dimensionality dependence of the exponents. The error bars represent the points for s and t. The dotted line is the De Gennes theory.

dimensions a dual relationship[12,13] forces u = 0.5 and s = t. (3) For d ≥ 6, the conduction exponents are independent of d, and take on the values given by the Cayley tree.[14] The value of u for d ≥ 6 has been supplied by the scaling relation. The values for the correlation exponent are calculated from the scaling relationship $d\nu = 2\beta + \gamma$, and values given by Kirkpatrick.[15] The values for s in three and four dimensions are due to an extension of previous work[12] using lattices of edge L = 39 and 15, respectively. I have taken the liberty of doubling the error estimate on Onizuka's calculation.[18]

De Gennes has given a relationship between t, ν, and the dimensionality. He proposes that the lattice be characterized as a collection of "active chains" (which are essentially one-dimensional conducting paths), which join at "nodes". The average geometrical distance between nodes varies with p like the correlation length, and we shall assume that the average end-to-end conductivity of a chain is also described by some power law:

$$\sigma_{chain} \sim (p - p_c)^x. \qquad (6)$$

Then in a uniform external field, power dissipated in a chain is of the order $\sigma_{chain} (\xi E)^2$, and the power dissipated per unit volume is

$$\Sigma E^2 \sim \sigma_{chain} \xi^{2-d} E^2 \sim E^2 (p - p_c)^{x+\nu(d-2)} \qquad (7)$$

giving the exponent relationship[20] $t = x + \nu(d-2)$. De Gennes further assumed that the chain conductivity is proportional to the reciprocal of the average distance between nodes as measured along the chain ($\overline{N}$), and that this quantity in turn is proportional to $(p - p_c)^{-1}$. Thus De Gennes' equality is $t = 1 + \nu(d-2)$, which is also shown in Figure 5.

The equality has in its favor that it is correct for the two cases d = 1 and d = 6, and reasonably close for d = 2 and d = 3. The result t = 1 for d = 2 is interpreted to mean $\Sigma \sim -(p_c - p_c)/\ln(p-p_c)$ or a similar logarithmic correction, which would tend to mistaken for t > 1 in approximate work. However, for the argument to be accepted there must be a similar reinterpretation of the correlation length exponent, since necessarily $\overline{N} \geq \xi$, implying[21] $1 \geq \nu$. Available estimates of ν considerably violate this inequality in two dimensions.

Levinshtein et al[22] gave a similar argument, but assumed that $x = \nu$, giving $t = \nu(d-1)$. This relationship does about as well as De Gennes' for the lower dimensionalities, but is wrong at d = 6 and thus can be ruled out.

APPENDIX. CONDUCTIVITY FLUCTUATIONS

Consider a lattice of edge 2L with conductances assigned at random. The specific conductivity σ of this particular sample will differ from the macroscopic conductivity Σ by an amount $\delta\sigma$ specific to the particular configuration.

We may consider this lattice as being composed of $N = 2^d$ sublattices of edge L; the specific conductivity of the ith sublattice is $\sigma_i = \Sigma + \delta\sigma_i$.

The sublattice conductivities determine the lattice conductivity through a functional $F(\{\sigma_i\})$, which is in general a nonlinear combination of its arguments, but which does have some special properties: it is homogeneous of first degree; it has the property that for the special case that all σ_i take on the same value $\sigma_i = X$, $F(\{X\}) = X$; and if σ_i is finite, then F is not singular as a function of σ_i.

If $L \gg L(p)$, the fluctuations $\delta\sigma_i$ will be small, and we may linearize F about $\sigma = \Sigma$. The properties just set out force the relationship

$$\Sigma + \delta\sigma = \Sigma + N^{-1} \Sigma_i \delta\sigma_i \qquad (8)$$

Since the variables $\delta\sigma_i$ are independent, they are uncorrelated, and so

$$\langle\delta\sigma^2\rangle = N^{-2} \langle(\Sigma_i \delta\sigma_i)^2\rangle = N^{-1} \langle\delta\sigma_i^2\rangle \qquad (9)$$

or equivalently, $\langle\delta\sigma^2\rangle \sim L^{-d}$. This relationship may then be written more completely as

$$\langle\delta\sigma^2\rangle \approx \Sigma^2 (\xi/L)^d \qquad (10)$$

which includes the estimate $\langle\delta\sigma^2\rangle \sim \Sigma^2$ for $L \sim \xi$ (i.e. at the beginning of the second step).

REFERENCES

1. S. Kirkpatrick, Rev.Mod.Phys. **45**, 570 (1973).
2. J.P. Straley, J. Phys. C **9** 783 (1976).
3. A.L. Efros and B.I. Shklovskii, Phys.Stat.Sol. **76b** 475 (1976).
4. M.E. Fisher, Rept. Prog. Phys. **30** 615 (1967).
5. P. Schofield, Phys. Rev. Lett. **22** 606 (1969).
6. B.P. Watson and P.L. Leath, Phys. Rev. B **9** 4893 (1974).
7. R.W. Cohen, G.D. Cody, M.D. Coutts, and B. Abeles, Phys. Rev. B**8** 3689 (1973).
8. O. Cheshnovski, U. Even, and J. Jortner, Solid State Comm. **22** 745 (1977).
9. M.E. Levinshtein, J. Phys. C **10** 1895 (1977).
10. T.J. Coutts, Thin Solid Films **38** 313 (1976).
11. V.N. Andreev, T.V. Smirnova, and F.A. Chudnovskii, Phys. Stat. Sol **77b** K97 (1976).
12. J.P. Straley, Phys. Rev. B **15** 5733 (1977).
13. A.M. Dykhne, Zh. Eksper. Teor. Fiz. **59** 110 (1970) [Sov. Phys. -JETP **34** 63 (1971)].
14. J.P. Straley, J. Phys. C **10** 3009 (1977).
15. S. Kirkpatrick, Phys. Rev. Lett. **36** 69 (1976).
16. A.G. Dunn, J.W. Essam, and D. Ritchie, J. Phys. C **8** 4219 (1975).

17. M.E. Levinshtein, B.I. Shklovskii, M.S. Shur, and E.L. Efros, Zh. Eksper. Teor. Fiz. 69 386 (1975) [Sov. Phys.-JETP 42 197 (1976)].
18. K. Onizuka, J. Phys. Soc. Japan 39 527 (1975).
19. P.G. DeGennes, J. Phys. Lett. (Paris) 37 L1 (1976).
20. A.B. Harris and S. Kirkpatrick, Phys. Rev. B 16 542 (1977).
21. B.I. Shklovskii and A.L. Efros, Usp. Fiz. Nauk 117 401 (1975) [Sov. Phys.-Usp. 18 845 (1976)].
22. M.E. Levinshtein, M.G. Shur, and A.L. Efros, Zh. Eksper. Teor. Fiz. 69 2203 (1975) [Sov. Phys.-JETP 42 1120 (1976)].
23. J.P. Straley, J. Phys. C 10 1903 (1977).

DISCUSSION

D. BERGMAN (Tel-Aviv Univ.): I would like to emphasize a point I made in my talk which is relevant to this paper. There is a singularity not only when $\Delta T = 0$ and when $A/B = 0$, but also for any non-zero value of ΔT; in this latter case, however, the singularity occurs at the negative value of A/B. There is thus a basic difference between this problem and the phase transition problem which warrants, in my opinion, greater recognition.

STRALEY: You have made a good point.

J. C. GARLAND (Ohio State Univ.): Is there a simple reason why the critical exponents stabilize at six dimensions?

STRALEY: The original explanation is attributed to Toulouse, although Scott Kirkpatrick convinced me it was correct. The renormalization group approach which worked for magnetic critical phenomena in four dimensions, has not yet been developed for this problem. For the percolation problem, however, Kirkpatrick showed numerically that exponents stablized at six dimensions. For this problem it is still something of a guess.

AN EXACT THEORY OF THE ELECTRICAL TRANSPORT AND OPTICAL PROPERTIES OF INHOMOGENEOUS MEDIA

P. DQ. Landau
Inst. for Physical Problems, Moscow

ABSTRACT

A general statistical mechanics is developed for the effects of randomly distributed inhomogeneities on the electronic properties of conducting and non-conducting media. The theory, valid for all concentrations, sizes, and shapes of the inhomogeneities, is based on an expansion of the system energy in a powers of a complex vector order parameter $\vec{\Delta k}$. The theory yields the conductivity and complex dielectric function for all frequencies, and in a trivial limit reduces to the results of the Effective Medium Approximation. The onset of metal-nonmetal transitions is accurately predicted. As a simple example, renormalization group predictions for critical exponents are shown to be a special case of the theory.

CHAPTER II: INVITED PAPERS:

ELECTRICAL TRANSPORT PROPERTIES

INHOMOGENEOUS SUPERCONDUCTORS*

M. Tinkham
Harvard University, Cambridge, Mass. 02138

ABSTRACT

The coherence length ξ and penetration depth λ set the characteristic length scales in superconductors, typically 100-5,000 Å. A lattice of flux lines, each carrying a single quantum, can penetrate type II superconductors, i.e., those for which $\kappa \equiv \lambda/\xi > 1/\sqrt{2}$. Inhomogeneities on the scale of the flux lattice spacing are required to pin the lattice to prevent dissipative flux motion. Recent work using voids as pinning centers has demonstrated this principle, but practical materials rely on cold-work, inclusions of second phases, etc. to provide the inhomogeneity. For stability against thermal fluctuations, the superconductor should have the form of many filaments of diameter ~10-100 μm imbedded in a highly conductive normal metal matrix. Such wire is made by drawing down billets of copper containing rods of the superconductor. An alternative approach is the metallurgical one of Tsuei, which leads to thousands of superconducting filamentary segments in a copper matrix. The superconducting proximity effect causes the whole material to superconduct at low current densities. At high current densities, the range of the proximity effect is reduced so that the effective superconducting volume fraction falls below the percolation threshold, and a finite resistance arises from the copper matrix. But, because of the extremely elongated filaments, this resistance is orders of magnitude lower than that of the normal wire, and low enough to permit the possibility of technical applications.

INTRODUCTION

To start talking about inhomogeneous superconductors, one must first establish the length scale on which the inhomogeneity must exist. The characteristic length scales in superconductors are set by the <u>coherence length</u> ξ, which describes the distance over which the superconducting electron density $|\psi|^2$ can vary, and the <u>penetration depth</u> λ, which is the distance for screening out magnetic fields by the supercurrent response. Both of these lengths diverge at the transition temperature T_c, but have values of the order of hundreds or thousands of Å in typical cases. Thus both might be called "semimicroscopic" lengths: small on a macroscopic scale, but very large on an atomic scale.

Because these lengths are of the same order, two possibilities exist, depending on which is larger. These correspond to the two types of superconductors, called type I and type II. [Both ξ and λ diverge as $(1-T/T_c)^{-\frac{1}{2}}$, keeping a constant ratio. However, they depend differently on the material parameters, so that materials with

*Research supported in part by the National Science Foundation.

high T_c and short electronic mean free path ℓ tend to be type II, while materials with low T_c and long ℓ are usually type I.] The significance of the sense of the inequality of ξ and λ is illustrated in Fig. 1. If we imagine a stable interface between superconducting and normal material in the presence of the critical field H_c, the profile looks as shown in Fig. 1(a) for type I ($\kappa \equiv \lambda/\xi < 1/\sqrt{2}$). $|\psi|^2$ cuts off slowly, but even a small $|\psi|^2$ gives a short λ. This gives rise to a <u>positive</u> interface energy, since the interface region fails to enjoy the superconductive condensation energy, but it pays the energetic price of excluding B.

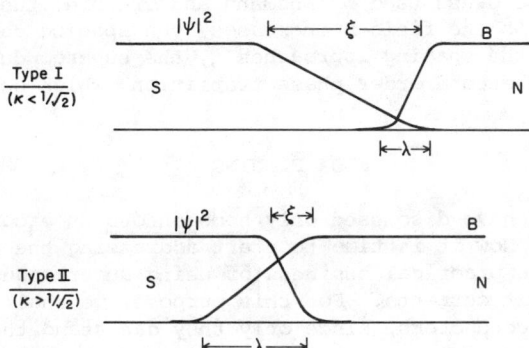

Fig. 1. Schematic diagram of interface between superconducting and normal material in presence of critical magnetic field. Superconducting order parameter $|\psi|^2$ dies out in coherence length ξ, while magnetic field B penetrates a distance λ. Type I materials have positive interface energy; type II have negative interface energy.

For type II materials, the situation is reversed: the interface benefits from $|\psi|^2$ without paying the price of flux exclusion. Hence the interface energy is <u>negative</u>, and in the presence of a strong magnetic field, the lowest energy state is one in which interface proliferates until it fills the sample volume. This is done most efficiently by going from a laminar structure to a normal spot structure, as shown by Abrikosov[1] in 1957.

Thus, while type I superconductors show the Meissner effect of excluding the flux, in type II superconductors flux penetrates in an array of flux tubes, each carrying a single quantum of flux

$$\Phi_o = \frac{hc}{2e} = 2.07 \times 10^{-7} \text{gauss-cm}^2$$

This quantization of flux is the exact macroscopic analog of the quantization of angular momentum in atomic systems, and it follows from the single-valuedness of $\psi(\vec{r})$, the macroscopic wavefunction describing the condensed Cooper pairs. This single-valuedness implies that the <u>phase</u> of ψ must change by $2\pi n$ in any complete circuit. The integer n is the number of flux quanta within the loop. If n=0 is the lowest energy state, one has the Meissner state. If n>0 has lower energy, as in type II materials above H_{c1}, one has flux penetration.

As shown by Kleiner, Roth, and Autler[2] (correcting a numerical error in Abrikosov's path-breaking paper), the lowest energy configuration is one in which the flux penetrates in a triangular array of vortices, each containing Φ_o. The predicted arrangement was confirmed experimentally by Essmann and Träuble[3] using a decoration and replica technique to make the flux line centers visible to the electron microscope. The spacing between vortices is set by the flux density, since the area of the unit cell is given by Φ_o/B. In the fields of ~100 Gauss used by Essmann and Träuble, the spacing is about ½ μm. As the field B increases, the spacing decreases accordingly. When the spacing approaches ξ, the superconducting density $|\psi|^2 \to 0$ in a second order phase transition, which occurs at $H_{c2} = \Phi_o/2\pi\xi^2$.

FLUX PINNING

So far I have discussed only homogeneous superconductors in equilibrium. Now it is time to start addressing the role of inhomogeneity in the practical business of using superconductors to carry large transport currents. For this purpose, we shall concentrate on type II superconductors, since only they can stand the large magnetic fields produced by large currents. Consider the effect of a tranport current passing at right angles to the flux-line-lattice (FLL). There will be a Lorentz force $\vec{J} \times \vec{B}$ tending to move the conductor (as in a motor). But if the conductor is fixed, what happens instead is that the FLL is driven to move relative to the material. If this motion occurs, the relative motion of flux and conductor turns it into a generator of an EMF which tends to oppose the current. In other words, the superconductor is resistive. In fact, one can show that if the material is perfectly homogeneous, so that the energy of the FLL is independent of position relative to the sample, there will be a resistance associated with the flux flow which is the order of the normal resistance times the volume fraction of the quasi-normal cores, i.e.,

$$\rho \approx \rho_n B/H_{c2} \qquad (1)$$

Such a material would be worthless as a practical conductor. Accordingly, the metallurgical processing of practical magnet wire material is designed to yield an inhomogeneous material, with the scale of inhomogeneity such as to provide the strongest possible "pinning" of the vortex lattice.

It is clear that the strongest pinning should result when the properties of the superconductor vary as strongly as possible in space in a pattern which exactly matches the FLL. To test this idea, Hebard, Fiory, and Somekh[4] at Bell Labs have recently reported experiments on a superconducting aluminum film in which a triangular array of holes had been perforated. They typically used 1μ diameter holes in a 3μ-spacing. With this structure, they did indeed find very strong pinning for the values of B for which the vortex array spacing matched their array of holes. In fact, they were able to reach critical currents (before flux motion and dissipation started) which approached the theoretical maximum current density for a film.

This experiment nicely confirms the theoretical basis for flux pinning to raise I_c, but what about the practical world? One must rely on a <u>random</u> pinning array, because one wants the material to work well for a wide range of field values and orientations. Moreover, the vortex line spacing at the higher fields of practical interest becomes very small. For example, at 50 kG, the spacing is about 200 Å, so that pinning sites will be most effective if they modulate the superconducting properties on that scale. Since this is well below the crystallite size obtained in conventional practice, <u>edges</u> between disparate crystallites are important. Also recourse is made to cold-work to introduce a high density of dislocations, giving distortions on a small scale.

Another approach, tried by Powell, et al.,[5] was to use a spray-drying technique to produce extremely tiny crystallites, which were then reacted in suitable atmosphere to make the NbN or NbC material desired, followed by compaction and sintering. This produced a void-impregnated solid, as shown in Fig. 2. But even these rather heroic procedures produced structure on a micron scale, rather than ~200 Å, and yielded a critical current density only modestly superior to the best sintered compacts produced by more conventional techniques.

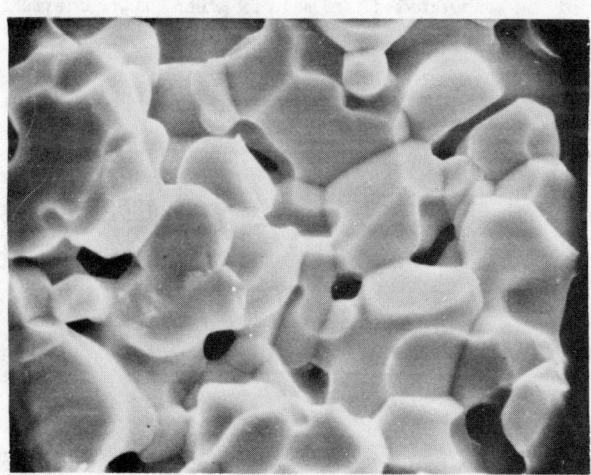

Fig. 2. Scanning electron micrograph of an NbN pellet sintered in a pressure of 10 atm of nitrogen. (After Powell, et al., Ref. 5)

MULTIFIMENTARY WIRE

Developing inhomogeneity on the scale of ξ or the inter-vortex spacing is necessary for high J_c, but it is not sufficient for making <u>stable</u> superconducting wire. The problem is that the pinning strength and hence critical current decrease strongly with increasing temperature, so that an upward fluctuation in temperature reduces the local critical current, causing a reduction in the maximum flux density gradients, which causes flux motion, which causes heating, which amplifies the initial fluctuation. This problem increases with the square of the diameter of the superconductor, since a given change in gradient is operative over a greater length. Figure 3 illustrates the essential point. For given J_c, the dissipation per unit volume scales with the time integral of the electric field induced by the flux change represented by the shaded area in the Figure. Clearly this increases as d^2. Carrying out the calculation, one finds[6] that the criterion for stability is

$$d < \left[\frac{3c^2 C}{\pi J_c (-dJ_c/dT)} \right]^{1/2} \sim 100 \text{ μm} \qquad (2)$$

where C is the specific heat per unit volume. In practice, to reduce losses in swept field applications and to give a margin of safety, filament diameters are usually taken in the range of 10-40 μm. These filaments must be embedded in a matrix with high thermal conductivity, usually copper.

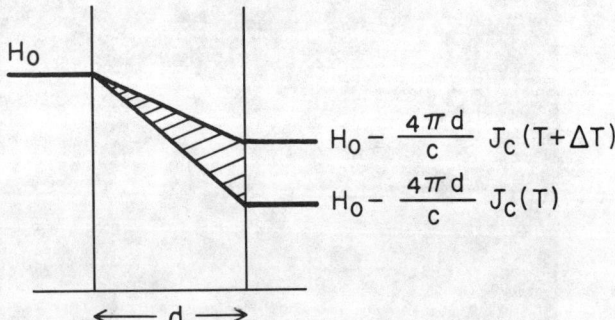

Fig. 3. Flux density profiles in a conductor of thickness d before and after a temperature fluctuation ΔT.

The current technology is to construct wires containing hundreds or thousands of such filaments by drawing down copper billets containing superconducting rods. So long as the final filaments are many microns in diameter, much greater than the coherence length, their properties are essentially the same as bulk material, and the composite conductor acts like parallel superconducting and normal conductors. In equilibrium, the superconducting filaments carry all the current, and the copper simply serves to provide thermal stability and fault protection, carrying the current if the superconductor quenches for some reason.

TSUEI WIRE

For the purpose of this conference, it is more interesting to move on to a discussion of "Tsuei wire" - a composite material[7] which resembles a commercial multifilamentary conductor except that the filaments are produced metallurgically, and hence can be made much finer, but they are <u>not continuous</u>. This raises non-trivial questions about conduction in a heterogeneous medium, in which the length scale is not too much greater than the coherence length, so that there is no longer a clearcut distinction between superconducting and normal regions.

Let me first briefly describe the process which produces the wire. If one wants Nb_3Sn filaments in Cu, one melts Nb, Sn, and Cu together (no mean feat to get a good uniform solution) and quenches the melt into an ingot. The Nb precipitates out in dendritic particles on a scale depending on the cooling rate. This material is then drawn into wire, which deforms the original Nb particles into long filaments. Finally, a heat treatment reacts the Sn out of the Cu matrix to form Nb_3Sn. To a first approximation, the deformation of the Nb particle follows that of the macroscopic ingot. If the cross-sectional area is reduced by a reduction factor Re, the length also increases by Re to conserve volume, so that the length to diameter ratio scales up as $Re^{3/2}$. For the representative value Re = 600, $Re^{3/2} \sim 10^4$, so the originally isotropic inclusions become very long and thin. Typical lengths are up to 1 cm, while diameters are typically only 1 μm. Thus, there will typically be 10^5 filaments passing through any cross section of a sample wire, while the current need make only a few jumps from filament to filament to get from one end of the sample segment to the other. A view of the end of a partially etched wire sample is shown in Fig. 4.

In assessing the potential usefulness of such materials, one is particularly interested in whether the resistance will be zero or at least close enough so that it could be used to make cheap magnet wire. Given the notion of a percolation threshold[8] at some volume fraction (~15% in 3-dimensions, ~50% in 2-dimensions) one might expect strictly zero resistance above that fraction, and finite (but small) resistance at smaller volume fractions. But the effective superconducting volume fraction will depend on temperature, current density, and magnetic field because of the "proximity effect", i.e., the fact that superconducting electrons can diffuse a normal metal coherence length

$$\xi_N = \left(\frac{\hbar v_F \ell}{6\pi kT}\right)^{1/2} \quad \text{dirty} \tag{3a}$$

$$\xi_N = \frac{\hbar v_F}{2\pi kT} \qquad \text{clean} \tag{3b}$$

which is about the same as the low-temperature coherence length $\xi(0)$ for a superconductor having $T_c \sim T$. Thus this length will be of the order of a fraction of a micron under typical conditions. But because the coherence energy of the wavefunction only falls exponentially, as e^{-x/ξ_N}, some superconductive coupling will continue until this energy falls below the thermal noise level. (See Fig. 5b.) Thus, at low temperatures and with small measuring currents, one expects to find true perfect conductivity, as in a bulk superconductor, and one does.

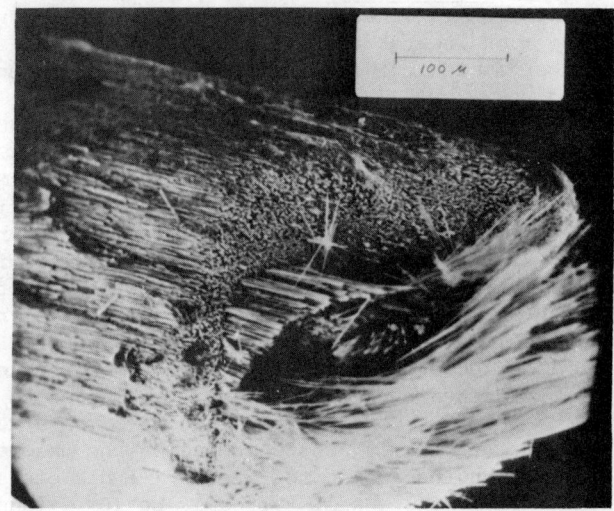

Fig. 4. End view of partially etched piece of Tsuei wire.

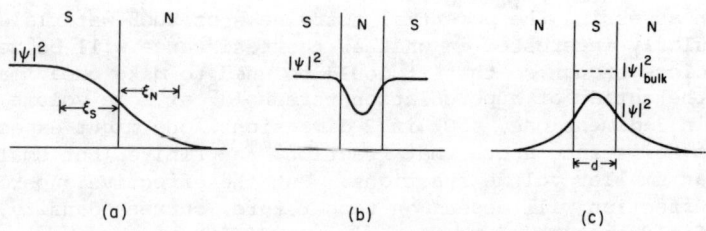

Fig. 5. Schematic illustration of the proximity effect. (a) Superconducting and normal state coherence lengths. (b) Superconducting coupling through a normal metal layer. (c) Depression of superconducting order parameter in small superconducting inclusion in a normal matrix. The T_c depression corresponding to (c) is given by Eq. (4).

But the technically interesting case is one of high current densities, $T \gtrsim 4^\circ K$, and in the presence of strong magnetic fields. All these influences reduce the proximity effect coupling, effectively shrinking the superconducting volume fraction. In fact, if the superconducting filament diameter is comparable to ξ, the diffusion of normal electrons into the superconductor will reduce its T_c, and the superconducting volume fraction may be even less than the nominal one, since the finest filaments will be normal. (See Fig. 5c.) Although the detailed variation of T_c depends on the geometry and the interface conditions, roughly one has

$$\frac{T_c}{T_{cb}} = 1 - \left(\frac{\pi \xi(0)}{d}\right)^2 \tag{4}$$

so that if $\xi(0) \sim 1000$ Å and $d \sim 1$ μm, a 10% reduction of T_c might be expected. This effect will be important for pure Nb in Cu, but in Nb_3Sn, $\xi(0)$ is so small that the effect should be negligible.

Because of the uncertainty about the details of this intermediate region, it is useful to consider first a simple limit, in which one assumes that the proximity effect is completely suppressed (by current or field), and that the superconductive volume fraction f_s consists of a collection of long thin cylinders of length L and diameter d formed by the drawing process, and hence aligned along the wire. The rest of the volume is assumed filled with a completely normal Cu matrix.

It is intuitively clear that these long aligned filamentary fragments will be much more effective in reducing the resistance of the wire than would an equal volume fraction in the form of isotropic inclusions, or of randomly oriented filamentary fragments. I gave an elementary treatment[9] of this problem several years ago, the result being that the remnant resistivity should be reduced below that of the matrix by the factor

$$\frac{\rho_{rem}}{\rho_o} \approx \frac{1}{1 + f_s L^2/d^2} \approx \frac{1}{f_s} \frac{d^2}{L^2} \approx \frac{1}{f_s} \frac{1}{R_e^3} \tag{5}$$

which can readily reach 10^{-7}. Since the matrix itself is Cu at low temperatures, the reduced resistance could be so small as to be detectable only with superconducting instrumentation. The physical reason for this result is that the resistance stems from the spreading resistance around the superconductive filaments as the current fans out to pass from one to an adjacent one. The spreading resistance per cell decreases as $1/L$, while the effective "Maxwell-Garnett cell" is deforming to a length L and diameter $\sim d/f_s^{1/2}$. A conclusion similar to (5) for extremely elongated filaments was reached by Callaghan and Toth,[10] apart from numerical factors. By contrast with this physically reasonable result, conventional effective medium theory applied to this case leads to a prediction of zero resistance for all f_s exceeding the depolarization factor of the inclusions. For the usual isotropic case, this would be the reasonable magnitude of 1/3, but for these filaments, it would be an unreasonable value of order $d^2/L^2 \approx 10^{-8}$!

Proceeding phenomenologically, it seems reasonable to build the percolation threshold concentration f_c for perfect conductivity into (5) by an additional factor $(1-f_s/f_c)^s$. As noted above, one expects $f_c \sim 0.15$ in 3-dimensions and $f_c \sim 0.5$ in 2-dimensions, while theoretical estimates[11] of the critical exponent s yield ~0.7 in 3-dimensions and range from 1 to 1.3 in 2-dimensions. (Values near the upper end of this range were favored by experimental results of Watson and Leath.[12]) Taking the long-filament limit of (5), we expect then that

$$\frac{\rho_{rem}}{\rho_o} \approx \frac{1}{f_s} \frac{d^2}{L^2} (1 - \frac{f_s}{f_c})^s \qquad f_s < f_c$$

$$= 0 \qquad f_s > f_c$$

(6)

where f_s will depend on current, temperature, and magnetic field.

What did the measurements show? Davidson[13] studied two samples of wire containing Nb_3Sn filaments in a Cu matrix. One containing 7½% of Nb_3Sn showed the expected drop of a factor of ~10^5 in resistance at the T_c of Nb_3Sn, followed by a long plateau in resistance, after which the resistance dropped again, to below our sensitivity. As shown in Fig. 6, the details of this plateau region are current-dependent, as might be expected since increasing current counteracts the proximity effect, while decreasing temperature enhances it. Another sample examined by Davidson contained ~15% Nb_3Sn. It showed no significant plateau region, with R dropping from its normal value to an unobservably small value in a small temperature range near the T_c of Nb_3Sn. These two results appeared to confirm the existence of a percolation threshold near 15% by volume.

Despite this apparent confirmation, we felt further measurements were required to test the threshold concentration more carefully, and to clarify the role of the proximity effect on f_s. Moreover, the interpretation of the Nb-Sn-Cu composite results were complicated by the presence in the core of the Nb_3Sn filaments of unreacted Nb, whose T_c would be near the temperature at which the resistance disappears in Fig. 6. These considerations motivated a second series of measurements now in progress by C. Lobb in our laboratory on samples containing simply Nb filaments in a matrix of Cu or of a Cu-rich alloy. Although only preliminary results are available, I shall now summarize the results to date.

On a sample containing 10% Nb in a pure Cu matrix, Lobb found no plateau region at all, just a smooth drop in resistance to unobservably small values in a current-dependent temperature interval somewhat below the T_c of Nb. This result indicated the presence of

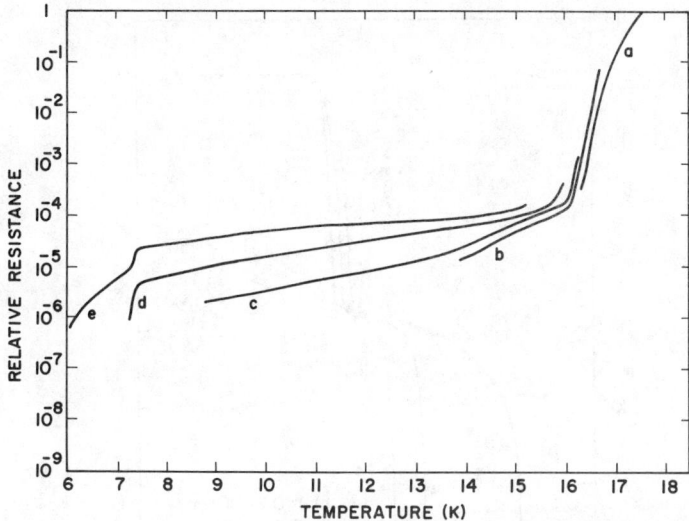

Fig. 6. Resistive transition of a 7½% superconductive wire made by Tsuei's process. Curves <u>a</u> through <u>e</u> are measurements made at different current levels, ranging in decades from 6×10^{-5} A for curve <u>a</u> to 0.6 A for curve e. The initial drop near 17 K marks the T_c of the Nb_3Sn surface layer on the Nb filaments in the copper matrix. (After Davidson and Tinkham, Ref. 13)

strong proximity effect coupling, giving a greatly expanded effective superconductive volume fraction. To test this interpretation, Lobb diffused Zn into the very same sample, to shorten the mean free path and weaken the proximity coupling. The result was a small plateau region at the expected resistance level $\sim 10^{-5}$ that of the normal state, before the final plunge to zero resistance at lower temperature. In an attempt at still further reducing the proximity effect, Lobb then made a series of samples with various concentrations of Nb filaments in a Cu matrix containing 3% Ni. (Magnetic impurities are specifically effective in breaking up the superconductive pairing.) In all cases, there was a drop in resistance of several orders of magnitude just below the T_c of Nb, as expected from (5), followed by a drop to zero resistance at a lower (current-dependent) temperature as implied by (6).

The data on a sample containing 7% Nb are shown in Fig. 7, and they appear sufficiently smooth and systematic to allow at least a speculative further level of interpretation using (6). There is no <u>intrinsic</u> critical temperature below the T_c of the Nb filaments, which is depressed to $\sim 7°K$ by proximity effect, finite current density, impurities, etc. Accordingly, we can interpret the disappearance of resistance at some lower T_{c1} as occuring when f_s, which increases roughly as $[d + \xi_N(T,I)]^2$ as T decreases, reaches f_c. More-

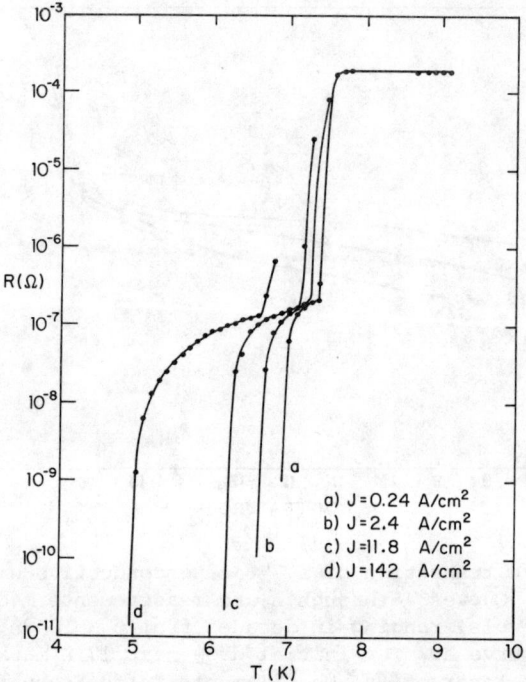

Fig. 7. Resistive transition of a wire sample made by Tsuei's process, containing 7% Nb filaments in a Cu matrix which has 3% Ni in solid solution. As indicated, the current density increases by almost three orders of magnitude from curve (a) to curve (d). Curve (d) is well fit by Eq. (7), with s = 1.29 and T_{c1} = 4.94 K, from T_{c1} up to the break in the curve at about 6.5 K, where presumably the Nb filaments themselves start to go normal in this high current density.

over, in the absence of any intrinsic critical temperature in this range of 5-6°K, it is reasonable to assume that f_s varies roughly linearly with T over this small interval, since ξ_N varies smoothly with T. In that case (6) implies that the resistance should vary as

$$R \propto (T-T_{c1})^s \qquad (7)$$

in this region. [The variation of f_s in the prefactor $f_s^{-1} d^2/L^2$ in (6) is cancelled by the variation in $d_{eff} \approx d + \xi_N(T,I)$.] If one fits the data on the curve for the highest current density in the range from 5-6.5K (ignoring the final drop which would be broadened by any gross inhomogeneity over the length of the sample), one finds quite a good fit for the critical exponent s ≈ 1.29 ± 0.1. A similar fit to the highest current data on another sample with higher Nb concentration yielded s ≈ 1.24 ± 0.1. These values are consistent with theoretical expectations for 2-dimensional percolation, but not with the value s ≈ 0.7 expected in 3-dimensions. At least in retrospect, this 2-dimensional exponent seems not unreasonable in view of the highly anisotropic sample. As is made graphically clear from Fig. 4, the huge number of long filaments assure conduction along the wire as soon as the threshold for 2-dimensional percolation between the filaments is exceeded. More work is clearly in order to see whether this preliminary interpretation will prove out. If so, this unusual system may offer some advantage for studying percolative conductivity because the superconducting volume fraction of a given sample can be varied continuously by varying temperature or current.

REFERENCES

1. A. A. Abrikosov, Soviet Physics-JETP 5, 1174 (1957).
2. W. H. Kleiner, L. M. Roth, and S. H. Autler, Phys. Rev. 133, A1226 (1964).
3. U. Essmann and H. Träuble, Phys. Letters 24A, 526 (1967).
4. A. F. Hebard, A. T. Fiory, and S. Somekh, IEEE Trans. Magnetics, Vol. MAG-13, 1977, p. 589; and to be published.
5. R. M. Powell, W. J. Skocpol, and M. Tinkham, J. Appl. Phys. 48, 788 (1977).
6. See, for example, M. Tinkham, Introduction to Superconductivity, McGraw-Hill, New York, 1975, p. 184.
7. C. C. Tsuei, Science 180, 57 (1973); J. Appl. Phys. 45, 1385 (1974).
8. See, for example, V. K. S. Shante and S. Kirkpatrick, Adv. Phys. 20, 325 (1971); H. Scher and R. Zallen, J. Chem. Phys. 53, 3759 (1970); R. Zallen and H. Scher, Phys. Rev. B4, 4471 (1971).
9. A. Davidson, M. R. Beasley, and M. Tinkham, IEEE Trans. Magnetics, Vol. MAG-11, p. 276, 1975.
10. T. J. Callaghan and L. E. Toth, J. Appl. Phys. 46, 4013 (1975).
11. J. P. Straley, Phys. Rev. B15, 5733 (1977), and references cited therein.
12. B. P. Watson and P. L. Leath, Phys. Rev. B9, 4893 (1974).
13. A. Davidson and M. Tinkham, Phys. Rev. B13, 3261 (1976).

DISCUSSION

S. NAM (University of Dayton): Have you studied the critical current in this wire?

TINKHAM: The figures I've presented are essentially a plot of critical current vs. temperature. The point where the resistance drops out is the critical current at that temperature. The materials that we are making are not intended to be practical because they contain just Nb in Cu. and nobody would use that for anything. The practical material is Nb_3Sn and Duke and Turnbull at Harvard have obtained very good current densities in this material.

G. CODY (Exxon Research and Engineering): Have you observed anisotropy in the critical currents or the flux flow resistances as a function of the angle of the magnetic field? There are interesting such effects in the literature and in your nice clean system you might be able to unravel them.

TINKHAM: We have no applied field in these experiments — only the self field of the current. It is an interesting problem.

LOW-FIELD AND HIGH-FIELD HOPPING CONDUCTION IN GRANULAR METAL FILMS

Ping Sheng
RCA Laboratories, Princeton, NJ 08540

ABSTRACT

Based on the assumption of composition homogeneity, a physical picture of hopping conduction in granular metals is developed which gives temperature and electric field dependences of the conductivity in excellent agreement with experiment. The same hopping conduction picture, when coupled with the assumption about spin-dependent tunneling, is found to explain the striking temperature variation of magnetoresistance in granular Ni-SiO$_2$. Two material characterization parameters, C and $\mathcal{E}_0$, are identified which respectively define the temperature and the electric field scales of granular metal systems. The relationship between C and $\mathcal{E}_0$, and the composition dependence of C are discussed.

Granular metals[1-4] are composite materials consisting of fine mixtures of metals and insulators. Depending on the volume fraction of metal, x, there are three possible structural regimes. When x is large, we have the metallic regime in which the metal grains touch and form a metallic continuum with dielectric inclusions (Fig. 1a). For x between 0.5 and 0.6, the dielectric particles become interconnected, and a labyrinth structure results (Fig. 1b). This is the transition regime. In this paper we will be mainly concerned with the electrical conduction mechanism of granular metals in the dielectric regime (x < 0.5), where we have small, isolated metal grains of the size 50-200 Å dispersed in a dielectric matrix (Fig. 1c).

In the dielectric regime, the temperature coefficient of resistivity is negative, which is usually taken as an indication that the charge carriers are temperature activated. The source of the activation energy in this case is not difficult to identify. Electrical conduction in granular metals results from electrons tunneling from one metal grain to the next across the dielectric barrier. In order to generate a charge carrier, an electron has to be removed from one neutral grain and placed on another neutral grain, thereby creating a pair of positively and negatively charged grains. Such a process of charge carrier generation requires a certain amount of charging energy[1-5] E_c, which can be pictured as the energy stored in the electric field of the charged grain pair. For 50 Å grains, E_c is of the order of 50 meV. From dimensional analysis, E_c has to have the form

$$E_c = \frac{e^2}{d} F\left(\frac{s}{d}\right), \tag{1}$$

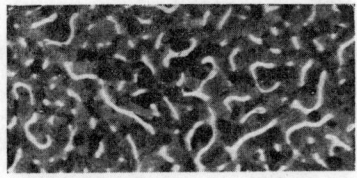

(a) 73 Vol % Au

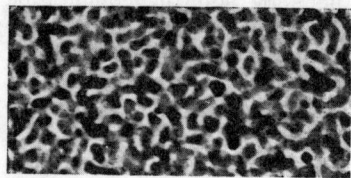

(b) 48 Vol % Au

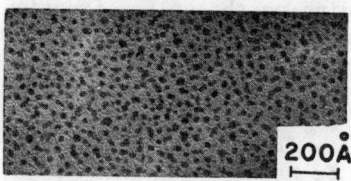

200Å

(c) 18 Vol % Au

Fig. 1. Electron micrographs of Au-Al$_2$O$_3$. After Ref. 1.

where e is the electronic charge, d is the size of a grain, s is the separation between grains, and F is a function whose form depends on shape and arrangement of the grains and on the interaction between the pair of charged grains. The existence of E_c directly implies that the thermal equilibrium number density of charge carriers, whose generation requires a charging energy E_c, is proportioned to the Boltzmann factor $\exp(-E_c/2kT)$.

However, the knowledge of the existence of E_c and its magnitude is not directly helpful in characterizing the conductivity of granular metals. This is because that in granular metals there is a wide variation in metal grain sizes and therefore a broad distribution in the values of E_c. The useful characterizing quantity, as it turns out, is the volume fraction of metal x. Since the metal grains are formed by surface diffusion of sputtered metal and insulator atoms, it is plausible to expect the composition x to be uniform when averaged over a few surface diffusion lengths. This condition of composition homogeneity[2] immediately implies a close relationship between the grain size d and the tunneling barrier thickness s. If we visualize each metal grain as surrounded by a shell of insulator, then in order for the relative volume fractions of metal and insulator to stay constant regardless of the grain size, the thickness of the insulator would have to be directly proportional to the size of the grain. In other words, the ratio s/d is a constant whose value depends only on x. From Eq. (1) and the condition s/d = constant it follows directly that[2] sE_c = constant for a given composition x.

Given the condition that s is inversely proportional to E_c, we can see that a charge carrier with generation energy E_c' would be inhibited from tunneling to grains with $E_c \leq E_c'$ because of the larger tunneling barriers associated with smaller charging energy. Since it is also improbable for the charge to tunnel to much smaller grains with $E_c \gg E_c'$ because of insufficient energy, the optimal path for the charge would follow regions with the least deviation of

E_c from E_c'. The corresponding tunneling barrier s' is therefore given by $C/\chi E_o'$, where

$$C = \chi s E_c \qquad (2)$$

is a constant and $\chi = [2m\Phi/\hbar^2]^{1/2}$ is the tunneling constant[6] with m denoting the electron mass, Φ the effective barrier height, and $\hbar$ Planck's constant. Since the conductivity σ is the product of mobility, charge, and number density of charge carriers, we can write

$$\sigma \propto \exp[-(E_c'/2kT) - 2\chi s'] = \exp[-(E_c'/2kT) - (2C/E_c')]. \qquad (3)$$

In Eq. (3) we have written the mobility as proportional to the tunneling probability $\exp(-2\chi s')$. It is easily seen that σ is a peaked function of E_c' with the maximum occurring at

$$E_c' = \sqrt{4kTC}. \qquad (4)$$

In granular metals where we have a distribution of values of E_c', the dominant contribution to σ would come from those charge carriers with E_c' close to the value which maximizes σ. Therefore, substituting Eq. (4) into Eq. (3), we get the dominant behavior (we say "dominant" because we have neglected the weaker effect of the percolation path density distribution on the temperature dependence of σ) of σ as given by [1,2]

$$\sigma = \sigma_o \exp[-2\sqrt{C/kT}], \qquad (5)$$

where σ_o is a constant. The temperature dependence of the conductivity predicted by Eq. (5) agrees very well with experiment as seen in Fig. 2. In fact, we have least-square analyzed all our data[7] in the form

$$\sigma = \sigma_o \exp[-A/T^n]. \qquad (6)$$

The result shows that the best value of n in every case falls between 0.4 and 0.6.

The physical picture of hopping conduction is therefore as follows. At low temperatures the conduction is by hopping between large grains with small charging energies. As the temperature is raised, the principal conduction path shifts to regions of smaller

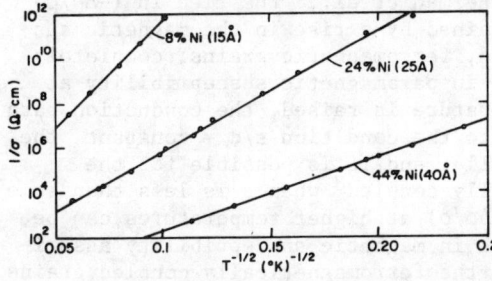

Fig. 2. Low-field resistivity $1/\sigma$ as a function of $\sqrt{T}$ for three compositions of Ni-SiO$_2$. The full lines represent the relation $\ln \sigma = -2\sqrt{C/kT}$. After Ref. 2.

grains to take advantage of the smaller tunneling barriers. An interesting consequence of this shift in conduction path with temperature is manifested in the magnetoresistance of granular Ni-SiO$_2$ films.[8,9] Magnetoresistance of granular Ni-SiO$_2$ arises because each Ni grain is ferromagnetic, and as a result the electron tunneling probability between two grains is sensitive to the relative orientation of their directions of magnetization.[8] Since the tunneling probability is larger between more parallel-aligned neighboring magnetizations, the application of a magnetic field would always increase the conductivity. For a given value of the magnetic field, the magnitude of the increase in conductivity varies directly with the magnetic susceptibility, which is a measure of the ease at which the magnetizations of the various grains can be aligned by an external magnetic field. Figure 3 shows a typical set of magnetoresistance data[8,9] taken as a function of temperature. The quantity

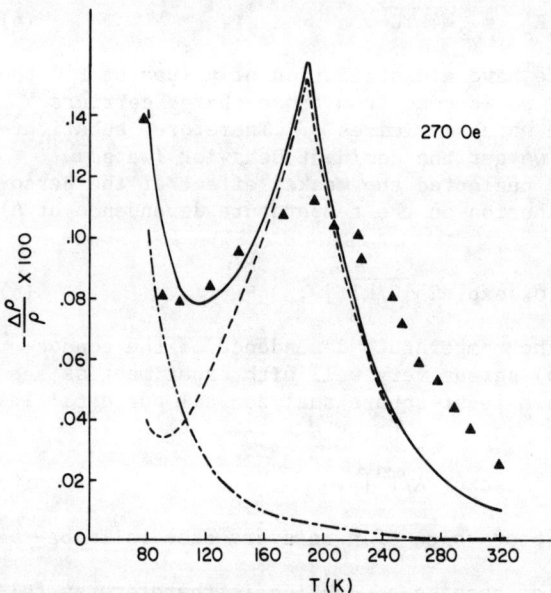

Fig. 3. The temperature variation of transverse magnetoresistance of a granular Ni-SiO$_2$ film with x = .48 . The applied magnetic field is 270 Oe. The triangles are experimental data, the solid curve is the theoretical result which contains a superparamagnetic contribution (dash-dotted line) and a ferromagnetic contribution (dashed line). After Ref. 9.

being plotted is the change in resistivity, $\Delta\rho$, resulting from the application of a magnetic field, divided by the resistivity ρ at zero field. As pointed out by Helman et al.,[9] the rise in $(-\Delta\rho/\rho)$ at low temperatures can be explained by a rise in the magnetic susceptibility of a set of isolated, ferromagnetic grains; completely analogous in nature to the rise in paramagnetic susceptibility at low temperatures. As the temperature is raised, the conduction path shifts to smaller grains. Due to the condition s/d = constant, the tunneling barriers are also smaller and it is possible for the grains to become ferromagnetically coupled[8] when s is less than some critical value. The peak in $(-\Delta\rho/\rho)$ at higher temperatures can be understood as caused by the peak in magnetic susceptibility associated with a phase transition in the ferromagnetically coupled grains from the ferromagnetic state to the superparamagnetic state[8,9] (the state in which each grain is ferromagnetic but intergrain magnetic ordering is lost). Therefore, by allowing the conduction path to shift with temperature, one can get the striking magnetoresistance behavior that is superparamagnetic in nature at low temperatures and ferromagnetic in nature at higher temperatures. In Fig. 3 it can be seen that the calculation[9] based on the above explanation reproduces all the main features and the order of magnitude of experimental results. The good agreement between the theory and the experiment on magnetoresistance gives support to the physical picture of hopping conduction responsible for the $\ln \sigma \sim 1/\sqrt{T}$ behavior.

In the above we have only considered temperature activation as the charge carrier generation process. However, charge carriers can also be generated by applying an external electric field.[3] The basic mechanism is illustrated schematically in Fig. 4. When the voltage difference ΔV between two neighboring grains is larger than

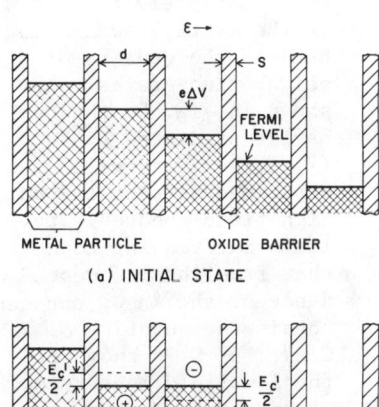

Fig. 4. Schematic illustration of effect of an electric field on the energy levels in the granular metal. For the sake of illustration the structure is taken as uniform. (a) Each grain is neutral before the tunneling takes place. (b) After field generation has occurred, a hole is left on one grain and an electron is added to the other grain. After Ref. 1.

E_c/e, a pair of positively and negatively charged grains can be generated. Due to the electrical interaction between the neighboring, oppositely charged grains, the energy threshold required for field generation can be substantially smaller than that required for generating a completely dissociated electron-hole pair. If we denote the energy required for field generation as E_c^1 and that for temperature activation as E_c^0, calculation[1] has shown that for a wide range of x values, $E_c^1 \simeq E_c^0/2$. At any given temperature, the relative importance of thermal activation and field generation is gauged by the ratio $kT/e\Delta V$. At T = 10 K, for example, field generation becomes the dominant process when $e\Delta V \geq 1$ meV. If the grains are separated by 50 Å on the average, $e\Delta V \geq 1$ meV implies an electric field $\mathscr{E} \geq 10^4$ V/cm.

In the regime $e\Delta V \gg kT$, we can derive a simple dependence of the conductivity on the electric field. Making the simplifying assumption that the electric field gives rise to equipotential layers of grains with equal voltage drop ΔV between neighboring layers, we get the result that at a given ΔV, there is a maximum value of charging energy $E_c^1 = e\Delta V$ above which field generation cannot occur. Due to the relation $2\chi s E_c^1 = C$ (we have used the relation $E_c^1 \simeq E_c^0/2$), the maximum threshold value of E_c^1 can be translated into a minimum value of tunneling barrier thickness $s = C/2\chi e\Delta V$. Since the rate of field generation depends on the tunneling rate $\exp(-2\chi s)$, it is expected that the dominant contribution would come from those pairs of grains with s close to the minimum value at any given ΔV. That is,

$$\sigma_H \propto \exp\left[-2\chi \frac{C}{2\chi e \Delta V}\right] = \exp[-C/e\Delta V] = \exp[-C/ew\mathscr{E}] \qquad (7)$$

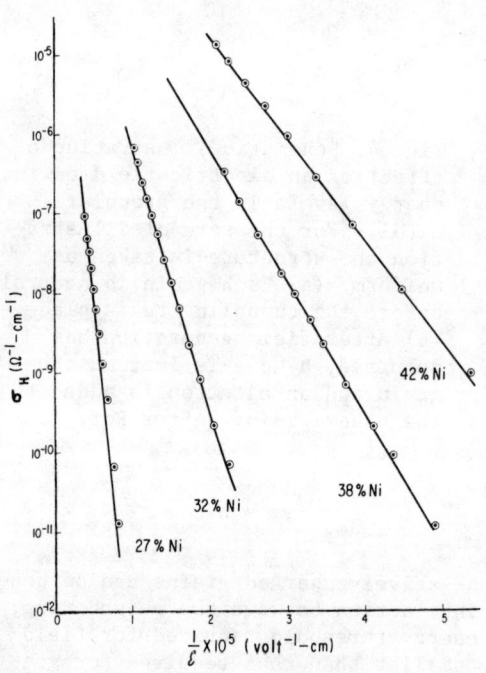

Fig. 5. Measured conductivity σ_H at 1.25 K plotted as a function of $1/\mathscr{E}$. The solid line represents the relation $\ln \sigma_H = -\mathscr{E}_0/\mathscr{E}$. After Ref. 3.

Here σ_H denotes the high-field conductivity, and w is the average separation between the centers of neighboring grains. As shown in Fig. 5, the behavior predicted by Eq. (7) agrees very well with experimental results[3] on high-field conductivity. It is interesting to note that from the $1/\mathscr{E}$ dependence of the $\ln \sigma_H$ one can obtain the quantity $\mathscr{E}_0 \equiv C/ew$. If C is known from the low field measurement of $\ln \sigma \propto 1/\sqrt{T}$, then the quantity w, which can be regarded as the "lattice constant" of granular metals, can be obtained as the ratio $C/e\mathscr{E}_0$. In Table I we compare the values of w calculated from $C/e\mathscr{E}_0$ with those determined from electron micrographs.[1] The good agreement between the two sets of values is a confirmation of the self-consistency of the theory.

When the thermal activation process is not negligible, it has been shown[1,2] that the high field conductivity can be expressed as

$$\sigma_H = \sigma_\infty f(kT/C, \mathscr{E}/\mathscr{E}_0). \qquad (8)$$

TABLE I. Values of $\mathscr{E}_o$, C, and w for Ni-SiO$_2$

x vol. fraction of Ni	$\mathscr{E}_o$ (10^6 V/cm)	C (eV)	w (Å) calc.	w (Å) el. micro.
0.44	0.22	0.13	60	50
0.34	0.48	0.22	45	49
0.24	1.0	0.35	44	35
0.14	1.8	0.68	38	25
0.08	3.4	1.10	34	20

The interesting property of Eq. (8) that σ_H/σ_∞ can be expressed as a universal function of the dimensionless variables kT/C, $\mathscr{E}/\mathscr{E}_o$ has been verified by experimental results.[1] The two constants C and $\mathscr{E}_o$ can therefore be regarded as defining the temperature scale and the electrical scale of the granular metal, respectively.

The structural constant C defined by Eq. (2) depends only on the metal fraction x. Thus, in a granular metal for a definite value of x the constant C is expected to be independent of grain size. This invariance of C with respect to grain size has been demonstrated experimentally in the W-Al$_2$O$_3$ granular system.[10] In these experiments it was found that by annealing the W-Al$_2$O$_3$ samples the grain size increased by a factor of 2 to 5 and the resistivities increased by several orders of magnitude, yet the value of C remained unchanged. The value of C as a function of x for several granular systems[1] are shown in Fig. 6. In order to obtain a theoretical expression for C(x), we note that from Eqs. (1) and (2) we have

$$C = \chi e^2 \frac{s}{d} F\left(\frac{s}{d}\right). \tag{9}$$

An approximate expression for F(s/d) can be obtained as follows. Consider the energy E_{es} required to remove an electron from a neutral grain to infinity. That energy is equivalent to the total energy stored in the electrostatic field of a positively charged metal grain. Since the electric field of the charged grain is shielded by its neighboring grains, for simplicity we will approximate the realistic geometry of Fig. 7a by that of Fig. 7b. Then elementary electrostatics yields

$$E_{es} = \frac{e^2}{\varepsilon d} \frac{s/d}{\left(\frac{1}{2} + \frac{s}{d}\right)}, \tag{10}$$

Where ε is the dielectric constant of the insulator. By comparing with Eq. (1) and noting that $E_c = 2E_{es}$ if there is no interaction

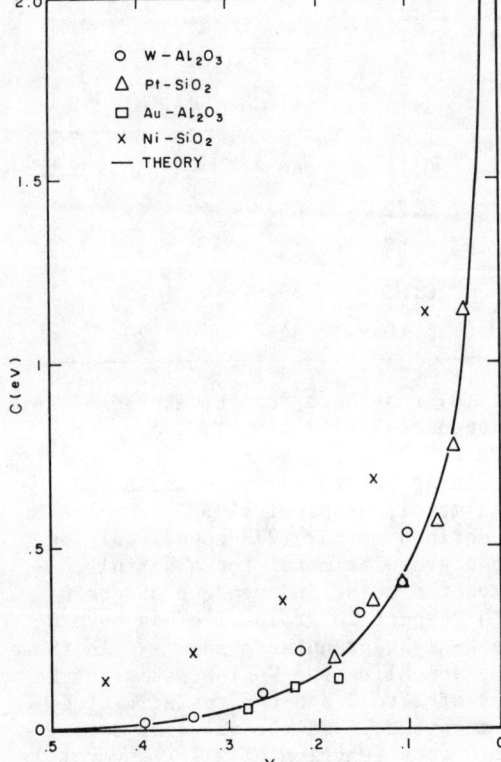

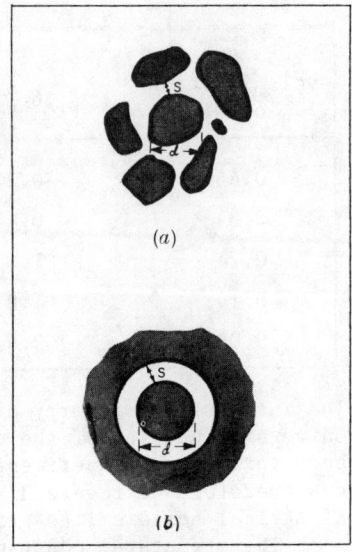

Fig. 7. For the purpose of calculating the function F(s/d), the realistic geometry in (a) is approximated by that of (b). After Ref. 1.

Fig. 6. Structural constant C as a function of x. The solid curve is obtained from Eq. (13) with η = 1 eV. After Ref. 1.

between the positively and negatively charged grains (i.e., completely dissociated), we get

$$F\left(\frac{s}{d}\right) = \frac{2(s/d)}{\varepsilon\left(\frac{1}{2} + \frac{s}{d}\right)} . \quad (11)$$

In order to relate s/d to x, we approximate the granular metal by a simple cubic lattice of metal spheres with lattice constant s + d and sphere diameter d. Then

$$\frac{s}{d} = \left(\frac{\pi}{6x}\right)^{1/3} - 1 . \quad (12)$$

Substitution of Eqs. (11) and (12) into Eq. (9) yields

$$C(x) = \eta \frac{[(\eta/6x)^{1/3} - 1]^2}{[(\eta/6x)^{1/3} - \frac{1}{2}]} \quad (13)$$

$$\eta = \frac{2\chi e^2}{\varepsilon}. \quad (14)$$

The solid curve in Fig. 6 is obtained from Eq. (13) with $\eta = 1$ eV (this value of η can be achieved, for example, by taking $\varepsilon \sim 10$ and $x \sim .5$ Å^{-1}). It can be seen that with the exception of Ni-SiO$_2$, where there is some uncertainty in the composition determination due to the presence of NiO, the agreement is very good.

In summary, we have shown that the temperature and electric field dependences of the hopping conductivity in granular metal films can be explained in a unified framework in which the most crucial elements are (1) the existence of charging energy, (2) the constancy of s/d for a given x, and (3) generation of charge carriers by field-induced tunneling. Furthermore, the analysis of hopping conductivity data in terms of the theory has yielded values of the material parameters C and E_o useful for the characterization of granular metal films.

REFERENCES

1. B. Abeles, P. Sheng, M.D. Coutts and Y. Arié, Adv. Phys. 24, 407 (1975).
2. P. Sheng, B. Abeles, and Y. Arié, Phys. Rev. Lett. 31, 44 (1973).
3. P. Sheng and B. Abeles, Phys. Rev. Lett. 28, 34 (1972).
4. B. Abeles, Appl. Sol. St. Sci. 6, 1 (1976).
5. C.J. Gorter, Physica 17, 778 (1951);
   E. Darmois, J. Phys. Radium 17, 210 (1956);
   C.A. Neugebauer and M.B. Webb, J. Appl. Phys. 33, 74 (1962).
6. L.I. Schiff, Quantum Mechanics, 3rd Ed. (McGraw-Hill, N.Y., 1968), p. 278.
7. P. Sheng and B. Abeles, Thin Sol. Films 41, L39 (1977).
8. J.I. Gittleman, Y. Goldstein and S. Bozowski, Phys. Rev. B5, 3609 (1972).
9. J.S. Helman and B. Abeles, Phys. Rev. Lett. 37, 1429 (1976).
10. B. Abeles, H.L. Pinch and J.I. Gittleman, Phys. Rev. Lett. 35, 247 (1975).

DISCUSSION

D. M. GINSBERG (University of Illinois): Could anything be learned from measurements of Hall effect in these samples?

SHENG: There have been some measurements. Mobilities are very low so the measurements are very difficult.

J. LASS (The Technical University, Munich): In ferromagnetic materials, such as Ni, the negative magnetoresistance goes as the square of the magnetization. Does a similar law hold in granular metals?

SHENG: We have not tested our results for this.

W. L. McClean (Rutgers University): Is there some correlation between the metals that obey this theory and the electron-phonon interaction in the metal? The systems you have talked about are not superconducting. What about Sn, Pb and Aℓ for example?

SHENG: The theory is concerned only with interparticle tunneling and should be insensitive to the electron-phonon interaction in the metal. We have not sputtered Sn or Pb because of the reactivity of these metals.

B. ABELES (Exxon): There has been some work on $Aℓ-Aℓ_2O_3$ which indicates that $1/T^{1/2}$ behavior is followed.

MAGNETORESISTANCE IN INHOMOGENEOUS METALS

A. B. Pippard

Cavendish Laboratory, Cambridge

ABSTRACT

Pure metals at a low temperature have an exceedingly anisotropic conductivity tensor in high magnetic fields, and one which is commonly strongly dependent on the orientation of the field relative to the crystal axes. A polycrystalline sample is therefore a mixture of very different species, and a general account of what is involved in calculating the bulk properties of polycrystalline copper serves to illustrate the characteristic features that recur in studies of magnetoresistance. It has probably been taken for granted too readily in the past that carefully prepared single-crystal samples are homogeneous, and such puzzling observations as the linear magnetoresistance of potassium are more likely to arise from material defects than from any intrinsic property. In such a case as this, defect size and free path may be comparable, generating a peculiarly difficult set of problems. The aim of the talk will be to set the scene for more specialized contributions rather than go into the details of any one aspect.

DISCUSSION

J. WOOLHAM (N.A.S.A.): In graphite, there is a longstanding problem of explaining a linear magnetoresistance. Graphite is a compensated conductor having no open orbits or extended orbits. Is it safe to assume that the current jetting effects you have described cannot occur in compensated materials?

PIPPARD: No, I do not believe you can make that assumption. The transverse magnetoresistance of graphite is very high, whereas the resistance along the magnetic field is quite low. Therefore the scaling process -- which is the origin of the jetting phenomenon -- is definitely present. I believe it would be perfectly reasonable to attempt to explain the linear magnetoresistance of graphite by invoking a current jetting argument.

J. C. GARLAND (Ohio State Univ.): For very small inhomogeneities, the classical, local analysis breaks down and the magnetoresistance problem becomes very complicated. Would you care to comment on non-local effects?

PIPPARD: Should it turn out, when all the sums are done, that my calculations of the linear resistance of copper are wrong, then I shall certainly take refuge in size effects. For example, the measurements on copper by Fickett were taken on samples whose resistance ratio was about 7,000 with a corresponding mean free path of about 0.1-0.2mm. The diameter of his wire samples was about 1.5mm. In order to satisfy the local conditions, the size of the crystallites must be large compared to the mean free path. I do not believe Fickett's samples satisfied that criterion, so that there would have been corrections due to size effects and non-locality. The most obvious corrections would result from open orbits and extended orbits having their trajectories terminated by scattering off of grain boundaries. For motion transverse to the magnetic field, the appropriate length is the cyclotron radius whereas along the magnetic field it is the mean free path. I do not know what happens to the magnetoresistance when the scale of inhomogeneities is small compared to a mean free path. This is a very important problem which has so far been remarkably untouched.

P. L. TAYLOR (Case Western Res.): As a theorist, I always have difficulty with linear magnetoresistance. All the ingredients that go into the problem are symmetric in B, yet I know that were I to do a power series the answer would come out in B^2.

PIPPARD: You theorists believe in power series. That's the trouble, isn't it!

M. YAQUB (Ohio State Univ.): Is it possible under any conditions for a conductor to have a negative magnetoresistance? I have heard that there is a thermodynamic argument which asserts that in bulk materials a negative magnetoresistance is not possible.

PIPPARD: In light of what we have heard today, I an not prepared to commit myself to the idea that a negative magnetoresistance is impossible; obviously it is possible. However, I will say that if there is no quantization of carrier levels, and if the scattering cross section is not changed by the magnetic field, then the magnetoresistance is always positive. In the quantum limit in semiconductors, for example, these sorts of things can happen: in ordinary metals and under ordinary conditions a negative magnetoresistance would be extremely unusual.

CALCULATION OF THE MAGNETORESISTANCE OF POLYCRYSTALLINE METALS

H. Stachowiak
Institute for Low Temperature and Structure Research,
Polish Academy of Sciences, Wrocław

ABSTRACT

A self-consistent effective medium method of computing the effective conductivity tensor of polycrystalline metals in a magnetic field is presented. The method is applied subsequently to the study of the influence of three different factors 1) fluctuations of the electron density 2) open orbits 3) extended orbits on the galvanomagnetic properties of polycrystalline samples. Formulas obtained by the author are compared with those obtained earlier by Herring and later by Dreizin and Dykhne.

It is concluded that a prolonged linear growth of the magnetoresistance cannot follow from a single relaxation time application of the LAK theory. Some possibilities of obtaining such a linear growth are pointed out, like small angle scattering and broken orbits.

I. INTRODUCTION

The computation of the effective magnetoresistance of polycrystalline metals is a particularly challenging problem among those concerning the electrical transport in inhomogeneous media. There are two reasons for that 1) the electrical conductivity of the particular crystallites has a nontrivially tensorial character, 2) the conductivity tensor in many cases varies strongly from one crystallite to another[1]. In this paper a self-consistent effective medium method of solving such a problem is presented and applied subsequently to metals like copper and other noble metals which are uncompensated and have an open Fermi surface. Such metals in polycrystalline form are known to exhibit a linear growth of the transversal magnetoresistance at high magnetic fields[2-5]. There has been several attempts to attribute this effect to the character of the electronic orbits in these metals, particularly to extended orbits[6-8] (cf. also other papers[9-11]). In Sec. III this problem is considered in detail using the method presented in Sec. II.

A more detailed account of the results presented here can be found elsewhere[9-11,13-17].

II. METHOD

The approach we use has been applied earlier by Bruggeman[18] and Odelevsky[19]. Its purpose is to compute the effective conductivity of polycrystalline mixtures and it works under the two following assumptions: 1) every crystallite of the mixture has a spherical shape and 2) every crystallite behaves like being embedded in a homogeneous medium of conductivity $\tilde{\sigma}$ equal to the effective conductivity of the sample.

The first assumption seems unnatural, since no polycrystal can be built of spheres. Nevertheless, allowing for more complicated shapes of the crystallites one would need additional information about shape distribution. If no such information is supplied, it is quite reasonable to assume a spherical shape. If one has some additional information about shape distribution this assumption can be generalized to needles or disks by allowing for spheroids, oblate or prolate instead of spheres. The solution of the boundary problems arising then would be facilitated owing to the existence of a solution for a very general boundary problem[20].

The true meaning of the second assumption is more difficult to elucidate. Some considerations concerning it can be found in another paper of the author[14]. It seems to be related to the problem of the relative dimensions of the crystallites (cf. the arguments exposed elsewhere[11]).

As concerns a regularly built polycrytsalline metal, the method has been tested in several ways. Here the approach elaborated by Herring[21] and applicable when the properties of the particular crystallites do not differ much from each other has provided a particular help. If the conductivity tensors of different crystallites tend to their average value, the conductivity tensor obtained using the present method should behave according to the appropriate Herring formula. This is really so in cases investigated[9]. When conductivity fluctuations are not small, the method has been particularly well tested for scalar conductivity, including experimental tests[19]. Moreover the following problem was considered:

Let the polycrystal be a mixture of conducting and insulating crystals. How large should be the concentration of conducting crystals in order to make the polycrystal conductive? The answer of the present method is one half in the two dimensional case and one third in three dimensions. Similar answer are obtained from direct statistical calculations[13].

The test lacking concerned the asymptotic behaviour of the effective conductivity tensor in the case of large fluctuations of the local conductivity, though the results obtained in this field by the present method[9,14] were reasonable and easily understandable on physical grounds.

This test has been executed by Dreizin and Dykhne[12]. The asymptotic qualitative formulas obtained in their paper show full agreement with the results of the effective medium method[11].

Thus, let us consider a spherical crystallite with conductivity tensor $\underline{\sigma}$ embedded in a medium of conductivity $\underline{\tilde{\sigma}}$ which we consider to be the effective conductivity tensor of the polycrystal.

An electric field is applied which is homogeneous and equal $\vec{E}^o$ at large distance from the sphere. The electric field inside the sphere $\vec{E}(\underline{\sigma}, \underline{\tilde{\sigma}}, \vec{r})$ is obtained by solving the appropriate boundary problem. The principle of continuity of the potential leads to the equation

$$\langle \vec{E}(\underline{\sigma}, \underline{\tilde{\sigma}}, \vec{r})\rangle = \vec{E}^o \tag{1}$$

where averaging is over the whole polycrystal. The effective conductivity tensor is then computed by solving Eq. (1).

The boundary problem for the electric potential ψ is the following:

$$\text{div } \vec{J} = 0 \tag{2}$$

where $\vec{J}$ is the electrical current with components

$$-J_i = \sum_j \sigma_{ij} \frac{\partial \psi}{\partial x_j} \, . \tag{3}$$

The boundary condition at infinity is

$$\vec{\nabla}\psi = -\vec{E}^o \, . \tag{4}$$

The boundary conditions on the surface of the sphere into consideration follow from the continuity of the electric field and the normal component of the electric current. Assuming $\underline{\sigma}$ and $\underline{\tilde{\sigma}}$ in the simplified form:

$$\underline{\sigma} = \begin{pmatrix} \sigma_{11} & \sigma_{12} & 0 \\ \sigma_{12} & \sigma_{22} & 0 \\ 0 & 0 & \sigma_{33} \end{pmatrix} \tag{5}$$

we obtain the following solution:

$$E_x = \frac{\alpha_2 \, a}{\alpha_1 \alpha_2 + \beta^2} E_x^o - \frac{\beta Q}{\alpha_1 \alpha_2 + \beta^2} E_y^o \, ,$$

$$E_y = \frac{\beta Q}{\alpha_1 \alpha_2 + \beta^2} E_x^o + \frac{\alpha_1 Q}{\alpha_1 \alpha_2 + \beta^2} E_y^o \tag{6}$$

where

$$Q = \tilde{\sigma}_{11} Q_1^1 - (\tilde{\sigma}_{11} \tilde{\sigma}_{33})^{1/2} Q_1^{1'} \, ,$$
$$\alpha_1 = \sigma_{11} Q_1^1 - (\tilde{\sigma}_{11} \tilde{\sigma}_{33})^{1/2} Q_1^{1'} \, ,$$
$$\alpha_2 = \sigma_{22} Q_1^1 - (\tilde{\sigma}_{11} \tilde{\sigma}_{33})^{1/2} Q_1^{1'} \, ,$$
$$\beta = (\sigma_{12} - \tilde{\sigma}_{12}) Q_1^1 \, . \tag{7}$$

Q_1^1 and $Q_1^{1'}$ are expressed by Legendre functions of the second kind

$$Q_1(z) = \frac{z}{2} \ln\frac{z+1}{z-1} - 1 \quad , \quad Q_1^1(z) = (z^2-1)^{1/2}\frac{dQ_1(z)}{dz} \qquad (8)$$

namely

$$Q_1^1 = Q_1^1(ish\,\eta_0) \quad , \quad Q_1^{1'} = \frac{d}{d\eta}Q_1^1(ish\,\eta)|_{\eta_0}. \qquad (9)$$

η_0 is obtained from the formula

$$\eta_0 = Ar\,tanh\left(\frac{\tilde{\sigma}_{11}}{\tilde{\sigma}_{33}}\right)^{1/2}. \qquad (10)$$

The solution of the boundary problem for a quite general conductivity tensor $\underline{\sigma}$ is given in another paper[9]. On the other hand, for symmetry reasons, the effective conductivity tensor $\underline{\tilde{\sigma}}$ has always the form (5) with $\tilde{\sigma}_{22} = \tilde{\sigma}_{11}$, unless polycrystal as such is anisotropic. The solution outside the sphere has also been given[9], while the electric field inside the sphere is constant as can be seen from (6).

III. APPLICATIONS

Three cases are particularly interesting, and all of them can be described by conductivity tensors of the form (5). These are 1) the Herring[21] case of fluctuations of the electron density, 2) the case when open orbits are present, 3) the case when extended (elongated) orbits occur i.e. when the Fermi surface contains sheets having the form of crimped cylinders[6-8].

These cases will be considered one by one.

In (1) averaging is over all crystallites, i.e. over all directions of the crystal axes with regard to the magnetic field. These directions are described by three Euler angles. The first two angles define the direction of the magnetic field with regard to the crystal axes while the third angle corresponds to a rotation of the crystal around the direction of the magnetic field. The z axis is defined by the magnetic field. If the components σ_{13}, σ_{23}, σ_{31} and σ_{32} of the conductivity tensor vanish, $\underline{\sigma}$ can be presented in the form (5) if a suitable coordinate system is chosen. Then $\underline{\sigma}$ depends only on the first two Euler angles. Of course, in a polycrystal the "suitable" coordinate axes take all possible orientations. So the components E_x and E_y in (6) must be averaged first over all directions of the xy axes with respect to the projection of the external electric field on the xy plane. In this way we get from (1) and (6) the following equations for $\underline{\tilde{\sigma}}$:

$$\tfrac{1}{2}\left\langle \frac{\alpha_1+\alpha_2}{\alpha_1\alpha_2+\beta^2}\right\rangle = \frac{1}{Q},$$

$$\left\langle \frac{\beta}{\alpha_1\alpha_2+\beta^2}\right\rangle = 0, \quad (11)$$

where averaging is over the first two Euler angles.

For a free electron gas the conductivity tensor is of the form (5) with

$$\sigma_{11} = \sigma_{22} = \frac{\sigma_{33}}{1+\omega_0^2 H^2},$$

$$\sigma_{12} = -\sigma_{21} = \frac{\omega_0 H \sigma_{33}}{1+\omega_0^2 H^2}, \quad (12)$$

σ_{33} and ω_0 independent of the magnetic field.

1. FLUCTUATIONS OF THE ELECTRON DENSITY

This case consists in variations of the σ_{12} component from one crystallite to another and has been considered first by Herring. σ_{11}, σ_{22} and σ_{33} will be considered constant over the polycrystal with $\sigma_{11} = \sigma_{22}$.

Under these conditions Eqs. (11) can be solved quite generally if one makes the assumption that

$$\langle(\Delta d)^2\rangle \ll 1 \quad (13)$$

where

$$d = \frac{\sigma_{12} Q_1'}{\alpha_1} \quad (14)$$

and

$$\Delta d = d - \langle d \rangle. \quad (15)$$

One obtains neglecting terms of order higher than the second in Δd

$$\tilde{\sigma}_{12} = \langle \sigma_{12} \rangle \quad (16)$$

and

$$(\tilde{\sigma}_{11} - \sigma_{11})\frac{Q_1^1}{Q} = \langle (\Delta d)^2 \rangle \quad (17)$$

At high magnetic fields when η_0 is very small (cf.(12))

one gets

$$\tilde{\sigma}_{11} = \sigma_{11} + \frac{\pi}{4} \frac{\langle (\Delta \sigma_{12})^2 \rangle}{(\tilde{\sigma}_{11} \tilde{\sigma}_{33})^{1/2}}, \qquad (18)$$

This equation is very similar to the formula

$$\tilde{\sigma}_{11} = \sigma_{11} + \frac{\pi}{4} \frac{\langle (\Delta \sigma_{12})^2 \rangle}{(\langle \sigma_{11} \rangle \langle \sigma_{33} \rangle)^{1/2}} \qquad (19)$$

obtained by Herring[21] for a similar case. The only difference is in the denominator. Replacing $\langle \sigma_{11} \rangle$ in the denominator by $\tilde{\sigma}_{11}$ would be equivalent to adding terms of order higher than the second in $\langle (\Delta \sigma_{12})^2 \rangle$ and such terms have been excluded by Herring from the very beginning.

From (18) one gets

$$\tilde{\sigma}_{11} - \sigma_{11} \sim \frac{1}{H} \quad \text{for} \quad \tilde{\sigma}_{11} - \sigma_{11} \ll \sigma_{11},$$

$$\tilde{\sigma}_{11} - \sigma_{11} \sim \frac{1}{H^{4/3}} \quad \text{for} \quad \tilde{\sigma}_{11} - \sigma_{11} \gg \sigma_{11}.$$

$$(20)$$

The magnetoresistance R is obtained from the formula

$$R = \frac{\tilde{\sigma}_{11}}{\tilde{\sigma}_{11}^2 + \tilde{\sigma}_{12}^2}. \qquad (21)$$

It grows like H at the beginning and like $H^{2/3}$ for very high magnetic fields. A similar results has been obtained by Dreizin and Dykhne[12] using diffusion arguments.

2. POLYCRYSTALLINE METAL CONTAINING OPEN ORBITS

Let us assume that the sample consists of two kinds of crystallites. The crystallites of the first kind contain open orbits and occur in concentration k. Their conductivity tensor $\underline{\sigma}_1$ has the form (5). We put

$$\sigma_{33} = 1, \quad \sigma_{11} = \frac{1}{1+H^2}, \quad \sigma_{22} = (1-c)\sigma_{11} + c,$$

$$\sigma_{12} = \frac{H}{1+H^2} \qquad (22)$$

where c is a constant. These formulas define at the same time the units in which the magnetic field is measured. The crystallites of the second kind contain only closed orbits and their conductivity tensor $\underline{\sigma}_2$ is equal $\underline{\sigma}_1$ except for the component σ_{22} which in this case is equal to σ_{11} as given by (22).

The effective conductivity tensor $\tilde{\sigma}$ of such a material can be computed from Eq. (11). Since the component σ_{12} can be put equal σ_{12} we can write

$$\left(1 - \frac{k}{2}\right)\frac{1}{\alpha_1} + \frac{k}{2}\frac{1}{\alpha_2} = \frac{1}{Q} . \tag{23}$$

This equation can be written in the form

$$\tilde{\sigma}_{11} = \sigma_{11} + \frac{k}{2}(\sigma_{22} - \sigma_{11}) \frac{\tilde{\sigma}_{11} Q_1' - \tilde{\sigma}_{11}^{1/2} Q_1'}{\sigma_{22} Q_1' - \tilde{\sigma}_{11}^{1/2} Q_1''} .$$

The above equation has been solved numerically and the results are shown in Fig. 1 for k = 8% and different values of λ where

$$\lambda = \frac{\pi}{4} c . \tag{25}$$

It can be easily seen that the effective conductivity σ_{11} reaches saturation at much lower magnetic fields than the magnetoresistance. This suggests an asymptotic formula for the magnetoresistance

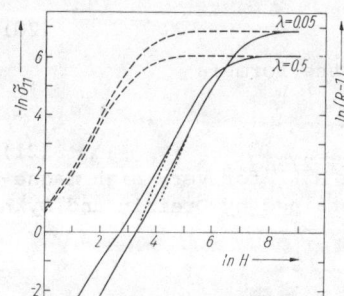

Fig.1. The curves $\ln(R-1)$ (full lines) and $-\ln \tilde{\sigma}_{11}$ (dashed lines) as functions of ln H in the presence of open orbits. The dotted lines were obtained by applying the asymptotic formula (26).

$$R(H) = \frac{\sigma_s}{\sigma_s^2 + \sigma_{12}^2(H)} \tag{26}$$

where σ_s is the saturation value of σ_{11}. In cases when σ_s is very small it obeys the formula

$$\sigma_s = \frac{\pi^2}{32} c^2 + \frac{k}{2} c$$
$$- \frac{\pi}{4} c \left(\frac{\pi^2}{64} c^2 + \frac{k}{2} c\right)^{1/2} . \tag{27}$$

How good this approximation is for different values of the parameters k and c can be seen from Fig.2.

Since some speculations have been made on the eventually linear increase of the magnetoresistance due to open orbits (cf. H. Stachowiak[9]), we show this effect in Fig.3 in some particularly favourable cases.

Expanding the square root in (27) into a Taylor series we get for very small c

$$\sigma_s = \frac{k}{2} c, \qquad (28a)$$

for very small k

$$\sigma_s = \frac{4k^2}{\pi^2}. \qquad (28b)$$

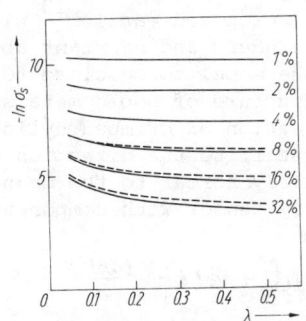 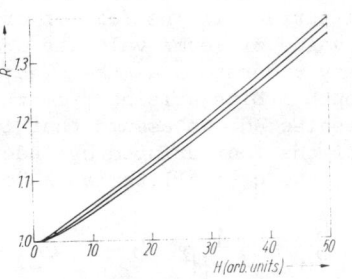

Fig.2. Comparison between asymptotic values of the conductivity σ_s obtained from (24) (full lines) and (27) (dashed lines) for different values of k.

Fig.3. The magnetoresistance R(H) for k=1% and λ =0.5 (upper curve), 0.3 (middle curve) and 0.2 (lower curve). The purpose of this figure is to demonstrate an eventual linear growth of the magnetoresistance.

A similar result was obtained by Dreizin and Dykhne and by Stroud[22].

The saturation of the magnetoresistance at high magnetic fields can be easily understood:

With increasing magnetic field σ_{11} decreases with respect to σ_{22}. The current flowing in the vicinity of crystals with open orbits is absorbed and flows inside of them in the direction of good conductivity (perpendicularly to the direction of the open orbits in k space). A further increase of the magnetic field leads to an increase in the z direction of the area from which the absorption occurs. This is possible owing to the high value of σ_{33}. At higher values of the magnetic field the area of absorption reaches other crystals with open orbits. At last σ_{11} becomes so small that in the xy plane the current flows uniquely in crystals with open orbits along the direction of easy conductivity, and the transition between such crystals takes place along the z axis. The conductivity of the circuit that settles in does not depend on the magnetic field. In such a way the magnetoresistance reaches saturation.

If the thickness of the sample in the z direction is too small, the circuit described above is not closed, because the current flowing in the z direction reaches the boundary of the sample before meeting the next crystal containing open orbits. This gives rise to specific size effects[15,12].

3. POLYCRYSTALLINE METAL CONTAINING EXTENDED ORBITS

We shall use as example the model proposed earlier[10] with Fermi surface in the form of crimped cylinders and constant absolute value of Fermi velocity. We wish the model to be close to reality understood as the electronic structure of noble metals. The open orbits arising from the intersection of crimped cylinders are neglected. We assume that if in a crystallite the direction of one of the four crimped cylinders is nearly normal to the magnetic field, the crystallite has a conductivity tensor with components

$$\sigma_{11} = \frac{1}{1+H^2}, \quad \sigma_{22} = (1-c)\sigma_{11} + c\left(1 - \frac{\tanh x}{x}\right) \tag{29}$$

Nothing particular happens with the σ_{12} component.
In the remaining crystallites σ_{10} has the free electron value. x is given by the formula[10]

$$x = \frac{\pi}{2} N \gamma_o \tag{30}$$

where

$$N = \frac{1}{\vartheta'} \frac{\Delta k}{k_o}, \quad \gamma_o = \frac{1}{H\tau}. \tag{31}$$

Here ϑ' is the angle between the cylinder axis and the direction perpendicular to the magnetic field;
Δk is the neck diameter and k_o the distance between opposite hexagonal faces of the Brillouin zone. τ is the relaxation time for small angle scattering as compared to the relaxation time equal to unity characteristic for ordinary resistivity. Only electrons in a thin layer in k space of thickness comparable to the neck diameter move along extended (elongated) orbits. Scattering even by a relatively small angle throws the electron out of the extended orbit region, while such a scattering hardly affects the contribution to conductivity of an electron on an ordinary closed orbit (cf. Pippard[23]). So, the effective relaxation time characteristic for electrons moving on an extended orbit is relatively short. c takes into account the fact that electrons can contribute either to extended or to closed orbit conductivities. We put

$$c = 1.5 \frac{\Delta k}{k_o} = 1.5\, dk. \tag{32}$$

The equation for $\tilde{\sigma}_{11}$ has the form

$$\tilde{\sigma}_{11} = \sigma_{11} + \frac{1}{2} \left\langle \frac{(\sigma_{22} - \sigma_{11})}{\alpha_2} \right\rangle Q \quad . \tag{33}$$

Taking into account that ϑ' cannot be larger than $\frac{\Delta k}{k_o}$ and that there are four cylinder axes we get

$$\tilde{\sigma}_{11} = \sigma_{11} + 2Q \int_0^{dk} d\vartheta' \frac{\sigma_{22} - \sigma_{11}}{\alpha_2} \quad . \tag{34}$$

Introducing (29) into (34) we can write this equation in the form

$$\tilde{\sigma}_{11} = \sigma_{11} + 2QcI \tag{35}$$

where

$$I = \int_0^{dk} \frac{d\vartheta'}{cQ_1' + \frac{S}{1 - \frac{\tanh x}{x}}} \quad , \tag{36}$$

$$S = \frac{1-c}{1+H^2} Q_1' - \tilde{\sigma}_{11}^{1/2} Q_1' \quad . \tag{37}$$

Since S is small for sufficiently high magnetic fields, it will play a role in I only for small values of $1 - \frac{\tanh x}{x}$, i.e. for small x. Expanding this expression into a Taylor series we get

$$1 - \frac{\tanh x}{x} \approx \frac{x^2}{3} \quad . \tag{38}$$

At the end we obtain

$$I = \frac{\pi}{2} \frac{dk}{H\tau \sqrt{3Sc\, Q_1^1}} \arctan\left(\frac{2}{\pi} H\tau \sqrt{\frac{3S}{cQ_1^1}}\right) \quad . \tag{39}$$

At high magnetic fields, when

$$S \approx Q \approx 2\,\sigma_{11}^{1/2} \tag{40}$$

we get the asymptotic formula

$$\tilde{\sigma}_{11} \approx \left(\frac{\pi}{2} dk\right)^2 \left(\frac{2}{H\tau}\right)^{4/3} \tag{41}$$

in agreement with the qualitative result of Dreizin and Dykhne.

The numerical solution of Eq.(35) presents no particular problem. The results of the computations for different values of τ are presented in Fig.4. Since dk takes the values 0.18 for copper 0.123 for silver and 0.159 for gold, we show results for dk = 0.159. It is easily seen that the magnetoresistance follows rat her closely the expected behaviour $R \sim H^{2/3}$. As concerns the famous problem of the linear growth of the magnetoresistance, it can be seen from Fig.5 that at least for τ as small as 0.1 a linear behaviour is observed.

Rahman Saad[16] has performed calculations of the magnetoresistance of polycrystalline copper allowing for the simultaneous presence of open and extended orbits. He used experimental parameters following from the measurements of Klauder et al.[24]. He obtained a

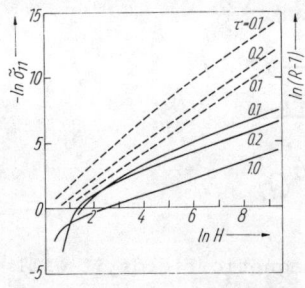

Fig.4. The functions $\ln(R-\frac{1}{\tau_0})$ (full lines) and $-\ln \sigma_{11}$ (dashed lines) in the presence of extended orbits.

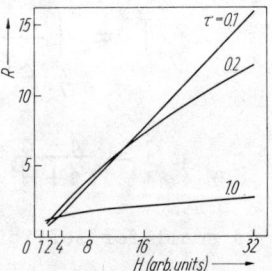

Fig.5. The magnetoresistance as a function of the magnetic field.

linear increase of the effective magnetoresistance at much lower values of τ than for extended orbits alone. These results, however, can only be considered as preliminary.

IV. CONCLUSIONS

The purpose of this paper was not to explain the linear growth of the magnetoresistance of polycrystalline noble metals observed experimentally, but to elaborate a method of computing the effective magnetoresistance starting from the knowledge of the properties of the particular crystallites. However it is visible that this effect is not a direct consequence of the LAK theory.

Looking around for a simple physical argument leading to a linear increase of the magnetoresistance we arrived to the problem of broken orbits[9]. Such orbits arise if scattering takes place mainly in particular regions of the Fermi surface. This leads to a relaxation time proportional to the inverse of the magnetic field and to a linear increase of the magnetoresistance. Such an effect should be observed even in single crystals. However tensions and deformations of the lattice occuring in polycrystals could lead to different conductivities of the crystallites than for isolated single crystals. Ideas supporting this point of view have been advanced by several authors[25-28].

REFERENCES

1. I.M. Lifshits, M. Ya. Azbel, M. I. Kaganov, Zh. Eksper. Teor. Fiz. 31, 63 (1956).
2. P. L. Kapitza, Proc. Roy. Soc. A123, 292 (1929).
3. B. Lüthi, Helv. Phys. Acta 33, 161 (1960).
4. N. E. Alekseyeysky, Yu. P. Gaidukov, Zh. Eksper. Teor. Fiz. 37, 672 (1959).
5. P. M. Martin, J.B. Sampsell, J. C. Garland, private communication.
6. J. M. Ziman, Phil. Mag. 3, 1117 (1958).
7. I. M. Lifshits, V. G. Pe-schansky, Zh. Eksper. Teor. Fiz. 35, 1251 (1958).
8. I. M. Lifshits, V. G. Peschansky, Zh. Eksper. Teor. Fiz. 38, 188 (1960).
9. H. Stachowiak, Physica (Utrecht) 45, 481 (1970).
10. H. Stachowiak, Acta phys. Polon. A40, 849 (1972).
11. H. Stachowiak, Phys. Stat. Sol.(a) 20, 707 (1973).
12. Yu. A. Dreizin, A. M. Dykhne, Zh. Eksper. Teor. Fiz. 63, 242 (1972).
13. H. Stachowiak, Acta Phys. Polon. 24, 749 (1963); 25, 211 (1964).
14. H. Stachowiak, Bull. Acad. Polon. Sci. sér. math. phys. astr. 15, 631, 637 (1967).
15. H. Stachowiak, Efektywne przewodnictwo elektryczne mieszanin polikrystalicznych oraz metali polikrystalicznych w polu magnetycznym (Institute for Low Temperature and Structure Research, Polish Academy of Sciences, Wrocław 1968, unpublished dissertation).
16. Abd el Rahman Ali Saad, Ph. D. Thesis, University of Wrocław 1976 (in English, unpublished).
17. H. Stachowiak, in Fizyka i Chemia Ciała Stałego. Wybrane Zagadnienia, ed. B. Staliński (Ossolineum, Wrocław 1977).
18. D. A. G. Bruggeman, Ann. Phys.(5), 24, 636 (1935).
19. V. I. Odelevsky, Zh. Tekhn. Fiz. 21, 667, 678, 1379 (1951).
20. V. Frank, Symp. Electromagnetic Theory and Antennas, Copenhagen 1962 (Pergamon Press 1963) p. 615.

21. C. Herring, J. Appl. Phys. 31, 1939 (1960).
22. D. Stroud, Phys. Rev. $\underline{B12}$, 3368 (1975).
23. A. B. Pippard, Proc. Roy. Soc. A305, 291 (1968).
24. J. R. Klauder, W. A. Reed, G. F. Brennert, J. E. Kunzler, Phys. Rev. $\underline{141}$, 592 (1966).
25. L. M. Falicov, H. Smith, Phys. Rev. Lett. $\underline{29}$, 124 (1972).
26. V. A. Gasparov, Proceedings of the 18th Soviet Low Temperature Conference, Kiev 1974, (in Russian) p. 349.
27. J. C. Garland et al., Phys. Rev. $\underline{B9}$, 1987 (1974).
28. T. Amundsen, P. Jerstad, private communication.

POTASSIUM: ARE THE MAGNETORESISTANCE ANOMALIES DUE TO INHOMOGENEITIES

R. S. Newrock and P. J. Tausch
University of Cincinnati, Cincinnati, Ohio 45221

ABSTRACT

The anomalous magnetoresistance of the simple metals is considered to be one of the outstanding unsolved problems in metal physics. These metals, of which potassium is the archetype, have a transverse electrical magnetoresistance which increases linearly with the applied magnetic field. This is in strong disagreement with the semiclassical theory of transport in metals, which, for these metals, predicts a saturating (field independent) electrical magnetoresistivity. Many different approaches have been investigated in an attempt to explain the linear magnetoresistivity, including both intrinsic and extrinsic theories. It has recently been shown that a linear electrical magnetoresistivity of the correct order of magnitude can be obtained by considering the conductivity to be inhomogeneous. These ideas are examined, insofar as they apply to potassium, in the light of the transverse and longitudinal electrical magnetoresistivity data and in the light of recent measurements of the other transport coefficients of potassium (the Hall and Righi-Leduc coefficients and the transverse and longitudinal thermal magnetoresistivity). We conclude that inhomogeneous conduction is not a likely cause of the magnetoresistance anomalies.

INTRODUCTION

The electrical magnetoresistance of pure, simple metals has seldom, if ever, shown the magnetic field dependence predicted by semiclassical transport theory.[1] In particular, the transverse electrical magnetoresistivity, which should saturate in strong fields ($\omega_c \tau \gg 1$) is instead linear in the magnetic field.[2] A variety of theoretical explanations[3] have been proposed but the issue is still very much unresolved and the problem of the electrical magnetoresistance of the simple metals has been called one of the most important unsolved problems in metals physics. Recent measurements[4,5] of the transverse thermal magnetoresistivity of one of the simple metals, potassium, show a still greater deviation from theory. It has been definitely shown that the problem is real - that is, it is not due to sample geometry and probe effects or to other problems associated with measuring techniques. Non-saturation has been observed in single and polycrystals, strained and "unstrained" specimens and by probe and probeless techniques. The problem is quite important; if the effects are indeed intrinsic, it is entirely within the realm of possibility that some fundamental process is being overlooked by the usual semi-classical theories.

It has recently been suggested that macroscopic sample inhomogeneities could be responsible for the magnetoresistance anomalies[6] and, in this paper, we wish to consider that possibility; i.e., the possibility that voids, inclusions and other macroscopic sample defects are responsible for the magnetoresistance anomalies of the simple metals. In this paper we will concentrate on potassium as it is the only simple metal in which all of the transport coefficients have been measured in very strong magnetic fields (up to 10 Tesla). In the next section we briefly review the results of the semi-classical theory and discuss the various experimental results. This is followed by a discussion of the contribution of sample inhomogeneities to the transport coefficients, and finally, in the last section we compare the field and purity dependence of each transport coefficient with the expected behavior to determine if inhomogeneities have an important effect on the transport properties of potassium.

THE SEMICLASSICAL THEORY AND THE EXPERIMENTAL RESULTS

The semi-classical theory of the magnetoresistivity of metals was derived by Lifshitz, Azbel and Kaganov[1], (LAK) in 1957. In solving the Boltzmann equation, they expand the deviation function in powers of $1/H$, and make the usual assumptions of semi-classical theory (semi-classical equations of motion, no field-induced interband transitions, etc.). By neglecting interband transitions, the band structure of the metal only enters via the topology of the cyclotron orbits. The type of orbit determines the leading term in the expansion and it is then possible to derive the field dependence of the various magnetotransport coefficients. These are shown in Table I. Note that these field dependences hold regardless of the number or type of electron scattering mechanisms present and that it does not depend on the existance of a well-defined relaxation time.

Table I Theoretical field dependence of the transport coefficients

Transport Coefficient $H \parallel z, J \parallel x$	Theory Compensated	Uncompensated Open	Closed	Experiment
ρ_{xx}	H^2	H^2	H^o	$\approx \rho_o + \Sigma_T H$
ρ_{zz}	H^o	H^o	H^o	$\approx \rho_o + \Sigma_L H^n$ $1 < n \leq 2$
$\rho_{xy} = R_H H$	no simple result	no simple result	$\frac{1}{ne}$	$R_H > \frac{1}{ne}$
$W_{xx} T$	H^2	H^2	H^o	$WT \approx W_o T + AH + BH^2$
$W_{zz} T$	H^o	H^o	H^o	$WT \approx W_o T + CH$
$W_{xy} T = R_{RL} TH$	no simple result	no simple result	$\frac{1}{L_o Tne}$	$R_{RL} \stackrel{<}{\sim} \frac{1}{L_o Tne}$

The field dependence of the various transport coefficients is shown for three cases: for an uncompensated metal with closed and with open orbits (in k-space perpendicular to the direction of the field), and for a compensated metal. We are interested in the simple metals; that is those metals which are uncompensated and only permit closed orbits perpendicular to the field. In particular, we are interested in potassium, an uncompensated metal with a Fermi surface known to be spherical[7] to a few parts in 10^4. Thus, for $\omega_c \tau \gg 1$, its transport coefficients should have the magnetic field dependences shown in the third column of Table I: Both the electrical and thermal longitudinal and transverse magnetoresistivities should saturate (become field independent) and both off-diagonal terms (the Hall coefficient and the Righi-Leduc (thermal Hall) coefficient) should be field and temperature independent and equal to their free electron values. (That is, independent of scattering and only dependent upon the electron density, a bulk property of the material). We also note that the longitudinal electrical and thermal magnetoresistivities should saturate regardless of the topology of the Fermi surface or the state of compensation.

These are the theoretical results to be compared with the experimental results, shown in the last column of Table I. Extensive measurements of the transverse electrical magnetoresistivity have shown it to be linear in the applied field to values of $\omega_c \tau_{el}$ in excess of 350 and fields in excess of 10T. The theory of course predicts strict saturation. The Kohler slope $S_{E,T}$ of the linear term is extremely purity dependent and has very little temperature dependence. (The Kohler slope is defined by $\frac{\Delta \rho}{\rho(H=0)} = S \omega_0 \tau$, where the τ appropriate to the experiment is used). The behavior of the electrical magnetoresistivity is very erratic for the first 1-2 Tesla of applied field, and strains and dislocations are known to have a very large effect on the magnetoresistivity. The anomalous electrical magnetoresistance is large; $\frac{\Delta \rho (H,T)}{\rho(0,T)}$ can be as large as 4 or 5 at 10 Tesla.

Compared to the large number in the transverse case, relatively few measurements have been performed on the longitudinal electrical magnetoresistance, and the results are not as clear-cut. The measurements extend to fields of about 4 Tesla and $\rho_{zz}(H,T)$ increases as H^n, where $1 \leq n \leq 2$ depending on the sample and the experiment[8,9]. When the data is linear the magnitude of the Kohler slope $S_{E,L}$ is comparable to the transverse slope. The last electrical coefficient to be considered is the Hall effect. The Hall coefficient is found to be field and temperature independent as predicted. However, recent measurements by three independent groups[10-12], using three distinctly different measurement techniques, all with a claimed accuracy of 3% or better, show the Hall coefficient to be 4-6% too large; that is, $|R_H|$ is greater than $1/ne$. One of these experiments[10] also indicates that the Hall coefficient may depend on crystal orientation.

Thus, as seen in the last column of Table I, none of the three electrical transport coefficients completely obeys the LAK theory. Table II summarizes the pertinent experimental facts for the various electrical transport coefficients.

Table II High field electrical transport coefficients

Transverse magnetoresistivity: $\rho_{xx}(H,T) = \rho(0,T) + S_{E,T}(\omega_c \tau_{el})$

$\rho_{xx}(H,T)$:   Wildly varying at low fields, including knees and plateaus.
                      Linear in field to at least $\omega_c \tau_{el} \sim 350$.
                      Does not cycle with temperature, essentially irreproducible.

$S_{E,T}$: 10^{-2} to 10^{-4}
                      Depends strongly on purity (RRR), stress, strain, pressure.
                      Very small temperature dependence.

Longitudinal magnetoresistivity: $\rho_{zz}(H,T) = \rho(0,T) + S_{E,L}(\omega_c \tau_{el})^n$; $1 \leq n < 2$.
$\rho_{zz}(H,T)$:   Similar to above, but limited data.
$S_{E,L}$:   Same as above (when linear).
Hall coefficient: $\rho_{xy}(H,T) = R_H H$

R_H: Independent of field and temperature.
            Larger than free electron value by 4 to 6%.
            Some orientation dependence.

We again note that the longitudinal magnetoresistivity should always saturate and that the Hall coefficient should be $\equiv \frac{1}{ne}$, where n, the electron density, is a property of the bulk material, that is, R_H does not depend on scattering. Thus, the observed small difference from the free electron Hall coefficient is very important. We note that this is not the only time problems have been noted with parameters determined from bulk properties of potassium. The Fermi surface parameters determined from measurements of the lattice parameter and those measured by de Haas-Van Alphen have long been known to differ by an amount outside experimental error and the pressure dependence of the de Haas-Van Alphen frequency varies considerably from the expected results[13].

Turning now to the thermal data, we find that the transverse thermal magnetoresistivity shows no sign of saturation to magnetic

fields in excess of 9 Tesla and values of $\omega_c \tau_{th}$ of 300 or more[4,5].
The transverse thermal magnetoresistivity is found to be essentially
a quadratic function of the applied field with a "small" linear term
present. The effects are huge; at 9T, $\Delta W(H,T)/W(0,T)$ can be as
large as 150. The transverse thermal magnetoresistance is strongly
temperature dependent and only very weakly purity dependent; its behavior is monotonic; it increases with field showing none of the erratic behavior so evident in the transverse electrical case. The
Kohler slope of the linear term $S_{T,T}$, is about four times as large
as S_{ET}.

We have recently obtained some longitudinal thermal magnetoresistance data. Although the data should be regarded as being <u>very</u>
preliminary, we have observed (Fig. 1) a longitudinal thermal magnetoresistivity that is linear in field to 9 Tesla ($\omega_c \tau \sim 150$). No
quadratic term appears. The Kohler slope of the linear term depends
on purity in a manner similar to that of the electrical case, although, as in the transverse case, the slopes are larger.

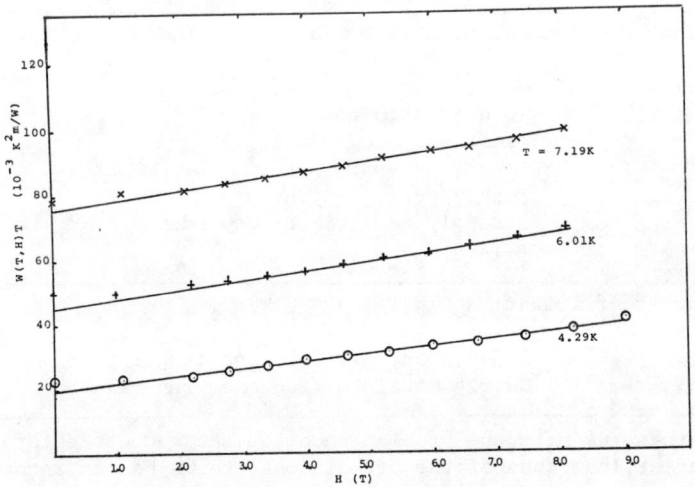

Fig. 1. The longitudinal thermal magnetoresistivity times temperature $W_{zz}(H,T)T$ as a function of the applied field.

Finally we have the Righi-Leduc or thermal Hall coefficient. As the LAK theory predicts, it is essentially independent of field and temperature but might be 3-5% less than the free electron value[14] (note however, that the geometry error was of that order).

The results of the thermal experiments are summarized in Table III.

Table III High-field electrical transport coefficients

Transverse magnetoresistivity: $W_{xx}(H,T)T = W(0,T)T + S_{T,T}(\omega_c \tau_{th}) + B(\omega_c \tau_{th})^2$	
$W_{xx}(H,T)T$:	No erratic low field behavior. Quadratic in field for large H. Reasonably reproducible from sample to sample.
$S_{T,T}$	Strongly temperature dependent ($\sim T^3$). Weakly purity dependent. Zero temperature value 4-5 times $S_{E,T}$.
B	Strongly temperature dependent. No discernable purity dependence.
Longitudinal magnetoresistivity: $W_{zz}(H,T)T = W(0,T)T + S_{T,L}(\omega_c \tau_{th})$	
$W_{zz}T$:	No erratic low field behavior. No quadratic term. Reasonably reproducible.
$S_{T,L}$:	Weakly purity dependent. Strongly temperature dependent. $\approx (4-5)S_{E,L}$
Righi-Leduc coefficient: $W_{xy}(H,T)T = R_{RL} T H$.	
TR_{RL}:	Independent of field and temperature. Nearly equal to free electron value.

In brief then, we have the problem: none of the magnetoresistances saturate and at least one of the off-diagonal terms has an incorrect magnitude.

VOIDS

Noting that the transverse electrical magnetoresistance of potassium and the other simple metals varies unpredictably with sample fabrication techniques, several authors have suggested that sample inhomogeneities may be the cause of the anomaly. Since the work of Herring[15] in 1960, it has been known that macroscopic sample inhomogeneities can have large effects on the magnetotransport properties of metals and semi-conductors; in fact, the distortions in the current streamlines created by such inhomogeneities may be shown to propagate a distance $\omega_c \tau R_o$ parallel to the field, where R_o is of the order of the size of the inclusion. The distortions create an excess concentration of current in the vicinity of the inhomogeneity which leads to a linear electrical magnetoresistance.[6]

We consider an inhomogeneity to be a macroscopic region where the local conductivity differs from that of the rest of the material. Such inhomogeneities could be caused by, for example, grain boundaries, crystallite orientation, local strain fields, dislocations, voids and inclusions. We are interested in macroscopic defects and for discussion purposes we consider two sizes: those inhomogeneities with linear dimensions greater than an electron mean-free-path, and those with linear dimensions considerably less than the mean-free-path, but still of a size sufficiently large so that ordinary electron scattering ideas are not applicable.(Presumably microscopic defects may be simply treated by the usual scattering techniques and there is mounting evidence that electron scattering mechanisms are not the cause of the anomalies[4,14,16]). Very little, if any, theoretical work has been performed for the latter case; a number of calculations have been done for the former. In that case, when $R_o > \lambda$ (the mean-free-path) the methods of continuum physics are applicable and the problem has been solved by many authors[6]. A particularly illuminating case is that of a single spherical void (that is, an area of zero conductivity) in a uniform, free-electron medium. The problem has been solved by Garland and Sampsell[6] and Stroud and Pan[17] among others. Fig. 2 shows some of their results; note the manner in which the current streamlines move around the void; in particular, note the current sheets formed at the edges of the "shadow" of the void. To see how these current sheets affect the dissipation, Fig. 3 shows $\vec{J} \cdot \vec{E}$ determined on a line perpendicular to the field passing near the void for various values of $\omega_c \tau$. Note the large increase in the dissipation at the edges of the void "shadow" where the current sheets are located. Finally, by integrating $\vec{J} \cdot \vec{E}$ over the volume of the specimen it is possible to calculate the magnetoresistivity, and,(Fig. 4) it may be shown that both the transverse and longitudinal electrical magnetoresistivity are linear in the applied field. The Kohler slope of the linear term may be obtained: $S=Af$, where A is a constant of order one and f is the volume fraction of voids. Stroud and Pan[17] also calculated the Hall coefficient and find a 0.1% change from the free-electron result; others find a Hall coefficient increasing roughly as f.

What does all this have to do with real metals? For purposes of this discussion, we adopt the following viewpoint: calculations such as these show that inhomogeneities can create magnetoresistance effects of the type necessary to explain some of the data. That is, small inhomogeneities produce long range fluctuations in the electric potential; these lead to a linear electrical magnetoresistance. We have reviewed many of the recent papers and observed that there are several conclusions common to the various calculations. It appears that, for macroscopic defects whose linear dimensions are greater than the mean-free-path of an electron,

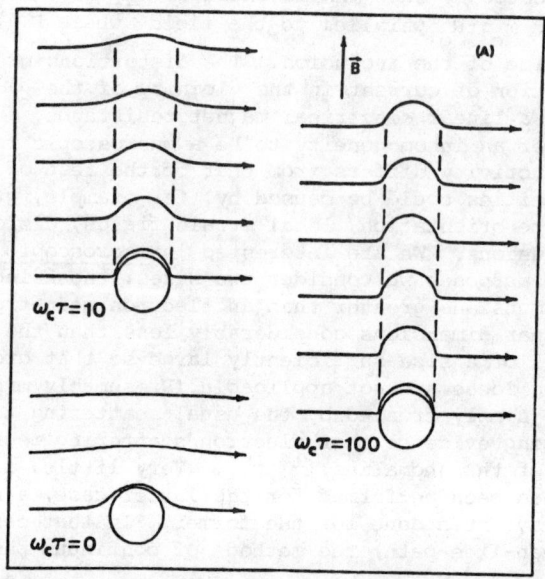

Fig. 2. Projection on the x-y plane of current lines (injected uniformly at $+\infty$) near a void, for different values of $\omega_c \tau$. (From Ref. 6)

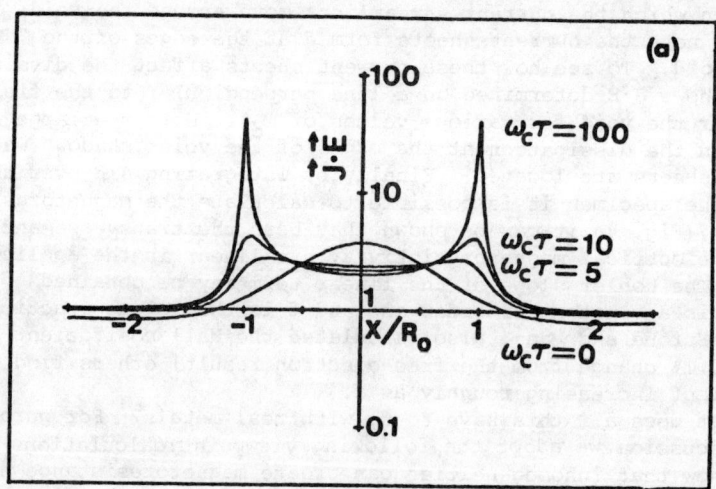

Fig. 3. Volume power density $\vec{J} \cdot \vec{E}$ for the current distribution in Fig. 1 plotted along a line parallel to the X-axis which touches the cylinder at $Z = R_o$. (From Ref. 6)

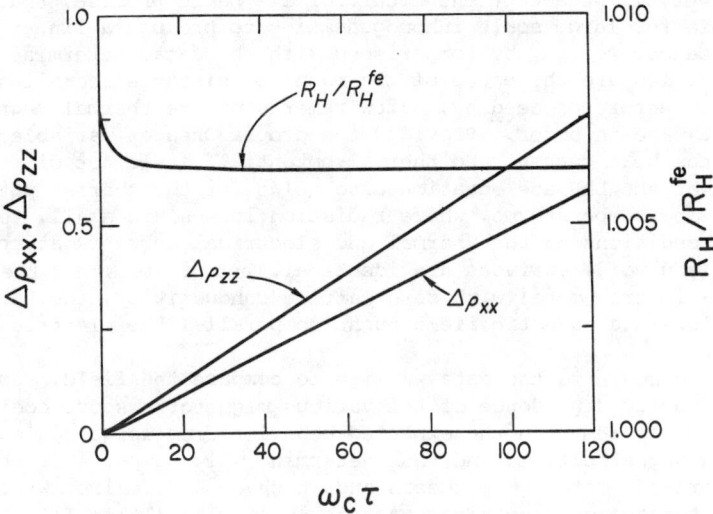

Fig. 4. The change in the transverse and longitudinal electrical magnetoresistivity due to the presence of voids and the Hall coefficient. (After Ref. 17)

    1. any spheroidal defect (and probably any shaped defect) leads to a linear transverse electrical magnetoresistance if, in the defect, either
        a. σ_{xx} and σ_{yy} are field independent, or
        b. $\sigma_{xy} \propto 1/H$,
    2. only a defect of strictly zero conductivity leads to a strictly linear longitudinal electrical magnetoresistance, otherwise it eventually saturates,
    3. the Kohler slope of the linear term is a constant of order one times the volume fraction of voids, and
    4. The Hall coefficient is essentially unchanged (it might increase about 1/2%).
In addition, Garland and Sampsell[6] point out that the contribution of surface defects, which may be shown to increase the linear term of both electrical magnetoresistances will eventually saturate in the longitudinal case. We also note the calculations assume the "shadows" of the defects don't overlap. The various calculations are essentially local in nature; defects which have linear dimensions less than the mean-free-path create an inherently non-local problem and careful calculations are necessary; to our knowledge this hasn't been done. It is likely however, that current distortions created by smaller voids and defects will be similar provided the void size is larger than a cyclotron radius.

However, if we accept the model for its basic premise, that it is possible for large scale inhomogeneities to produce a linear magnetoresistance, we can, by comparisons with the data, determine if inhomogeneities are the cause of the magnetoresistance anomalies in potassium. Before proceeding, a few remarks on the thermal magnetoresistivity are in order. Provided the proper Onsager variables are used for the heat current and thermal potential, it is not difficult to show that the LaPlace equation also holds for the thermal potential. At low temperatures, where radiation losses are small, the boundary conditions on the thermal and electrical currents at the specimen (and void) surfaces are identical. Thus, at least for the case of a free-electron metal with zero lattice conductivity, the thermal current flow in a magnetic field ought to parallel the electric current flow.

Turning again to the data we wish to compare the field, temperature and purity dependence of the various magnetotransport coefficients of potassium with the expected behavior from large scale defects in a magnetic field and thus determine which aspects of the data support the defect hypothesis and which do not. First we consider the transverse electrical magnetoresistivity (Table II). If we apply the results above, a Kohler slope of 10^{-2} implies a volume fraction of voids (or perfectly conducting inclusions) of about 1%. Not only is such a large concentration of voids unlikely, but Schouten and Swenson[18] have measured the lattice parameter of pure potassium at 4K using a volumetric technique and they find a lattice parameter that agrees with x-ray determinations to about a part in 10^3; that is, no large-scale finite volume voids are present. But the other properties noted in Table II are consistent with a defect theory - especially the wildly unpredictable low-field behavior, the strong effects of strain, pressure, etc., and the lack of a strong temperature dependence.

Considering the longitudinal electrical magnetoresistivity, the fact that the slope of the linear term has the same magnitude as the slope of the transverse resistivity supports a defect hypothesis. However, no real conclusions can be based on this data until the differences between the experiments are resolved and the actual power of the field dependence determined. Note that if it is indeed strictly linear, it eliminates the possibility of highly-conducting large-scale inclusions.

Lastly, we have the Hall coefficient. None of the void or inclusion theories can account for the roughly 5% difference between the measured value and the free electron value: the maximum change in R_H calculated by various authors appears to be about 1/2 to 1%. It is remotely conceivable that some form of small-scale defect might cause such an effect. The orientation dependence of R_H is not inconsistent with a defect theory, particularly if the defects are assymetrically shaped and oriented along certain crystallagraphic directions.

Thus portions of the electrical data are explainable by a defect model but it appears that large voids are definitely ruled out by the Schouten-Swenson work, and highly conducting inclusions are improbable because of the non-saturating longitudinal magnetoresistance. The electrical data do not really allow us to reach any conclusions about small-scale inhomogeneities, however their total volume fraction must be less than 0.1%.

Turning to the thermal data, as discussed the thermal and electrical currents should flow in a similar manner; thus if a defect is of sufficient size (so that we are not considering a scattering problem) some sort of Wiedemann-Franz law should hold and the defect ought to affect the thermal and electrical resistivities in the same manner.

We note from Fig. 1 that the longitudinal data is strictly linear over a very wide field range. The observed magnitude of the Kohler slope would require a larger density of voids than in the electrical case and the strictly linear field dependence and lack of any sign of saturation implies there are no conducting inclusions.

However, it is when we consider the transverse thermal magnetoresistivity where we observe the greatest difficulties with the inhomogeneity hypothesis. The semi-classical theory predicts saturation and large scale inhomogeneities should produce a linear field dependence; the data are quadratic in the applied field.

Before continuing, it should be pointed out that a non-zero lattice conductivity can create a thermal magnetoresistivity with a quadratic field dependence[19]. Lattice conduction also results in a field dependence to the Righi-Leduc coefficient, and by comparing thermal magnetoresistivity data with Righi-Leduc data taken on the same specimen it is possible to extract a value for the lattice conductivity and use it to correct the measured thermal magnetoresistivity and obtain the electronic contribution to the thermal magnetoresistivity. It was found that the contribution of the term due to the lattice conduction was far too small to account for all of the quadratic field dependence of the thermal magnetoresistivity.

Since large scale inhomogeneities should influence the thermal and electrical magnetoresistances in a similar manner, the transverse thermal data appear to make any theory based on large-scale inhomogeneities untenable; additionally, unless small-scale inhomogeneities can be shown to affect electrical currents differently from thermal currents, it would appear that small-scale inhomogeneities are also an unlikely cause of the magnetoresistance anomalies of potassium.

In addition, we note that the temperature dependent portion of the Kohler slope and the coefficient of the quadratic term (B) of the transverse thermal magnetoresistivity are, within the geometry errors, essentially independent of purity (as measured by the RRR)[4,16]. None of the erratic low field behavior that is so evident in the electrical case is observed. This may be taken as a further indication that inhomogeneities are not the prime cause of the anomalies.

One possible problem must be noted: it is not clear what the full effect of the lattice conductivity is on the flow of thermal current near a void; we are currently investigating that problem to see if the quadratic term might be considerably enhanced. We do not think that the effects will be large.

The conclusion is straight-forward: inhomogeneities, large or small, cannot be responsible for the magnetoresistance anomaly in potassium. In sum, the main reasons for reaching that conclusion are:

1. the data of Schouten and Swenson,
2. the quadratic field dependence of the transverse thermal magnetoresistivity,
3. the overly large Hall coefficient,
4. the non-saturating, strictly linear longitudinal magneto-resistances, and
5. the assumption of similar behavior for large and small inhomogeneities.

It is conceivable that small scale voids (perhaps several hundreds of missing atoms) might exist without significantly altering the volume of the metal. It is not clear what effect such defects would have; we are currently investigating that problem, both theoretically and experimentally.

REFERENCES

1. I. M. Lifshitz, M. Ya. Azbel, and M. I. Kaganov. Zh. Eksp. Teor. Fig. 30, 200 (1955), (Sov. Phys. - JETP 3 143 (1956)) and Zh. Eksp. Teor. Fig. 31, 63 (1956), (Sov. Phys. - JETP 4, 41 (1957)).
2. For a list of pertinent references see H. Taub, R. L. Schmidt, B. W. Maxfield and R. Bowers, Phys. Rev. B4, 1134 (1971).
3. For a list of pertinent references see references 2, 4, 16.
4. R. S. Newrock and B. W. Maxfield, J. Low Temp. Phys. 23 119, (1976).
5. R. Fletcher, Phy. Rev. Lett. 32 930, (1974).
6. For a list of work in this area see J. B. Sampsell and J. C. Garland, Phy. Rev. B13 583 (1976).
7. D. Shoenberg and P. J. Stiles, Proc. Roy. Soc. A281, 62 (1964).
8. A. M. Simpson, J. Phys. F3, 1471 (1973).
9. J. S. Lass, J. Phys. C3, 1976 (1970).
10. D. E. Chimenti and B. W. Maxfield, Phys. Rev. B7 3501 (1973).
11. P. A. Penz and T. Koshida, Phys. Rev. 176, 804 (1968).
12. S. A. Werner, K. Hunt and G. W. Ford, Sol. St. Comm. 14, 1212 (1974).
13. Z. Attounian, C. Verge and W. R. Datars, to be published.
14. P. J. Tausch and R. S. Newrock, to be published, Phy. Rev. B, Aug. 1977.
15. C. Herring, J. Appl. Phys. 31, 1939 (1960).
16. P. J. Tausch and R. S. Newrock, to be published.
17. D. Stroud and F. P. Pan, Phys. Rev. B13, 1434 (1976).
18. D. R. Schouten and C. A. Swenson, Phys. Rev. B10, 2175 (1974).
19. For a discussion of this point see Ref. 14.

DISCUSSION

J. LASS (Technical Univ. Munich): I would like to comment on the longitudinal magnetoresistance. The data you showed was taken when I was a research student in 1966; I should have left out of my dissertation those curves which showed a non-linear longitudinal magnetoresistance since I found them subsequently to be highly non-reproducible. On the other hand, those curves showing a linear magnetoresistance with a slope slightly greater than the transverse slope were taken somewhat later and are reproducible.

NEWROCK: I would like to point out that Simpson has also reported a quadratic longitudinal magnetoresistance for potassium.

LASS: That is another kettle of fish which I would also criticize.

A. B. PIPPARD (Cavendish Lab.): I wasn't happy with the easy way you dismissed the lattice conductivity. It seems to me that all the other discrepancies you have discussed between the electrical and thermal conductivity are irreproducible and, on the whole, rather small. But in the transverse magnetoconductance there is an enormous discrepancy between the electrical and thermal behavior. And this is a reproducible discrepancy that must have a genuine explanation quite different from the others. The lattice conductivity can explain a quadratic magnetoresistance and therefore should be very carefully examined. However, we do not know what the magnitude of the lattice thermal conductivity ought to be inasmuch as all theories of the thermal conductivity of metals are quite inexact. I wish you had analyzed these results to show what magnitude and temperature dependence of lattice thermal conductivity take away all that discrepancy.

NEWROCK: In order to explain the discrepancy one needs a lattice conductivity about four times as large as predicted by theory.

PIPPARD: I wouldn't have thought that a discrepancy of that amount to be particularly serious for this subject.

NEWROCK: I might also mention that an explanation based on the lattice conductivity is not compatible with my measurements of the Righi Leduc effect in potassium.

PIPPARD: But the "hump" in your data is strongly reminiscent of phonon drag. I don't want to bring phonon drag into the discussion except to stress that it is not immediately obvious that the two transport coefficients are independent; could there be a coupling between them which is analogous to phonon drag coupling between the electron current and the phonon current?

NEWROCK: To the best of my knowledge there is no phonon drag influence on the thermal resistivity of potassium.

PIPPARD: But are there not uncertainties in the theory about the relationship between the thermal and electric currents? I am not very certain of my ground here, but I would suggest that a good model would assume that the phonon gas is always in equilibrium with the electron gas, and vice-versa.

LASS: I am afraid I must disagree with both of these points-of-view. I am sure Professor Newrock is aware that in analyzing his data to include the thermal conductivity of the lattice he is using equations which are valid for a homogeneous sample without inclusions or other inhomogeneities. What is clearly needed is a theory like that of Garland and his colleagues which includes the effects of inclusions. Further, I would disagree with Professor Pippard about the lack of agreement between measured and calculated values of the thermal conductivity of pure metals. In indium, at least, the theoretical values of the thermal conductivity are "dead on" with the measured values. As far as phonon drag is concerned, one can argue fairly simply that it should not be a major factor in the thermal conductivity.

NEWROCK: I neglected to mention earlier that we are now doing the calculations you suggest of the thermal magnetoresistance of a metal with inclusions; we are assuming in the calculation the presence of a field independent lattice conductivity. Although we have not yet completed the calculations, I do not believe the lattice conductivity will have a major impact on the quadratic resistivity.

MAGNETORESISTANCE OF POTASSIUM

J. S. Lass

Physik-Department, Technische Universität München

8046 Garching

ABSTRACT

Current jetting and current channelling are discussed as representative models of magnetoresistance due to inhomogeneities. Induced torque experiments on the linear magnetoresistance of potassium show that its orientational dependance has uniaxial symmetry; and that the axis of symmetry is determined by the direction of growth of the single crystal. These, and a variety of other contactless measurements are consistent with a boundary-conditions model for the current flow around non-spherical inhomogeneities.

The phenomenon of linear magnetoresistance in potassium needs no long introduction. It still has no accepted explanation, and is refered to as an anomaly, despite the fact that it has been a research subject for about 10 years. The list of authors in the field includes 50 names by the last count, and I will not attempt to give any kind of balanced credit to the various groups or approaches. I will concentrate on the electrical magnetoresistance, since the thermal case will be treated separately at this conference.

In discussing the role inhomogeneities might play in the (roughly) linear increase of the resistivity with magnetic field, I will be assuming that the source of the effect is not inherent to ideal potassium. I wish to focus on the problem of cleaving some information about the character of the inhomogeneities from the magnetoresistance itself. Let me start with two hypothetical models.

CURRENT JETTING

A drastic inhomogeneity would be a void (a bubble, an inclusion) in the metal, i.e., a well defined region with zero conductivity. Such a void clearly poses a definite boundary condition, $j_n = 0$; there can be no current flowing normal to the walls of the void. This clear case of boundary-condition-induced magnetoresistance has been named "current jetting"[1].

It has been established experimentally in cases where particular boundary conditions have been created intentionally. The effect is based on the large difference between transverse magnetoconductivity $\sigma_{xx} \propto B^{-2}$ and longitudinal magnetoconductivity $\sigma_{yy} \propto B^0$. Current distortions introduced by the boundary conditions extend along the magnetic field for distances that increase as B, and cause an apparent increase in the resistivity.

There have been exact calculations for spherical and cylindrical voids which show a linear magnetoresistance, both transverse and longitudinal. These calculations have been based on purely macroscopic considerations, assuming that the void dimension is large compared to the mean free path, typically 100μ.

This should probably be relaxed to void size larger than the cyclotron radius r_c. If we assume that these size considerations are important, then we would expect a change in resistivity as a function of magnetic field, when B is such that void size is equal r_c. Since experimentally $\rho(B)$ is non-linear for B less than about 1 Tesla, the implication would be that the minimum void size is 10μ.

One would like to know if the current distortions should really be all that different in the case of a void for which one, two, or three of its dimensions are smaller than r_c. It can be shown that even in the non-local limit, in which the electric field around a void varies considerably over distances smaller than r_c, the magnetoconductivity tensor, although a function of position, will have qualitatively the same magnetic field dependence as a local conductivity defined for a homogeneous electric field.

Does this mean that the effect of a very small void will be similar to that of a larger one, apart from being somewhat smeared out due to averaging over distances comparable to r_c? It might very well not be so, due to another aspect of non-locality: the scattering of electrons on the void walls. This will be mentioned under the next heading.

CURRENT CHANNELLING

Let us first consider a surface parallel to B. Approximately 50 % of all electrons within a "skin layer" of a thickness r_c (see Fig. 1a) are scattered within a time $\tau_{eff} \lesssim 1/\omega_c$. For electrons in the bulk, $\omega_c \tau \gg 1$, and the large electric field (E_{Hal},) is nearly orthogonal to the transverse current j_y^{Hal}. Nevertheless, near the surface their relation will be more typical for $\omega_c \tau \approx 1$,

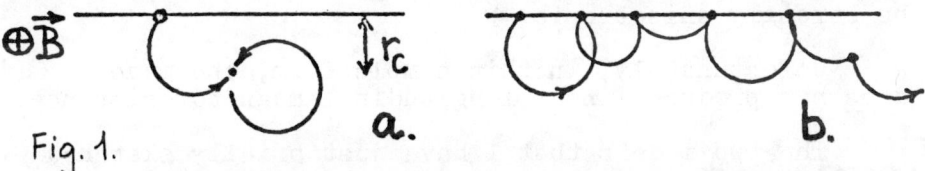

Fig. 1.

i.e. , a large component of the electric field will be parallel to the surface. It is interesting to note that these electrons are nearly trapped (Fig. 1b); only by additional scattering (on impurities or phonons) will current parallel to the surface be destroyed. As a result, the conductivity parallel to the surface will more probably be proportional to B^{-1}, rather than B^{-2} as in the bulk[3]. I would like to call this picture "current channelling".

Though reduced, this effect will be present even if the surface is extended mostly in one direction. Let us now imagine a needle-like internal surface or region of concentrated scattering (Fig. 2). Within an

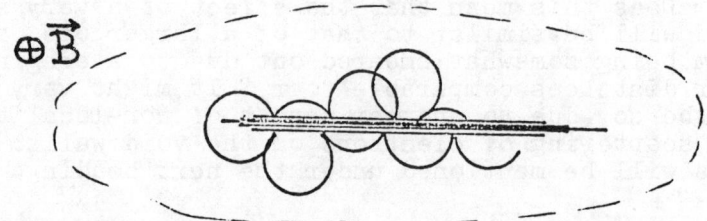

Fig. 2.

effective volume around this needle, the transverse conductivity will be similar to that of a bulk metal with a narrow band of open orbits. Unlike real open orbits, these are limited to a narrow, needle-like, region in real space- and not in k-space - and their contribution to the conductivity is $\propto B^{-1}$ (not B^0). If for a system of overlapping effective volumes the conductivity is $\sigma_{xx} = aB^{-2} + bB^{-1}$, by inversion the resistivity will have a term linear in B: $\rho_{yy} \propto a + bB$. Should the needles be oriented, we would expect the magnetoresistance to be strongly anisotropic.

Unfortunately, in this simple form, the model does not produce linear longitudinal magnetoresistance.

The two models that I have just briefly sketched for linear magnetoresistance are both due to inhomogeneities. There is obviously considerable overlap possible between the two cases, but what are the main differences?

"Current jetting" involves a mental picture of non-conducting voids. The physical basis of such voids are possibly clustered impurities, embedded oxides, or inert gases which are soluble in liquid potassium, but are probably trapped on vacancy clusters during solidification. A rather large fractional volume of voids is required to explain the observed magnitudes of linear magnetoresistance, and the importance of strain is not easily understood.

In the case of "current jetting" it is the boundary conditions acting as variables that control the appearance of the magnetoresistance in potassium with a nearly-free-electron magnetoresistivity tensor.

The "current channelling" model brings to mind regions of intense scattering extended in one direction - perhaps dislocation lines with precipitated impurities.

This would possibly involve a smaller volume of defects, and is clearly linked to strain. The magnetoresistance appears as an average property for a medium containing such extended defects, therefore this model is an example of a case described by a rather unusual magnetoresistivity tensor.

INDUCED TORQUE

An experiment that might help us distinguish between boundary conditions and the magnetoresistance tensor is the induced torque. The technique has been used in low magnetic fields to study inhomogeneous semiconductors [4], but here I will be discussing effects in high magnetic fields.

In the induced-torque experiment currents are excited by a rotating magnetic field rather than by contacts. It has in common with other contactless methods (eddy current decay, helicons) that it involves rotating electric fields, i.e., such for which curl $\vec{E} = -\partial \vec{B}/\partial t$, contrary to the curl $\vec{E} = 0$ for fields applied in measurements with contacts. For such rotating electric fields the boundary condition on the current, $j_n = 0$, has quite different solutions in high magnetic fields than in the case of curl-free fields. Very low frequencies are assumed, for which screening of $\partial B/\partial t$ is negligable.

One solution that is known [5] is the case of a conducting ellipsoid. The dissipation in such an ellipsoid in high magnetic fields has a) terms proportional to the magnetoresistivity (which are present also in a conducting sphere), but also b) terms proportional to B^2 that depend upon ellipsoid parameters and orientation.

In the geometry of the induced torque, and of the related effect soft helicons [6], $\vec{B}$ is perpendicular to $\partial \vec{B}/\partial t$, see Fig. 3a ($\vec{\xi}$ is the direction of the shortest semiaxis). In such a case the extra terms (referred to as b) above) are proportional to $B^2 \times \sin^4\varphi \cdot \cos^2\theta \cdot \sin^2\theta$. Therefore, only when $\varphi = 0$, or $\theta = 0$ or $\pi/2$, will the normal terms (referred to as a) above) proportional to $\rho_{zz} + \rho_{yy}$ be detectable, see Fig. 3b.

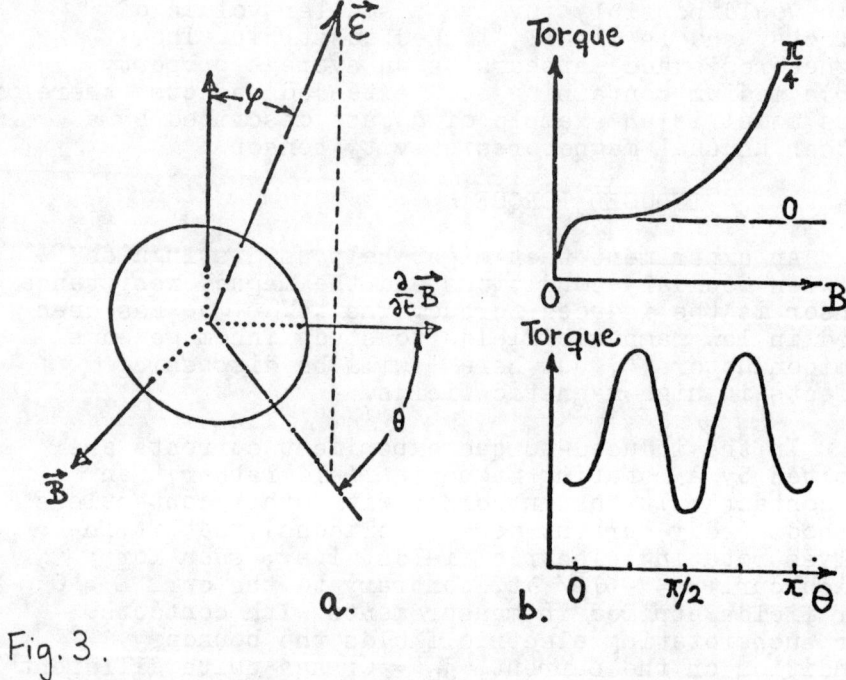

Fig. 3.

The induced torque experiments of Schaefer and Marcus[7] and of Holroyd and Datars[8] showed exactly such behaviour of the torque as expected for an ellipsoid (Fig. 3b), although the later[8] showed that this was not due to non-spherical samples. The axis of symmetry could be linked to the direction along which the single crystal sample was grown. This uniaxial symmetry and the quadratic field dependence suggest that some boundary conditions of the ellipsoidal-surface type are present.

Let us assume that there are internal surfaces of the kind I mentioned earlier, in the form of oblate ellipsoidal voids, all oriented along a direction that was determined during the growth of the single crystal sample. The sample itself is a sphere. It is plausible that currents around the ellipsoidal voids consist of two orthogonal and therefore additive solutions. One will be the usual "current jetting", the other will be circulating currents with the same symmetry properties as determined by the boundary conditions in conducting ellipsoids.

Then, even without a serious calculation, it is possible to make, on the basis of symmetry arguments, a consistent picture of the contactless measurements on potassium using the theory of the induced torque on an ellipsoid, applied to the ellipsoidal voids.

The results for induced torque are consistent not only with the experiments already mentioned, but also with those of Lass[1], and with the soft helicon measurements of Simpson[9], depending on the orientation of the growth axis.

In the geometry of the eddy current decay, and of longitudinal helicons, $\vec{B}$ and $\partial \vec{B}/\partial t$ are parallel; in which case the extra terms are only small and of the same power of B as the normal terms. This is consistent with the measurements of Simpson[9], and of Werner, Hunt and Ford[10].

That the results of 5 independent groups of authors[11] can be made consistent adds weight to the boundary-conditions approach in which potassium is regarded as having a nearly-free-electron magnetoresistance tensor.

Theories in which unusual magnetoresistivity tensors for potassium are assumed (due to inhomogeneities or otherwise) have not been ruled out. However, to explain the experimental data, such theories would have to reproduce, apart from linear magnetoresistance, a large and highly anisotropic longitudinal magnetoresistance quadratic in B.

REFERENCES

1. J.S. Lass, J. Phys. C 3, 1926 (1970).
2. J.B. Sampsell and J.C. Garland , Phys.Rev. B 13, 583 (1976); D. Stroud and F.P.Pan, Phys.Rev. B 13, 1434 (1976); K.D. Schotte and D. Jacob, phys, stat. sol. (a) 34, 593 (1976).
3. The situation is similar to the static skin effect discussed by M. Ya. Azbel', JETP 44, 983 (1963) - Sov. Physics JETP 17, 667 (1963).

4. U. Strom and P.C. Taylor, J. Appl. Phys. 45, 1246 (1974).
5. J.S. Lass, Phys.Rev. B 13, 2247 (1976).
6. J.A. Delaney and A.B. Pippard, Rep.Prog.Phys. 35, 677 (1972)
7. J.A. Schaefer and J.A. Marcus, Phys.Rev.Lett. 27, 935 (1971)
8. F.W. Holroyd and W.R. Datars, Can.J. Phys. 53, 2517 (1975).
9. A.M. Simpson, J.Phys. F 3, 1471 (1973).
10. S.A. Werner, T.K. Hunt and G.W. Ford, Sol. St. Comm. 14, 1217 (1974).
11. Contactless measurements using helicon standing wave resonances in disks, P.A. Penz and R. Bowers, Phys. Rev. 172, 991 (1968), were made at frequencies too high for the present concepts to be applicable.

DISCUSSION

M. H. COHEN (Univ. of Chicago): I have a question about your current channeling model. Is the effective volume interaction around each inclusion field dependent or field independent?

LASS: The interaction might well be field dependent, but it probably would suffice to treat it as field independent and to suppose that the effect arises from the changing proportions of current carried by either the surface or the bulk. Although I did not emphasize this point, there must be a certain amount of "communication" between the effective volumes.

D. L. MITCHELL (N.S.F.): In speaking of voids, dislocations, and other defects, I have heard no mention of the common metallurgical techniques which would reveal their presence: small angle x-ray scattering, electron microscopy, etc. Are these techniques impossible to use to characterize these samples?

LASS: I think it is clear that these techniques will not be easy to apply. For example, one cannot cut a very thin slice off of a potassium sample and look at it under an electron microscope. Other techniques could be applied, however, such as small angle x-ray scattering or mosaic structure studies using gamma rays. In fact, there is a richness of this kind of data on the unusual properties of potassium near the melting point which may bear on the magnetoresistance problem. Perhaps not enough effort has been expended in this direction.

CHAPTER III: INVITED PAPERS:
OPTICAL PROPERTIES

PHYSICAL AND OPTICAL PROPERTIES OF SMALL METAL PARTICLE COMPOSITES

H. G. Craighead and R. A. Buhrman
Dept. of Physics, Cornell University, Ithaca, N. Y. 14850

ABSTRACT

Small metal particle composite systems have been produced from a number of metallic elements both by co-evaporation of a metal and dielectric and by the evaporation of a metal in an inert gas atmosphere. The physical properties of these systems will be discussed. The optical response of several of these composite systems has been measured in the visible and near-infrared spectral regions. Comparison between the data and predictions of effective medium theories will be made, and possible causes for the cases of disagreement discussed.

DISCUSSION

D. R. McKENZIE (University of Sydney): Did you have any problems in determining the optical constants from the measured reflectance and transmittance? I find that sometimes the optical constants cannot be unambiguously assigned from the reflectance and transmittance.

BUHRMAN: You also must know the thickness. The equations can then be solved for the optical constants.

D. BERGMAN (Tel-Aviv University): If the Bruggeman equation (effective medium approximation) doesn't work for the reasons which you gave, then the Maxwell-Garnet theory shouldn't work either. Does you data agree with the Maxwell-Garnett theory?

BUHRMAN: At low enough fill fraction, tunneling effects should be small and some effective medium theory should work. I think our data agrees with the Maxwell-Garnett theory, but I can't prove that this agreement is not coincidental. We have some data on W which is also in agreement. The work should be done on materials like Ag and Cu.

D. L. MITCHELL (National Science Foundation): You mentioned that the medium displayed semiconducting behavior. What were the activation energies seen in the optical constants and how did these energies change as the fill fraction increased?

BUHRMAN: At dc the activation energy is tenths of electron volts. This energy decreases with increasing fill fraction, in agreement with the work at RCA. If we try to fit the optical data to an activated behavior, we must use hundreths of electron volts.

B. ABELES (Exxon Research and Development): The high absorption in the mid-infrared is observed in other systems, such as Au-MgO, and seems to be a quite general effect. This absorption is bad for solar energy applications, but is interesting physically. Could the mechanism recently put forward by D. Stroud, namely eddy current losses, account for this absorption? Regarding the mechanism you have proposed, photon assisted tunneling, we at RCA and others elsewhere have looked for photoconductivity in these systems. Photoconductivity has never been observed and so I doubt that tunneling is the answer.

BUHRMAN: How big the photoconductivity effect is depends upon the tunnelling length. In these systems I think this length is very short. We have also looked for photoconductivity and have observed none. We have also considered the eddy current effect, which should increase as the particle size goes up, giving more absorption in the infrared and less in the visible. But the effect of large infrared absorption is also seen in cases where the grain size and the electronic mean free path are both very small, for example Ni-MgO.

ABELES: To test the validity of the Maxwell-Garnett or Bruggeman theories, one should choose a system which exhibits a plasma resonance in a convenient wavelength region because that resonance gives rise to an absorption peak. The peak is not predicted by the effective medium theory, while the Maxwell-Garnett theory does predict such a peak, as Garnett showed many years ago.

W. T. DOYLE (Dartmouth University): It is no surprise that the Maxwell-Garnett theory works well in the low density regime because of the asymmetry between the metal and the suspending material. It is well documented that this theory works well under those circumstances. As the concentration increases the particles start to interact. It becomes necessary to consider all the multipoles. The incoming wave excites dipoles in the particle and so it is generally assumed that only dipoles enter, which leads to the Maxwell-Garnett theory. This assumption is wrong because as soon as the particles get near each other, each particle will feel the spatially inhomogeneous field of its neighbors. This inhomogeneous field causes many multipoles to be excited and they all have to be taken into account.

C. D. MITESCU (Pomona College): Have you varied the thickness of your films significantly? Your samples have particles of 100-200Å diameter in a 2000Å layer. I wonder whether an effective medium theory which is derived under the assumption that the medium extends quite considerably in all directions can be applied to what is in effect ten to twenty layers.

BUHRMAN: The range of thickness is limited. If the films are too thick there is no transmission, and if they are too thin there is no absorption.

McKENZIE: I have also observed extra infrared absorption in cermets and, along with Doyle, I think it can be accounted for by correct inclusion of interaction between particles. This interaction also broadens the resonance peak in the visible in the way which is observed.

OPTICAL PROPERTIES OF ULTRAFINE GOLD PARTICLES

C.G. Granqvist
Physics Dept., Chalmers University of Technology,
Fack, S-402 20 Gothenburg, Sweden

ABSTRACT

This paper reports on optical transmittance through samples comprised of ultrafine gold particles. These were prepared by two techniques: gas evaporation and island growth in discontinuous films. The measured data could be brought into detailed quantitative agreement with computed data based on the Maxwell-Garnett theory provided an accurate model for the particle morphology was employed and the local-field effects were accounted for properly.

I. INTRODUCTION

Ultrafine noble metal particles may exhibit optical properties which are very different from the bulk behaviour. This fact is illustrated in Fig. 1 where the dotted curve refers to a continuous gold film having bulk properties. By going to smaller wavelengths the transmittance is first enhanced, as expected from free electron (Drude) theory, but at wavelengths smaller than 0.51 μm there is a sharp decrease owing to the onset of interband transitions. A strikingly different behaviour is encountered for ultrafine gold particles as seen from the solid and dashed curves obtained respectively for a discontinuous gold film comprised of extremely small islands and minute gold spheres prepared by evaporation in a noble gas atmosphere and collecting the soot-like deposit on a glass substrate. Both these curves display a pronounced minimum - i.e. an anomalous absorption which is not present in the bulk - in the visible range.

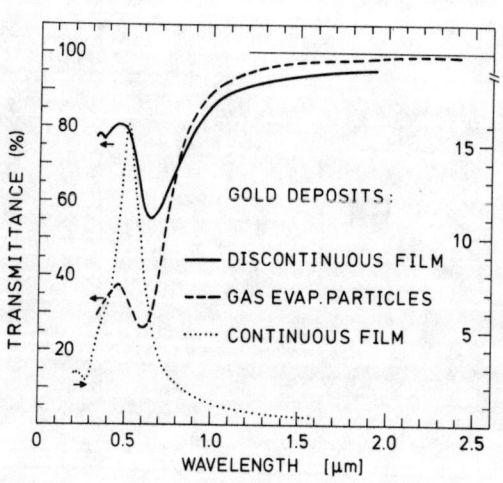

Fig. 1. Transmittance vs. wavelength for gold deposits being a discontinuous film (average thickness 3 nm), gas evaporated particles (mass density 0.6 g/m^2), and a continuous film (thickness 48 nm).

ISSN: 0094-243X/78/196/$1.50 Copyright 1978 American Institute of Physics

The main purpose of this paper is a quantitative analysis of optical transmittance through coatings of gas evaporated gold particles and through discontinuous gold films. The first step towards this goal is given in Sec. II where the measured data are reported together with a detailed characterization of the specimens prepared by the two experimental techniques. The second step (in Sec. III) is the formulation of an effective medium theory which is sufficiently general to allow the optical properties of the inhomogeneous media to be computed accurately - including size distributions, randomly oriented non-spherical islands etc. The rest of the paper is devoted to the exploitation of this theory to achieve theoretical data which can be compared with the measurements for gas evaporated particles (Sec. IV) and for discontinuous films (Sec. V). The most crucial point is the proper handling of clustering of adjacent particles which is significant for both types of samples. This effect can be treated only by introducing adjustable parameters. For the gas evaporated samples we need two parameters: the fraction of the spheres aggregated into linear chains resp. closepacked clusters. For the discontinuous films one parameter suffices: the width in the distribution of island-to-island separations. Reasonable values of these parameters yield theoretical data in excellent agreement with the measurements.

II. EXPERIMENTS AND SAMPLE CHARACTERIZATION

(a) Gas evaporated particles

The ultrafine particles were prepared by evaporation of gold (purity 99.99 %) from a tungsten boat in a conventional bell jar system containing 2 torr of air[1]. The metal atoms, which are effused from the heated vapour source, lose their energy rapidly by collisions with the gas atoms, so that a highly supersaturated state is reached from which stable clusters of metal atoms are produced by homogeneous nucleation. These embryonic particles then grow by liquid-like coalescence to form larger grains whose ultimate size is governed by several experimental conditions like evaporation rate, atomic or molecular weight of the gas, its pressure etc. The presence of some oxygen in the evaporator makes the gold particles electrically decoupled from each other, presumably by the formation of an oxide.

The particles were collected on a glass substrate positioned 8 cm above the vapour source. By use of a shutter we exposed the substrate to the particles for 1/4 to 1 minute, which was sufficient to produce an appropriate soot-like coating. Its thickness (t) was determined by optical microscopy and its mass per unit area (W/A) by weighing. Experimental data of the latter quantity are shown in Table I for four samples. The volume fraction of particles in the coatings (i.e. the "filling factor") was obtained from

$$f = (W/A) (\rho t)^{-1} \tag{1}$$

Table I. Data for the gas evaporated Au particles

Sample	W/A (g/m^2)	f (%)	$\bar{x}$ (nm)	σ
18G	0.29	0.32±0.05	3.5	1.37
19A	0.32	0.37±0.06	3.4	1.35
20A	0.36	0.27±0.05	4.0	1.34
20B	0.55	0.49±0.08	3.3	1.40

where ρ denotes the density of bulk gold. As seen from Table I the f's are only part of a percent. This is in agreement with the results from several other determinations of f for similarly prepared gold particles[2-5].

An electron microscope grid covered with an amorphous carbon layer was placed in the immediate vicinity of the substrate and was exposed to the particles simultaneously with the glass surface. Fig. 2a shows a typical bright field micrograph taken at a magnification of 130 000 x with a Philips EM 300 electron microscope operated at 100 kV. The spherical particles are seen to be aggregated into complex chains and clusters; this is a characteristic feature of gas evaporated particles[1-4] which may be caused by van der Waals interaction as well as by electrostatic attraction.

Fig. 2(b) reproduces an electron diffraction image of the same sample. The Debye rings signify that the main constituent is f.c.c. Au. In addition there is one diffuse ring, which we ascribe to WO_3, which must originate from the vapour source.

Size distributions were evaluated from the electron micrographs. It was shown that the fractional number of particles Δn per logarithmic diameter interval $\Delta(\ln x)$ could be approximated accurately by

$$\Delta n = \frac{1}{(2\pi)^{1/2} \ln \sigma} \exp\left[-\frac{1}{2} (\frac{\ln(x/\bar{x})}{\ln \sigma})^2 \right] \Delta(\ln x), \qquad (2)$$

i.e. by a log-normal distribution function characterized by a median diameter $\bar{x}$ and a geometric standard deviation σ. This particular distribution is expected for particles growing by binary collisions accompanied by liquid-like coalescence. Table I gives $\bar{x}$ and σ for four samples.

Optical transmittance in the wavelength interval $0.33 < \lambda < 2.5$ μm was recorded at normal incidence with a UNICAM SP 700 double beam spectrophotometer. In order significantly to decrease the influence from the glass we always put a substrate with particle coating in one of the light beams and an uncoated substrate in the other and monitored the ratio between the two transmitted intensities.
Fig. 3 shows wavelength-dependent transmittance for four samples with different W/A's. The most noteworthy feature is the dip in the

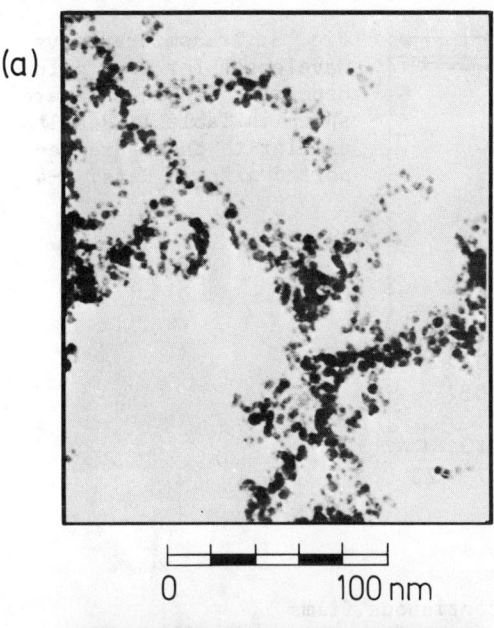

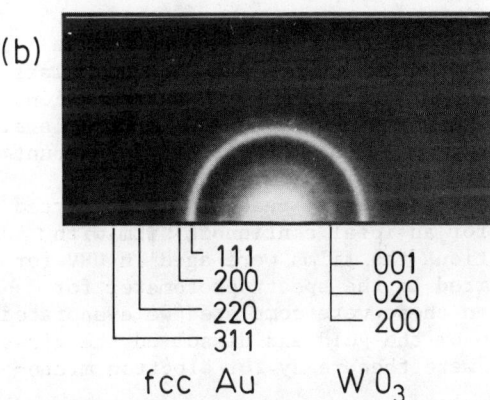

Fig. 2. Part (a) depicts ultrafine gold particles prepared by evaporation in air. Part (b) shows electron diffraction from the same sample. Calculated positions of the four innermost reflections for f.c.c. Au are indicated. The weak line closest to the center is due to three overlapping reflections from triclinic and/or orthorombic WO_3.

transmittance at a wavelength in the range 0.56 to 0.60 µm. It should be noted that the overall reproducibility of the curves is excellent. This is not the case of deposits which are much thicker than those of Fig. 3, but for such samples the transmittance minimum is smeared out considerably. This is caused by coagulation of the particles, presumably by excessive heating, which is difficult to avoid for the thick coatings.

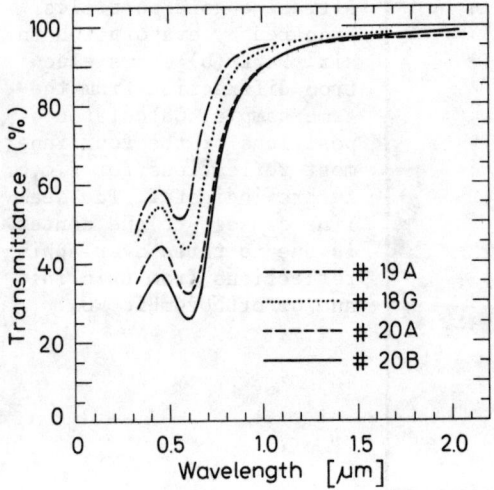

Fig. 3. Transmittance vs. wavelength for four gold deposits. Sample data are shown in Table I. Results similar to these are reported also in Refs. 2-4.

(b) Discontinuous films

The discontinuous gold films were produced by evaporation from a resistively heated alumina crucible onto Corning 7059 glass substrates under UHV-conditions ($<4 \times 10^{-9}$ torr during evaporation). An electric field of 20 V/cm was maintained in the substrate plane. The substrate temperature was kept constant at 290±5 K. The amount of material impinging towards the substrate was recorded on a quartz crystal oscillator microbalance whose reading was converted to an effective thickness (t) for an ideal continuous film with bulk density. After the depositions the films were aged in UHV for > 20 h. They were then transferred to the spectrophotometer for transmittance measurements. When these were completed we evaporated an amorphous carbon film on top of the gold and dissolved the glass in 5 % HF acid. The free films were then ready for electron microscopy.

Representative electron micrographs for films with four mass thicknesses are shown in Fig. 4. It is seen that when t is increased the average size of the islands as well as their irregularity is enhanced. Also the number of islands per unit area is decreased which proves that coalescence growth is dominating.

It is clear from the electron micrographs that the islands do not look round but that a better model for their two-dimensional images is to treat them as being ellipses. Empirically, lognormal distribution functions yield good representations of the major axes (a) as well as of the minor axes (b). Thus both quantities obey Eq. (2) with $\bar{a}$ and σ_a resp. $\bar{b}$ and σ_b signifying the median axes and the geometric standard deviations. Evaluations for eight discontinuous films are given in Table II.

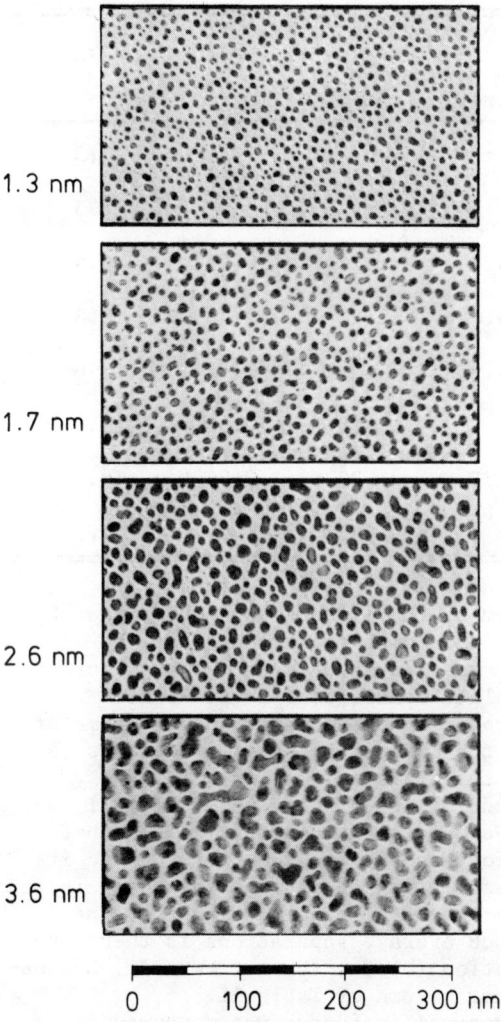

Fig. 4. Electron micrographs for discontinuous gold films with four mass thicknesses.

A quantification of the average deviation from circularity is convenient for characterizing the islands. This is achieved by defining a mean eccentricity ($\bar{e}$) according to

$$\bar{e}^2 = 1 - \left[N^{-1} \sum_{K=1}^{N} (a_K/b_K) \right]^{-2} . \qquad (3)$$

This puts most weight on the islands with largest elongation, which turns out to be appropriate for calculations of optical properties of non-spherical islands.

Table II. Data for the discontinuous Au films

t (nm)	$\bar{a}$ (nm)	σ_a	$\bar{b}$ (nm)	σ_b	$\bar{e}$	D (nm)
1.1±0.2	5.7	(1.5)	–	–	–	7.3
1.3±0.2	6	(1.35)	–	–	–	8.3
1.5±0.3	8	1.44	5.8	1.22	0.79	9.9
1.7±0.3	8.8	1.38	6.8	1.22	0.74	10.8
2.6±0.2	11.8	1.36	8.5	1.2	0.78	13.6
3.0±0.3	16	1.5	9.8	1.28	0.86	15.8
3.6±0.3	19	1.5	11.5	1.22	0.86	18.1
4.0±0.3	25	(1.6)	13	(1.36)	0.91	21.5

The three-dimensional form of the islands can be obtained by combining electron microscopic data with measurements of, for example, their number density. As discussed in Ref. 6 an accurate model is provided by whole prolate spheroids ("cigars") with symmetry axis along the substrate. Thus the distribution of major and minor axes evaluated from the two-dimensional micrographs is sufficient for describing the three-dimensional islands.

The final parameter needed for the discontinuous films is a mean interisland separation. We obtain this quantity – somewhat arbitrarily – by determining for each island the distances to its four nearest neighbours and associating an average island-to-island separation with the arithmetic mean of these four lengths. The mode in the distribution of such average separations is then taken as the mean interisland separation (D) for the entire film. Evaluations of D are shown in the last column of Table II.

Optical transmittance at normal incidence was measured for the discontinuous gold films in exactly the same way as for the gas evaporated samples. Fig. 5 depicts transmittance vs. wavelength for the discontinuous films of Table II. The most salient feature is, again, the transmittance minimum occurring at 0.56 to 0.70 μm, i.e. in the visible range. The minimum is seen to shift towards larger wavelengths in the thicker films, which is in agreement with the results of several studies[7-11] of discontinuous gold films.

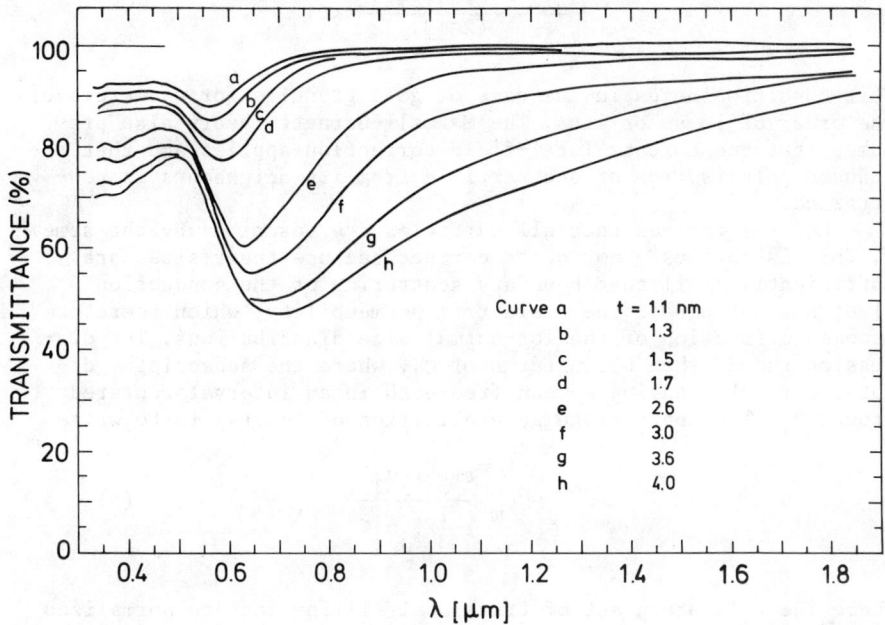

Fig. 5. Transmittance vs. wavelength for discontinuous gold films of different thicknesses.

III. EFFECTIVE MEDIUM THEORIES

The gas evaporated samples and the discontinuous films are both comprised of particles which are orders of magnitude smaller than the wavelengths of the light. This indicates that the optical properties of the deposits can be described in terms of an effective medium, being a spatial average over the dielectric permeabilities for the metal islands (ε) and for their surrounding material (ε_m).

The simplest effective medium theory is that of Maxwell-Garnett[12,13] which yields an effective dielectric permeability ($\bar{\varepsilon}^{MG}$) for spherical particles given by the well known expression

$$\frac{\bar{\varepsilon}^{MG}-\varepsilon_m}{\bar{\varepsilon}^{MG}+2\varepsilon_m} = f \frac{\varepsilon-\varepsilon_m}{\varepsilon+2\varepsilon_m} \quad . \tag{4}$$

It can be shown from the Mie[14,15] theory that the Maxwell-Garnett approach is valid when

$$\frac{4\pi^2 x^2}{15\lambda^2} \frac{(\epsilon+4)(\epsilon+2)}{(2\epsilon+3)} \ll 1 \quad .$$

This condition holds in the case of gold granules for diameters of the order of 10 nm or less. The Maxwell-Garnett theory also presumes that the Lorentz local-field correction applies and that the induced polarisation of one particle from its neighbours is non-retarded.

Eq. (4) implies that all particles are described by the same ϵ. This is obviously not quite correct because their sizes are sufficiently small that boundary scattering of the conduction electrons influences the dielectric permeability, which therefore becomes a function of the log-normal size distributions. The discussion should then be in terms of ϵ_j, where the subscript j denotes particles having a mean free path in an interval centered around ℓ_j. The appropriate generalization of Eq. (4) is to write

$$\bar{\epsilon}^{MG} = \epsilon_m \frac{1+\frac{2}{3}\sum_j f_j \alpha_j^-}{1-\frac{1}{3}\sum_j f_j \alpha_j^-} \quad , \tag{5}$$

where the f_j's are a set of fractional filling factors normalized by

$$\sum_j f_j = f \tag{6}$$

and α_j^- is proportional to the polarisability of the jth particles. Spheres – with depolarisation factor $\frac{1}{3}$ – have

$$\alpha_j^- = \frac{\epsilon_j - \epsilon_m}{\epsilon_m + \frac{1}{3}(\epsilon_j - \epsilon_m)} \quad . \tag{7}$$

For randomly oriented ellipsoids with depolarisation factors L_1, L_2, and L_3 the corresponding relation is[16]

$$\alpha_j^- = \frac{1}{3} \sum_{i=1}^{3} \frac{\epsilon_j - \epsilon_m}{\epsilon_m + L_i(\epsilon_j - \epsilon_m)} \quad . \tag{8}$$

Finally, ellipsoids with one of their axes perpendicular to a plane but with otherwise random orientations are characterized by

$$\alpha_j' = \frac{1}{2} \sum_{i=1}^{2} \frac{\varepsilon_j - \varepsilon_m}{\varepsilon_m + L_i(\varepsilon_j - \varepsilon_m)} \quad . \tag{9}$$

The latter expression, which holds only for normal incidence, was derived by Hunderi[17] in a treatment of the influence of grain boundaries and lattice defects on the optical properties of metals.

Eq. (7) is clearly applicable in a treatment of the transmittance through coatings of spherical gold particles prepared by gas evaporation. As we will find shortly Eq. (8) allows a discussion of local field effects for such samples. Eq. (9) is well suited for treating the optical transmittance of discontinuous films. In this case the depolarisation factors L_1 and L_2 can be connected with the evaluated mean eccentricity for the prolate-spheroidal islands by[18]

$$L_1 = \frac{1-\bar{e}^2}{2\bar{e}^3} \left[\ln\left(\frac{1+\bar{e}}{1-\bar{e}}\right) - 2\bar{e} \right] \tag{10}$$

$$L_2 = \frac{1}{2}(1-L_1). \tag{11}$$

From its derivation the Maxwell-Garnett model is valid only for very small filling factors and it is not clear how well this theory works in general at larger f's. One would expect that - at least in principle - a better theory could be formulated in a self-consistent manner. There exist several self-consistent effective medium theories, the most well known being the one by Bruggeman[19] which has attracted much interest recently[20-22]. It gives an effective dielectric permeability

$$\bar{\varepsilon}^{BR} = \varepsilon_m \frac{1-f+\frac{1}{3}\sum_j f_j \alpha_j}{1-f-\frac{2}{3}\sum_j f_j \alpha_j} \tag{12}$$

with

$$\alpha_j = \frac{\varepsilon_j - \bar{\varepsilon}^{BR}}{\bar{\varepsilon}^{BR} + \frac{1}{3}(\varepsilon_j - \bar{\varepsilon}^{BR})} \tag{13}$$

for spherical particles. Expressions analogous to Eqs. (8) and (9) can be obtained simply from the substitution $\varepsilon_m \to \bar{\varepsilon}^{BR}$. It is important to notice that Bruggeman's theory, as well as other self-consistent formulations, is a mean field theory which takes account of interactions among neighbouring particles only in so far as these can be represented by a constant far-field, whereas the specific near-field effects are suppressed. And, as we will see shortly, a proper description of the near-field is essential for predicting the transmittance quantitatively.

The calculations of $\bar{\varepsilon}^{MG}$ or $\bar{\varepsilon}^{BR}$ implies averaging over sets of size dependent dielectric permeabilities, ε_j. These quantities were derived by starting from an accurate ellipsometric determination of the dielectric function for bulk gold - ε_{exp}, as measured by Winsemius[23] - and modifying these data to account for size dependent scattering in the Drude part according to the construction

$$\varepsilon_j = \varepsilon_{exp} - \varepsilon_{exp}^{Drude} + \varepsilon_j^{Drude} . \tag{14}$$

The two free electron terms are given by

$$\varepsilon_{exp}^{Drude} = 1 - \frac{\omega_p^2}{\omega(\omega+i/\tau_b)}, \tag{15}$$

$$\varepsilon_{exp}^{Drude} = 1 - \frac{\omega_p^2}{\omega(\omega+i/\tau_j)}, \tag{16}$$

where ω_p is the bulk plasma frequency, τ_b is the mean electron lifetime for bulk gold and

$$\tau_j^{-1} = \tau_b^{-1} + v_F/\ell_j, \tag{17}$$

where v_F is the Fermi velocity and ℓ_j is the mean free path. It should be noted that this construction leaves the interband part of ε_{exp} unchanged and also neglects possible size quantization effects for the conduction electrons.

From Winsemius[23] the bulk properties are

$$\hbar\omega_p = 8.55 \text{ eV}$$

$$\hbar/\tau_b = 0.108 \text{ eV}$$

$$v_F/c = 4.7 \times 10^{-3};$$

these values were used throughout our computations.

IV. DISCUSSION OF GAS EVAPORATED GOLD PARTICLES

The gas evaporated gold particles look round under the electron microscope which allows us to write[24] $\ell_j = x_j/2$, i.e. the mean free paths are equal to the particle radii. The particles are surrounded by air so that $\varepsilon_m = 1$. The expressions of the preceding section then allow a quantitative calculation of $\bar{\varepsilon}^{MG}$ and $\bar{\varepsilon}^{BR}$ from which we derive the corresponding optical transmittance by the standard equations[25] for a thin film on a substrate. The filling factor is $\lesssim 1\,\%$ which is sufficiently small for the two effective medium theories to yield practically indistinguishable results.

A preliminary comparison of theory and experiments is given in Fig. 6, where the dash-dotted curve was calculated with parameters appropriate to sample 20 B (cf. Table I). The measured transmittance is shown by the solid curve. It is evident that the <u>qualitative</u> shapes of the theoretical and experimental graphs are similar, but that there exist two important <u>quantitative</u> discrepancies: the location of the transmittance minimum in the calculated curve occurs at too small wavelength, and the overall magnitude is approximately a factor twice too high. The differences are much too

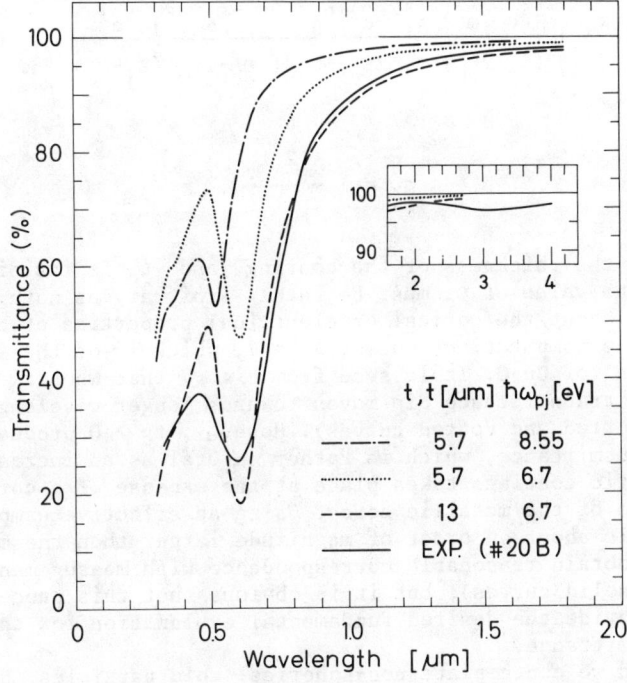

Fig. 6. Transmittance vs. wavelength for sample 20 B. The solid curve denotes experimental data. The other curves represent calculations based on the Maxwell-Garnett theory.

large to be blamed on an inadequacy in the determination of $\bar{x}$ or σ. If we choose a plasma frequency for the particles (ω_{pj}) which is lower than in the bulk the transmittance dip is shifted towards longer wavelengths, and for $\hbar\omega_{pj} \approx 6.7$ eV agreement is found for the positions of the minima as seen from the dotted curve in Fig. 6. Furthermore, if we take an effective sample thickness ($\bar{t}$) which is larger than the bulk value we can obtain a rough agreement with the measurements (cf. the dashed curve). Clearly the use of $\omega_{pj} < \omega_p$ and $\bar{t} > t$ is very much ad hoc and a more fundamental explanation for the apparent differences between measured and calculated results is needed. To this end we will now consider the roles of dielectric coatings, non-spherical particle shapes, and dipole-dipole interactions. The latter effect will be seen to be the most important one.

The individual spheres are electrically decoupled from each other, and it is therefore obvious that some kind of dielectric coating (presumably an oxide) must be present on their surfaces. Such pellicles can be treated simply by replacing Eq. (7) with an expression which is proportional to the polarisability of a coated sphere, viz.[15]

$$\alpha_j{'} = 3 \frac{(\varepsilon_c-\varepsilon_m)(\varepsilon_j+2\varepsilon_c)+q_j^3(2\varepsilon_c+\varepsilon_m)(\varepsilon_j-\varepsilon_c)}{(\varepsilon_c+2\varepsilon_m)(\varepsilon_j+2\varepsilon_c)+q_j^3(2\varepsilon_c-2\varepsilon_m)(\varepsilon_j-\varepsilon_c)} \qquad (18)$$

with

$$q_j = 1 - \frac{2 t_c}{x_j}, \qquad (19)$$

where t_c is the thickness of the coating, and ε_c is its dielectric constant. The value of ε_c must be rather arbitrary as nothing seems to be known about the optical or electrical properties of any gold oxide. In the computations we set $\varepsilon_c = 7$, which is of the same magnitude as for Cu_2O. It is seen from Fig. 7 that when t_c is increased the transmittance dip moves towards longer wavelengths (cf. dash-dotted and dotted curves). However $t_c > 0$ produces a higher transmittance, which is rather natural as an increase of the dielectric coatings takes place at the expense of a corresponding decrease of the metallic cores. Using an effective sample thickness which is about an order of magnitude larger than the measured one we can obtain reasonable correspondence with measurements (cf. dashed and solid curves), but it is obvious that this procedure does not provide the desired fundamental explanation for the optical transmittance.

Instead we contemplate non-spherical gold particles. No evidence for such shapes was gained from the electron micrographs, but the attainable resolution does not preclude some modest deviations from spherical forms. In our calculations we considered

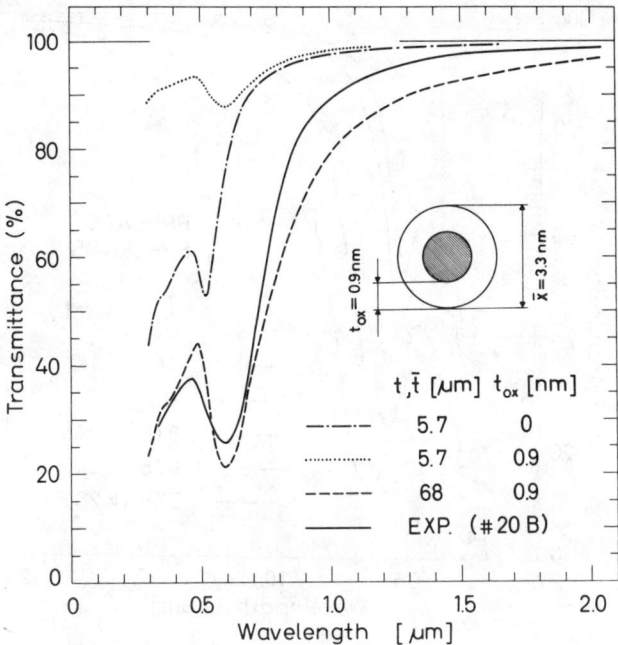

Fig. 7. Transmittance vs. wavelength for sample 20 B. The solid curve denotes experimental data. The other curves were calculated for coated spheres from the Maxwell-Garnett theory.

randomly oriented prolate spheroids ($L_1 < L_2 = L_3$) as well as oblate spheroids ($L_1 = L_2 < L_3$). The depolarisation factors were introduced into the effective medium formalism via Eq. (8). Fig. 8, pertaining to prolate spheroids, shows that when the difference between L_1 and L_2 is increased (i.e. for increased eccentricity) the overall transmittance is decreased and the minimum displaced towards longer wavelengths. A fair agreement is obtained for $L_1 = 0.12$ and $L_2 = L_3 = 0.44$. However, these values are equivalent to a ratio between major and minor axes of the spheroid of about 3, which is much too large to be reconciled with the electron microscopic evidence. The same conclusion was reached from calculations on oblate spheroids. Hence non-spherical shapes cannot explain the optical data in spite of the good agreement which can be achieved by use of appropriate depolarisation factors.

We finally consider dipole-dipole interactions. Previously it was tacitly presumed that the spatial distribution of particles was rather uniform throughout the sample, in which case the large particle-particle separations allow us to write the local field as the sum of the external field and the Lorentz field. From a direct inspection of typical electron micrographs (cf. Fig. 2a) it can be concluded that an assumption of large particle-particle

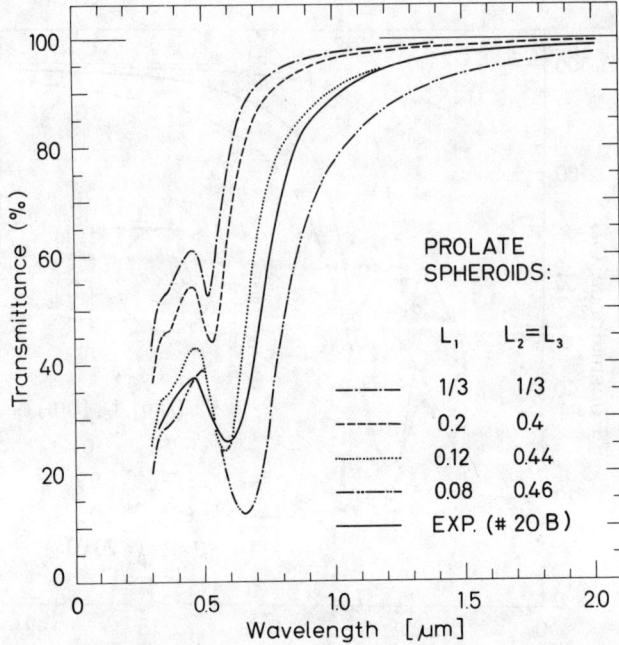

Fig. 8. Transmittance vs. wavelength for sample 20 B. The solid curve denotes experimental data. The other curves were calculated for randomly oriented prolate spheroids from the Maxwell-Garnett theory.

distances must be a very crude one, to the extent that it can be maintained at all, and it is evident that the individual particles stick together to form complex chains, clusters etc. Consequently dipole-dipole interactions must be important. Our semiquantitative description of the non-retarded coupling is founded on a recent paper by Clippe, Evrard, and Lucas[26], who computed the resonance frequency for several geometrical configurations of identical touching spheres. From their results we extracted the effective depolarisation factors listed in Table III. These make the approach formally identical to the one we used previously to treat non-spherical particles. It is important to note, though, that the present depolarisation factors are, in a sense, fictitious quantities which are not related to any ellipsoidal forms.

To get a final comparison of theory and experiments we considered the sample to be a mixture of aggregates of spheres with several geometrically well-defined configurations. Specifically, we chose single spheres, linear infinite single-strand chains, and close packed clusters with f.c.c. lattice. The existence of such structures was documented by the micrographs. Double spheres and double-strand chains were not included owing to the small differences with single spheres resp. single-strand chains. We hence

Table III. Equivalent depolarisation factors

Geometrical configuration	Equivalent depolarisation factor		
	L_1^*	L_2^*	L_3^*
Single sphere	1/3	1/3	1/3
Double sphere	0.250	0.375	0.375
Single-strand chain	0.133	0.435	0.435
Double-strand chain	0.139	0.342	0.518
fcc lattice	0.0865	0.0865	0.827

required that the filling factor would be

ξf for single spheres;

ζf for infinite chains;

$(1-\xi-\zeta)f$ for fcc clusters;

where $\xi, \zeta \geq 0$ and $\xi + \zeta \leq 1$. It should be pointed out that ξ and ζ do not necessarily represent exactly the "true" fractions of single spheres or infinite chains in the samples, because we have not attempted to invoke any relative absorption strengths for the different aggregates. The two parameters should then be interpreted as, loosely speaking, the product of an "oscillator strength" and a "true" fractional filling factor. In practice this does not play any role as the "true" fractions are not experimentally accessible quantities.

Fig. 9 depicts calculated and measured transmittance for two samples. With the shown percentages of single spheres, infinite linear chains, and close-packed clusters it is evident that we can bring the computed results into very good quantitative agreement with the measured data over the entire wavelength interval. The fits are not absolutely unique as a small increase of the single sphere fraction and a corresponding decrease of the fcc portion do not lead to any significant modifications. The slight deviations between measured and computed data which still remain are probably caused by the strong simplification in regarding non-retarded dipole-dipole coupling for only three different configurations, whereas in reality there must be a wide spectrum including chains, rings, and clusters of all shapes with, in general, retarded multipole-multipole interaction.

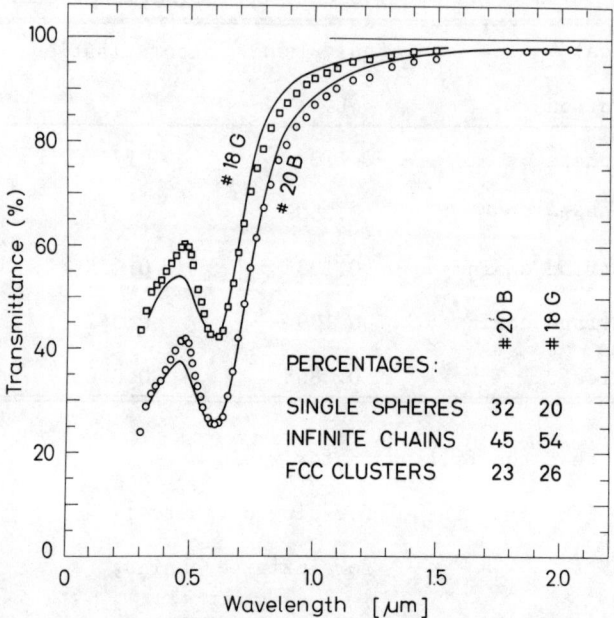

Fig. 9. Transmittance vs. wavelength for samples 20 B and 18 G. The solid curves denote experimental data. The squares and circles represent calculations based on the Maxwell-Garnett formalism with effective depolarisation factors taken to describe the dipole-dipole interaction.

V. DISCUSSION OF DISCONTINUOUS GOLD FILMS

When we want to use an effective medium theory for the optical properties of discontinuous films we encounter several problems: the proper meanings of filling factor and film thickness are not obvious and, furthermore, the average distance between the islands is so small that it must be assumed that different effective medium formulations yield different results. Fortunately, these three entangled problems can be be resolved - at least partially - by reference to the elaborate calculations of the dielectric properties of discontinuous metal films by Bedaux and Vlieger[27]. They showed that the Maxwell-Garnett theory would be applicable provided the discontinuous film was described by an "optical" thickness

$$t_{opt} = \frac{4}{3} <\Lambda>, \qquad (20)$$

which is related to an average distance $<\Lambda>$ between the islands. Eq. (20) applies to islands in an "amorphous" configuration, and somewhat different numerical factors are obtained under other conditions. The filling factor is given by

$$f = t/t_{opt}. \qquad (21)$$

This approach is completely analogous to the previous introduction of effective depolarisation factors to treat dipole-dipole interaction, but we find the use of $<\Lambda>$ to be more transparent in the present case. The proper selection of $<\Lambda>$ is not straight-forward, though. Bedaux and Vlieger[27] do not indicate any weighting in favour of the smaller or larger separations and therefore we feel that $<\Lambda> = D$ (cf. Table II) should be most in line with their approach. This definition is used in the calculations reported on in Figs. 10 and 11 below, but we return to the choice of $<\Lambda>$ towards the end of this section.

The mean free path is equal to the radius for spherical particles[24] but no analytic expression for ℓ_j is known for prolate spheroids. It is obvious that $b_j < 2\ell_j < a_j$, and it must be supposed that ℓ_j lies closer to the lower limit than to the upper. There are also some grain boundaries for the larger islands which tend to decrease ℓ_j. Hence $\ell_j = b_j/2$ is expected to be a good approximation; this relation was used in all of our computations.

A preliminary comparison of theory and experiments is given in Fig. 10 where we investigate the role of island shape for discontinuous films with two mass thicknesses. In these calculations we set $\varepsilon_m = 1$. Spherical islands (dash-dash-dotted curves) are seen to display much higher transmittance than obtained experimentally (solid curves). The computed transmittance is decreased when we use the measured eccentricity ($\bar{e}$, dashed curves), but there is still a large discrepancy with the experimental data. This conclusion is similar to the one we reached for the gas evaporated particles from Fig. 6. The best agreement is obtained when the eccentricity is as large as 0.95 (dash-dotted curves), but this value is much too high to be reconciled with the island shapes. The origin of these differences between theory and experiment is the topic of the rest of this section where we discuss the roles played by ε_m and $<\Lambda>$.

The use of $\varepsilon_m > 1$ can be motivated if we note that the islands must have a certain region of contact with the substrates. Hence one expects ε_m to fall in the range $1 < \varepsilon_m < \varepsilon_s$, where $\varepsilon_s = 2.34$ is the dielectric constant for the Corning 7059 glass substrates. The prolate-spheroidal model for the islands, introduced in Sec. II (b), implies that this contact area must be small, and therefore ε_m ought to lie much closer to the lower limit than to the upper. Fig. 11 shows our calculated results for $\varepsilon_m = 1$, $\varepsilon_m = \varepsilon_s$, and $\varepsilon_m = (\varepsilon_s+1)/2$. The latter definition was introduced by David[28] and has since been used in, for example, Ref. 8. When ε_m is increased we find that the transmittance is diminished, and for $\varepsilon_m = \varepsilon_s$ we obtain a fair agreement with the measurements. However, this value for ε_m is certainly much larger than what we expect, so

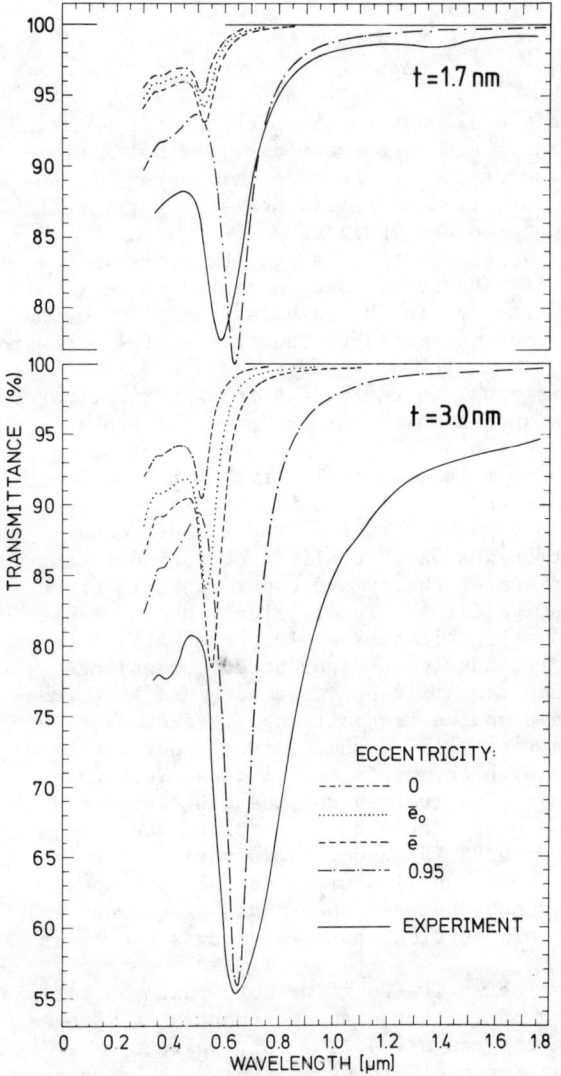

Fig. 10. Transmittance vs. wavelength for two discontinuous gold films. The solid curves denote measurements. The other curves represent calculations for prolate-spheroidal islands with different mean eccentricities. $\bar{e}$ is defined by Eq. (3) and $\bar{e}_o \equiv 1 - (\bar{b}/\bar{a})^2$.

the good fit between theory and experiments – though striking – should nevertheless be regarded as irrelevant.

Instead we return to Eq. (20) where the local field effects were simulated by identifying the effective film thickness with an unweighted average island separation. This cannot be quite correct, though, because the dipole-dipole coupling implies a strong weighting in favour of the smallest separations. Hence the width in the distribution of center-to-center distances must enter the problem. It is difficult to obtain an unambiguous measure of this

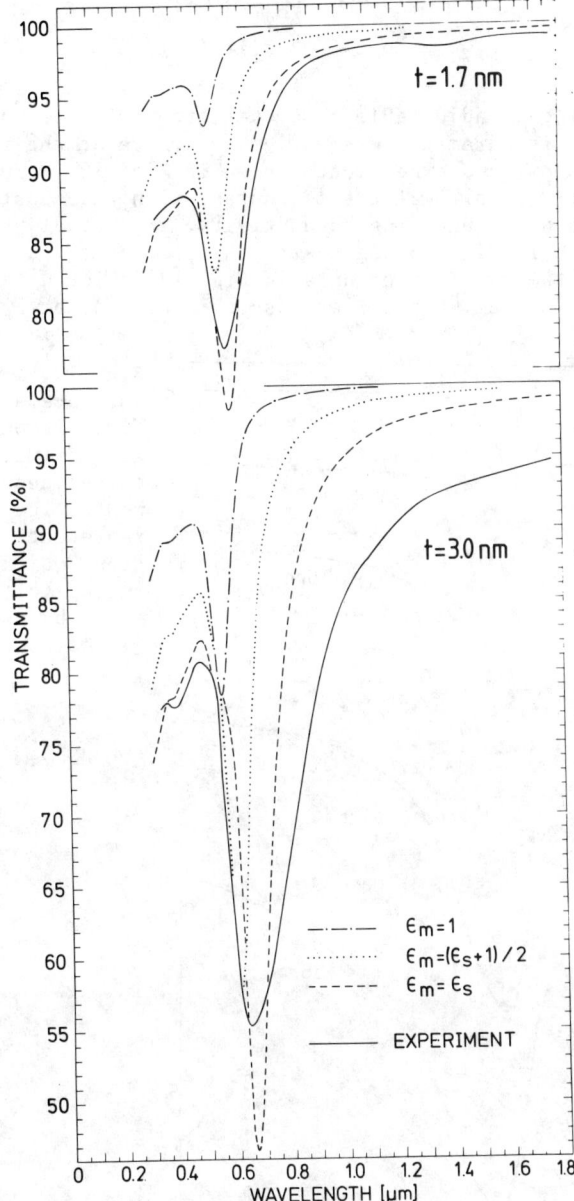

Fig. 11. Transmittance vs. wavelength for two discontinuous gold films. The solid curves denote measurements. The other curves represent calculations with different values of the dielectric constant for the medium which surrounds the islands.

quantity, and therefore we write

$$t_{opt}' = \frac{4}{3} \xi_{fit} D, \tag{22}$$

with ξ_{fit} being an adjustable parameter (the only one in our computations). This parameter was varied to arrive at the final comparison of theory and experiments shown in Fig. 12. We used $\varepsilon_m = 1.2 + i0.12$ to allow for some contact with the substrate in addition to a weak tunneling conductivity. The actual values of ξ_{fit} are given in Fig. 13 together with $f_{fit} \equiv t/t_{opt}'$. The agreement between the two sets of data in Fig. 12 is found to be excellent for $t < 3$ nm. For larger mass thicknesses the measured

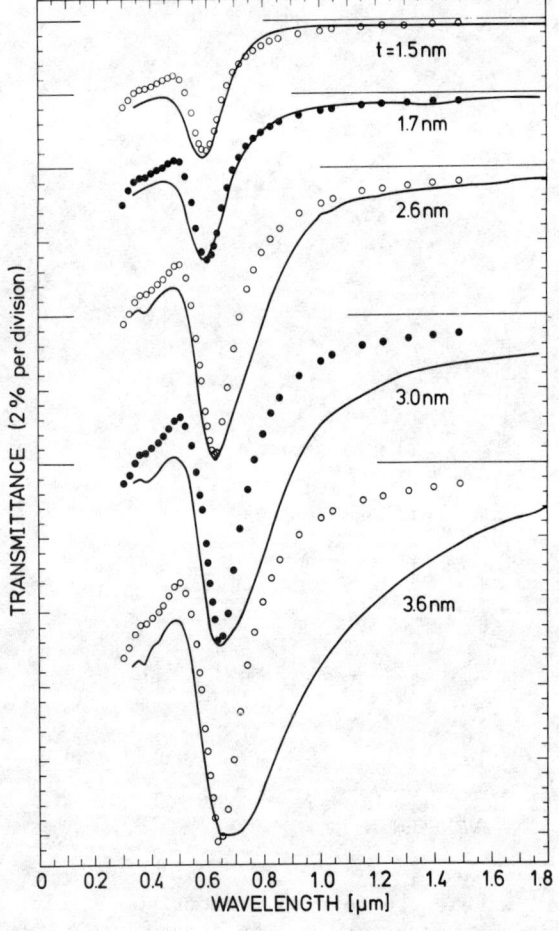

Fig. 12. Transmittance vs. wavelength for five discontinuous gold films. The solid curves denote measurements. The circles represent computations which were fitted to the experiments by use of the ξ_{fit}'s shown in Fig. 13.

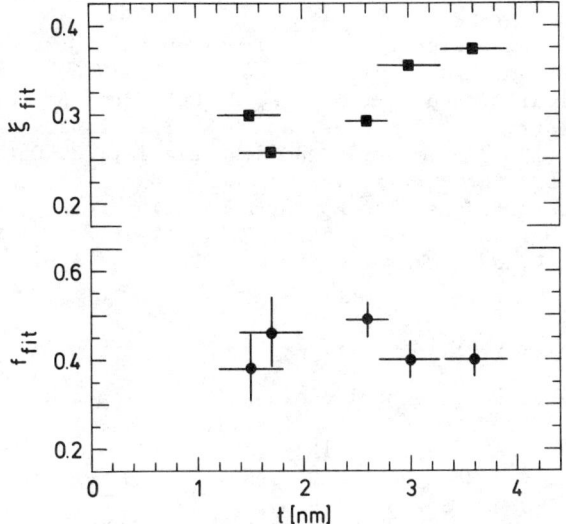

Fig. 13. ξ_{fit} and f_{fit} derived from the fittings in Fig. 12.

minima seem to be too broad. This is not an unexpected result, because in these films the islands look rather irregular so the spheroidal approximation becomes questionable.

VI. CONCLUSION

The most important result of the work reported here is the good agreement between theory and experiments documented for gas evaporated gold particles in Fig. 9 and for sufficiently thin discontinuous gold films in Fig. 12. We conclude that the optical properties for both kinds of ultrafine particles are governed entirely by the shapes of the islands and their separation. The band structure for gold appears to be virtually unaffected for diameters down to 3 nm. No noticeable effects seem to be caused by quantization of electronic levels. A more complete description of our experiments and calculations will appear elsewhere[29,30].

Acknowledgements. The results presented here were obtained in collaboration with Thorwald Andersson and Sten Norrman at Chalmers University of Technology, and with Ola Hunderi at Dept. of Physical Metallurgy, Norwegian Institute of Technology, Trondheim, Norway. Financial support was received from the Swedish Natural Science Research Council.

REFERENCES

1. A survey of the gas evaporation technique is given in C.G. Granqvist and R.A. Buhrman, J. Appl. Phys. $\underline{47}$, 2200 (1976).
2. L. Harris, R.T. McGinnies, and B.M. Siegel, J. Opt. Soc. Am. $\underline{38}$, 582 (1948); L. Harris, D. Jeffries, and B.M. Siegel, J. Appl. Phys. $\underline{19}$, 791 (1948); L. Harris and J.K. Beasley, J. Opt. Soc. Am. $\underline{42}$, 134 (1952); L. Harris, The Optical Properties of Metal Blacks and Carbon Blacks (The Eppley Foundation for Research, Monograph Series No. 1, Newport, R.I., 1967).
3. V.N. Sintsov, Zh. Priklad. Spektrosk. $\underline{4}$, 503 (1966) [J. Appl. Spectrosc. $\underline{4}$, 362 (1966)].
4. D.R. McKenzie, J. Opt. Soc. Am. $\underline{66}$, 249 (1976).
5. C. Doland, P.O'Neill, and A. Ignatiev, J. Vac. Sci. Technol. $\underline{14}$, 259 (1977).
6. The experimental technique is described in T. Andersson and C.G. Granqvist, J. Appl. Phys. $\underline{48}$, 1673 (1977).
7. S. Yamaguchi, J. Phys. Soc. Jpn. $\underline{15}$, 1577 (1960).
8. G. Rasigni, Rev. Opt. $\underline{41}$, 383, 566, 625 (1962); G. Rasigni and P. Rouard, J. Opt. Soc. Am. $\underline{53}$, 604 (1963).
9. R.H. Doremus, J. Appl. Phys. $\underline{37}$, 2775 (1966).
10. V.V. Truong and G.D. Scott, J. Opt. Soc. Am. $\underline{66}$, 124 (1976).
11. D.N. Jarrett and L. Ward, J. Phys. D: Appl. Phys. $\underline{9}$, 1515 (1976).
12. J.C. Maxwell-Garnett, Phil. Trans. R. Soc. Lond. $\underline{203}$, 385 (1904); $\underline{205}$, 237 (1906).
13. L. Genzel and T.P. Martin, Surf. Sci. $\underline{34}$, 33 (1973).
14. G. Mie, Ann. Phys. (Leipzig) $\underline{25}$, 377 (1908).
15. H.C. van de Hulst, Light Scattering by Small Particles (Wiley, New York, 1957).
16. D. Polder and J.H. van Santen, Physica (Utrecht) $\underline{12}$, 257 (1946).
17. O. Hunderi, Phys. Rev. $\underline{B7}$, 3419 (1973).
18. L.D. Landau and E.M. Lifshitz, Electrodynamics of Continuous Media (Pergamon, New York, 1969), §4.
19. D.A.G. Bruggeman, Ann. Phys. (Leipzig) $\underline{24}$, 636 (1935).
20. R.J. Elliott, J.A. Krumhansl, and P.L. Leath, Rev. Mod. Phys. $\underline{46}$, 465 (1974).
21. D. Stroud, Phys. Rev. $\underline{B12}$, 3368 (1975).
22. D.M. Wood and N.W. Ashcroft, Phil. Mag. $\underline{35}$, 269 (1977).
23. P. Winsemius, Ph.D. Thesis (Rijksuniversiteit te Leiden, the Netherlands, 1973) (unpublished).
24. J. Euler, Z. Physik $\underline{137}$, 318 (1954).
25. See, for example, O.S. Heavens, Optical Properties of Thin Solid Films (Dover, New York, 1965).
26. P. Clippe, R. Evrard, and A.A. Lucas, Phys. Rev. $\underline{B14}$, 1715 (1976).
27. J. Vlieger, Physica (Utrecht) $\underline{64}$, 63 (1973); D. Bedaux and J. Vlieger, Physica (Utrecht) $\underline{67}$, 55 (1973); $\underline{73}$, 287 (1974); see particularly Sec. 6 in the latter paper.
28. E. David, Z. Physik $\underline{114}$, 389 (1939).

29. C.G. Granqvist and O. Hunderi, Phys. Rev. B, in press.
30. S. Norrman, T. Andersson, C.G. Granqvist, and O. Hunderi, Solid State Commun., 23, 261 (1977); to be published.

DISCUSSION

J. FURDYNA (Purdue University): We have made microwave transmission measurements on loosely pakced semiconductor powders in which the grains are touching. We found the concept of an effective depolarization factor unavoidable. It is a single adjustable parameter which accounts for the local fields and for particle shapes in a phenomenological fashion.

GRANQVIST: I can obtain a rather good fit to my data for the gas evaporated particles with only one parameter. I think that, in reality, one must account for both the clusters and the chains and so I have introduced two parameters.

R. W. CHANG (Rockwell International): I have made a fit to your discontinuous film data using the effective medium theory. It seems to be not too bad a fit.

GRANQVIST: I think that the detailed analysis of J. Vlieger indicates that one should use the Maxwell-Garnett theory.

M. H. COHEN (University of Chicago): The effective medium theory is either identical to the Maxwell-Garnett theory or better than the Maxwell-Garnett theory under the conditions in which the effective medium theory holds. In the conditions of your experiment, the effective medium theory does not hold simply because the assumption of randomness is not met. As in the cermet films, the probability of finding vacuum or dielectric outside a particle is unity. The probability is not unity in situations to which the effective medium equations are usually applied. There is a variation of the effective medium theory, however, which would hold in your case and which would be better than the Maxwell-Garnett theory.

W. T. DOYLE (Dartmouth College): Because of the very complicated way in which all of the parameters enter the theory, the optical coefficients of the discontinuous films are quite different for non-normal incidence. If the same parameters hold for normal and for non-normal incidence, one can begin to have some confidence in the theory.

GRANQVIST: We are planning to do this experiment. The computation is rather difficult because the samples consist of prolate spheroids (looking like little cigars) lying on the substrate. This geometry is easy to treat at normal incidence; although away from normal incidence the computation is more difficult, it can be done.

K. OKUMURA (Xerox Corporation): You might consider small angle x-ray scattering as a means of characterizing your films. From the size of the particles and clusters in your experiment, I believe it would be possible to obtain the second order correlation function by this technique.

GRANQVIST: It would be very interesting if the particle-particle correlation function could be found, although I do not know if an experimental correlation function could be built into the formalism. Perhaps it could be done.

D. R. McKENZIE (University of Sidney): Why was scattering neglected in your analysis? According to Mie theory, scattering effects can broaden the resonance in a way similar to the experimental data. If the particle size is on the order of 1000Å, scattering is very important. If there are a dozen or so particles in your samples electrically in contact, the effective particle size would be about 1000Å. To neglect scattering one must assume that every particle is electrically insulated from every other particle.

GRANQVIST: Particle sizes in our specimens were much smaller than 1000Å, and even clusters of particles would not be that large. However, we did implicitly assume that particles were electrically isolated from one another.

T. YOUNG (University of Houston): The gold smokes have a very low reflection coefficient. What was the reflection coefficient of the discontinuous gold films?

GRANQVIST: We have not measured the reflectance. The films appear greyish, so they do reflect a little bit.

OPTICAL PROPERTIES OF COMPOSITE MATERIALS

B. Abeles
Exxon Research and Engineering Co., Linden, N. J.

ABSTRACT

Optical properties of composite materials, such as aggregated metal films, rough surfaces, cermets (metal-insulator mixtures), semiconductor-insulator mixtures and other multiphase systems have been of continuing interest since the beginning of the century. A review is given of their optical properties in the infrared and the visible. When the wavelength of light is large compared to the grain size, the composite material can be described in terms of a dielectric constant which is a function of the dielectric constants of the component materials and their relative concentrations. The applicability of theoretical models to the various composite materials is discussed. The importance of composite optical materials in solar energy applications is pointed out.

DISCUSSION

R. A. BUHRMAN (Cornell University): Were the measurements of efficiency of solar collectors done at 500°C?

ABELES: No. This was calculated from measurements at room temperature and at normal incidence. We assumed that the properties do not change with temperature. This assumption is very far from reality. What should be done is to make the measurement over a hemispherical reflectance and at the operating temperature.

R. CHANG (Rockewell International): What is the meaning of the negative efficiency which you showed?

ABELES: It is ficticious. In order to maintain the collector at a particular temperature, heat in addition to the sun must be put into the collector. In some rooftop collectors, by the way, there is an electric heater standing by to help the sun.

BUHRMAN: H. Craighead, of Cornell, has been working with Ni-Al_2O_3. He has graded from a 20% concentration at the back, or metal, surface to zero, pure Al_2O_3, on the front. Most of the interference effects in the visible can be eliminated and he obtains 94-95% efficiency at room temperature.

D. LAMBETH (Eastman Kodak): You alluded to surface roughness as being a desirable quality. Is the roughness made by the substrate?

ABELES: No. It is the MgO which causes the roughness. We do not understand the effect, except that it is related to sputtering. One suggestion is that MgO has a strong electron emission and gets charged. A lot of resputtering goes on and Au resputters faster than MgO and so this faceted structure is found.

LAMBETH: Did you investigate bias sputtering?

ABELES: No. That would be a worthwile thing to do.

W. T. DOYLE (Dartmouth College): High multipoles make important contributions to the oscillations of spheres. Both D. McKenzie and I have made calculations of these multipoles for various kinds of materials. I've looked at the real and imaginary parts of the dielectric constant of cermets made of Drude metals (with various assumptions about the quality of the metal) and also of Au and Ag. You show, in the case of Ag, the position of the peak shifting from the wavelength characteristic of spheres to what is characteristic of cylinders. In my interpretation you are looking at a mixture of modes - the dipole mode and all the higher modes. When higher order modes come into effect, new resonances pop up in a high quality sample - one with very small damping. If there is a lot of damping the new modes will not be resolved. The presence of these other modes shifts the peak very strongly towards the infrared. This is a very general effect which is always present and it will be present in random mixtures. (I have done the calculation in ordered mixtures as it is easier). There is nothing in the calculation except the optical constants of the metal and Maxwell's Equations.

A. J. SIEVERS (Cornell University): Your data for W particles showed a simple interference effect. Are you sure you still have

W metal there? I did a calculation on all the transition elements with Maxwell-Garnett and W was actually one of the best for absorbtion in the visible.

ABELES: I agree with you. We studied the whole composition range and about 40 volume percent W, just about at the metal-insulator transition, was optimal. We were very disappointed.

J. GITTLEMAN (RCA Laboratories): We may not have made it thick enough.

D. R. McKENZIE (University of Sydney): I'm a bit troubled about your use of room temperature emissivity in calculating the conversion efficiency. The emissivities usally increase dramatically with temperature.

ABELES: We don't know what the temperature dependence is. It's certainly worth measuring. As you say, things only get worse and not better.

OPTICAL PROPERTIES OF A MICROSCOPICALLY TEXTURED SURFACE

G. D. Cody and R. B. Stephens
Corporate Research Laboratories
Exxon Research and Engineering Company
Linden, New Jersey 07036

ABSTRACT

Current interest in optimum surfaces for solar collectors has focussed on rough surfaces. We consider surface texture on a scale less than the wavelength of incident light. We utilize effective medium theories to relate a spatial variation in mass density to a spatial variation in dielectric constant. An iterative solution to Maxwell's equations for the above configuration is stated in terms of reflectivity, transmission and absorption for the surface grade as a function of angle of incidence, polarization and wavelength. Comparison is made to an exact analytic solution. Experimental data are presented for some examples of microscopic texturing. Finally we consider within the framework of the effective medium theory the contribution of polarization phenomena.

INTRODUCTION

Current interest in solar technology has led to a revival of research on the optical properties of surfaces. The motivation is apparent. Although the commercial viability of solar energy depends on many factors, it is hard to disagree with the assumption that the optimization of the optical properties of all solar devices will always be a significant contributor to performance and hence life cycle cost. Of course, this assumption predicates cost-effective processing and materials.

In the solar thermal application a variety of useful surfaces have been obtained in an attempt to realize the selective surface concept of Tabor[1,2]. Recent success in understanding and controlling the optical properties of composite materials has given a new direction to this research and one that shows great promise. In particular, the various effective medium theories for the optical properties of composite materials have proven to be useful conceptual models for the design of selective surfaces[3,4].

In the photovoltaic application where optical optimization is particularly important due to marginal efficiency, the majority of devices are prepared with some form of antireflection film for control of reflectivity. The reduction of reflectivity by a macroscopically textured surface and multiple reflections has also received some recent attention.[5]

This paper will not go further into the solar justification for controlling the optical properties of surface and will not focus on the performance of new optimum surfaces for solar technology. We will rather present a systematic approach to the optical properties of surfaces which is rigorous within the limitations of certain simplifying assumptions. We will demonstrate the utility of appropriate effective medium theories in modeling "graded" surfaces which should possess optimum optical properties. While emphasizing the utility of the effective medium and Maxwell Garnett theories[6] for the design and interpretation of experiments, we will also indicate areas of uncertainty for the application of these models.

Perhaps the major feature of our approach is that it sets a bench mark against which the optical performance (reflection, absorption and transmission coefficients) of any real surface can be compared. Within the framework of a rigorous optical calculation, but an approximate physical model, we will present optical properties of surfaces which delineate the advantages of a "graded optical" surface. Finally we will present data on a real surface which exhibits some of the features of the model systems.

THEORY: CLASSICAL OPTICS

As noted, it is obvious that control of the reflectivity is crucial for any solar device. Anti-reflection films are one attempt to solve this problem, but they are expensive to apply, are delicate, suffer in performance unless the solar radiation is incident normal to their surface, and are not effective over the broad solar spectrum. A more desirable route to low reflectivity could be a low reflecting surface formed directly on the materials of interest.

Since reflectivity is due to the abrupt transition at the surface, between the dielectric constant of air and that of the material, it is appealing to consider a graded transition in the dielectric constant in order to reduce the reflectivity. In 1880, Rayleigh[7] considered the modification in reflectivity "caused by the substitution of a gradual for an abrupt transition" from one media to another. He remarked that the interesting question was not the existance of the effect, but rather the magnitude of the reduction in reflectivity as well as the characteristic distance that distinguishes an abrupt from a graded junction. Rayleigh chose to consider a stretched string for analysis, but was undoubtedly interested in the optical problem as well.

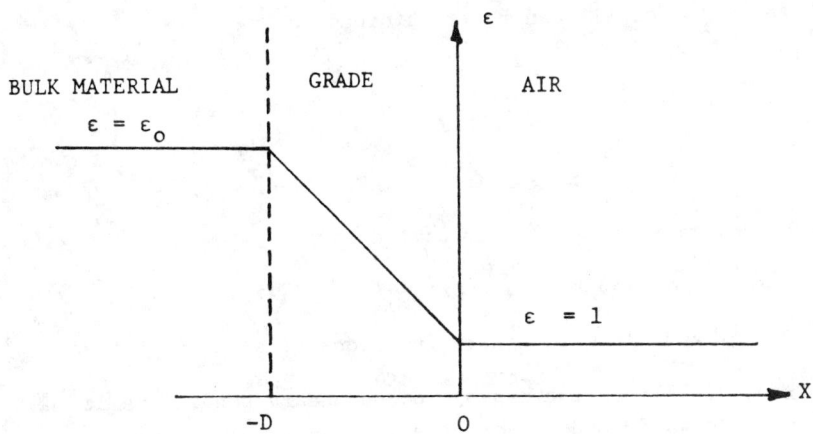

Fig. 1. Variation of dielectric constant with depth for a linear grade line of depth D from air ($\varepsilon = 1$) to its bulk value ($\varepsilon = \varepsilon_0$)

Fig. 1 sketches an experimental configuration. A plane wave traveling in the negative "X" direction is incident on a medium where the dielectric constant $\tilde{\varepsilon}(x)$ varies with position from its value in air ($\tilde{\varepsilon}(0) = 1$), to its value ($\tilde{\varepsilon}$) in the bulk of the material. If the grade occurs over a distance D, $\tilde{\varepsilon}(-D) = \tilde{\varepsilon}$. For a plane wave of unity power density, we are interested in the reflectivity R at $x = 0$, the transmission coefficient T at $x = -D$, and the absorbed power, A, in the layer. In general, $R + T + A = 1$. If $\tilde{\varepsilon}(x)$ is real, $A = 0$.

For an abrupt junction $R = R_o$ where

$$R_o = \left| (1 - \tilde{\varepsilon}^{1/2})/(1 + \tilde{\varepsilon}^{1/2}) \right|^2$$

For a graded junction the optical properties are determined by solving the appropriate Maxwell equations subject to the usual boundary conditions at $x = 0$ and $x = -D$.

For an E field polarized in the Z direction we have,

$$d^2E/dx^2 + k^2 \tilde{\varepsilon}(x) E = 0 \qquad (1)$$

where $k^2 = (2\pi/\lambda)^2$ and where λ is the free space wavelength of the incident radiation.

An analytic solution to Eq. (1) was first shown by Albini and Jahn[8] for a linear variation in $\tilde{\varepsilon}(x)$ i.e.

$$\tilde{\varepsilon}(x) = (1 - (\tilde{\varepsilon} - 1)(x/D)) \qquad (2)$$

From Eq. (2), Eq. (1) can be transformed to

$$d^2E/dV^2 - VE = 0 \tag{3}$$

where

$$V(x) = -(\tilde{\varepsilon}(x))(kD/\tilde{\varepsilon}-1)^{2/3} \tag{4}$$

The solution to Eq. (3) is of the form

$$E(x) = A\ Ai(V(x)) + B\ Bi(V(x)) \tag{5}$$

where $Ai(V)$ and $Bi(V)$ are Airy functions of argument V^9.

From Eq. (5) and the appropriate boundary conditions it is straightforward to determine R and T, and A from $A = 1 - T - R$.

The transmission coefficient T is given by

$$T = (4n/\pi^2)\ (\gamma^2)\ (\alpha b - a\beta^2) \tag{6}$$

In Eq. (6),

$$\tilde{n} = (\tilde{\varepsilon})^{1/2} = n + i\kappa$$

$$\gamma = (-i)(kD/\tilde{\varepsilon}-1)^{1/3}$$

$$a = Ai'(V(0)) + \gamma Ai(V(0))$$

$$\alpha = Ai'(V(-D)) - \gamma\tilde{n}\ Ai(V(-D))$$

$$b = Bi'(V(0)) + \gamma Bi(V(0))$$

$$\beta = Bi'(V(-D)) - \gamma\tilde{n}\ Bi(V(-D))$$

The prime in the above denotes the derivative of the appropriate Airy function. A considerably more complicated algebraic form for R is obtained. In the following calculation, we will only consider real $\tilde{\varepsilon}$, hence $A = 0$ and $R = 1-T$. For a given material, the above equations are only a function of kD.

Table I summarizes the calculation of T and hence R, for Si where $\lambda > 0.5\mu$, $\tilde{\varepsilon} \approx 16$. The grade distance D is $1.0\mu m$. R_o is the abrupt reflectivity.

TABLE I
VALUES OF T, R AND R_0 FOR $\tilde{\varepsilon} = 16$, D = 1.0μm
AS A FUNCTION OF FREE-SPACE WAVELENGTH λ

λ (μm)	D/λ	T	R	R_0
0.5	2.0	0.99	0.01	0.36
1.0	1.0	0.96	0.04	0.36
5.0	0.2	0.88	0.12	0.36
10.0	0.1	0.78	0.22	0.36

Several interesting features emerge from Table I. First, the reflectivity is quite small even for D/λ $\tilde{\sim}$ 1. Second, interference structure in the reflectivity is small even though no absorption has been included in ε(x). Third, at $2\pi D/\lambda \tilde{\sim} 1$, or $kD \tilde{\sim} 1$, the reflectivity falls to about one half its abrupt value R_0. This result suggests a criteria for distinguishing between an abrupt and graded transition in the dielectric constant.

The lowering of the reflectivity over a broad band of wavelength illustrated in Table I can be contrasted with the narrow band response of single layer anti-reflection films. Finally, at λ = 0.5μm the reflectivity can be expressed as $R \tilde{\sim} R_0^{4.5}$. The high value of the exponent should be compared to the R_0^2 dependence expected for macroscopically rough surfaces where multiple geometric reflections are required before the incident radiation can be reflected.[5]

Clearly a graded surface presents advantages for all solar devices over either macroscopically rough surfaces or anti-reflection coatings. Furthermore, the approach to abrupt reflectivities for $2\pi D/\lambda \ll 1$, suggests possibilities for "Tabor" selective absorbers with low reflectivity in the visible and high reflectivity in the infrared. All that is required is sufficiently high R_0 and an appropriate grade distance D. As will be seen, this is true for good conducting metals.

A major question remains as to how to realize such a graded surface in real materials. We are also interested in the inherent limitations of the classical optical approach due to scattering and polarization phenomena when we consider the construction of such graded surfaces. Before we consider such questions we will present a more practical computational approach than that represented by Eqs. (2) - (6).

THEORY: COMPUTER CALCULATION

The approach represented by Eqs. (2) - (6) to the solution of Eq. (1) has the advantage of mathematical precision, but that is all. Even for real $\tilde{\epsilon}(x)$, the equations are complex and laborious to evaluate. Non-normal angles of incidence would lead to yet more algebra. However, the major problem in the use of the analytic solution is the unphysical form of Eq. (2). The most natural approach to a grade in $\tilde{\epsilon}(x)$ is through either a variation in composition for a composite material as shown in Fig. 2a or through a microscopic roughening or texturing of a surface as shown in Fig. 2b to produce a mass variation. Both grades can be related to $\tilde{\epsilon}(x)$ through either the Maxwell Garnett or Effective Medium models. There is by now sufficient experimental evidence to suggest that these theories of the dielectric constant of composite materials are good starting points for modeling $\tilde{\epsilon}(x)$[6].

When we do apply these theories for a plausible linear variation of mass or composition, the spatial dependence of $\tilde{\epsilon}(x)$ is by no means as simple as that represented by Eq. (2). In Fig. 3 the concentration dependence of the dielectric constant for gold spheres in air is shown within the effective medium approximation at a wavelength of 2 µm. Clearly Eq. (2) is not even a good approximation for such a variation. Additional complication is produced by the geometrical structure shown in Fig. 2b which leads to a tensor formulation for $\tilde{\epsilon}(x)$ through the grade, although the bulk material is isotropic. Again Eq. (2) would be inadequate. Finally, multiple discontinuous changes in $\tilde{\epsilon}(x)$ through the grade are not easily handled by the analytic approach.

We have described in a previous paper[3] an iterative approach to the solution of Eq. (1). A versatile computer program has been developed which divides the grade into an arbitrary number of layers in each of which R, T and A are calculated by the usual equations. Recursion relations then determine the stacked films' R, T and A to a set precision for arbitrary polarization, angle of incidence and form of $\tilde{\epsilon}(x)$. The calculation agrees with the analytic solution represented by Table I for normal incidence for $\tilde{\epsilon} = 16$.

For Fig. 4 and Fig. 5 we show calculated results for glass and gold where a linear compositional variation has been related to $\tilde{\epsilon}(x)$ through either the effective medium theory for symmetrical systems or through the Maxwell Garnett theory where the particles can be distinguished from the medium.[6] While Fig. 4 is a model calculation, a recent publication[10] presents data on a multi component glass whose surface has been graded by selective leaching and the results are similar. Fig. 4 also shows the optical reflectivity produced by an anti-reflection film.

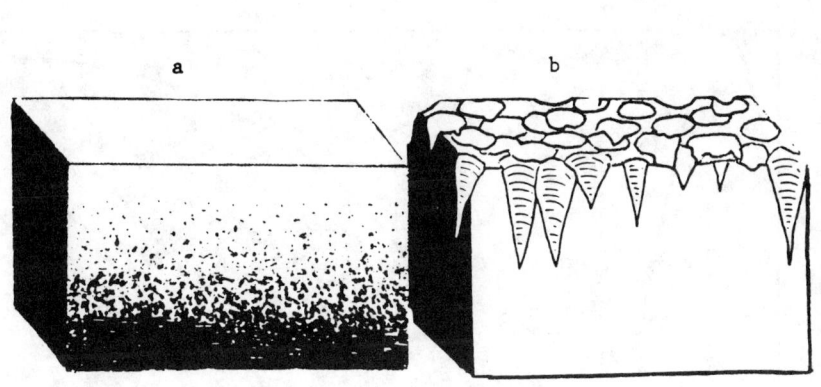

Fig. 2. Graded Surface Topologies.

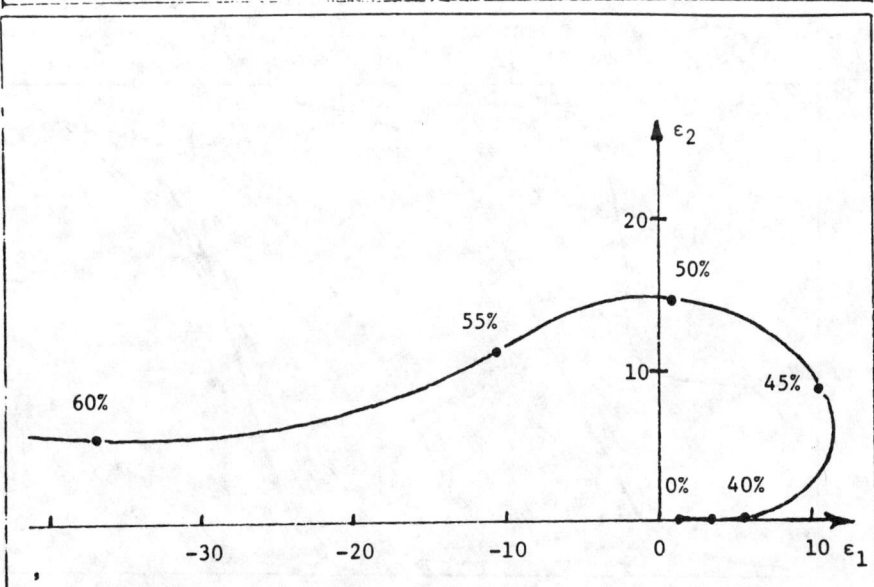

Fig. 3. Complex dielectric constant ($\varepsilon_1 + i\varepsilon_2$) of gold spheres in air as a function of their concentration for $\lambda = 2\mu m$ and using a mean field approximation.

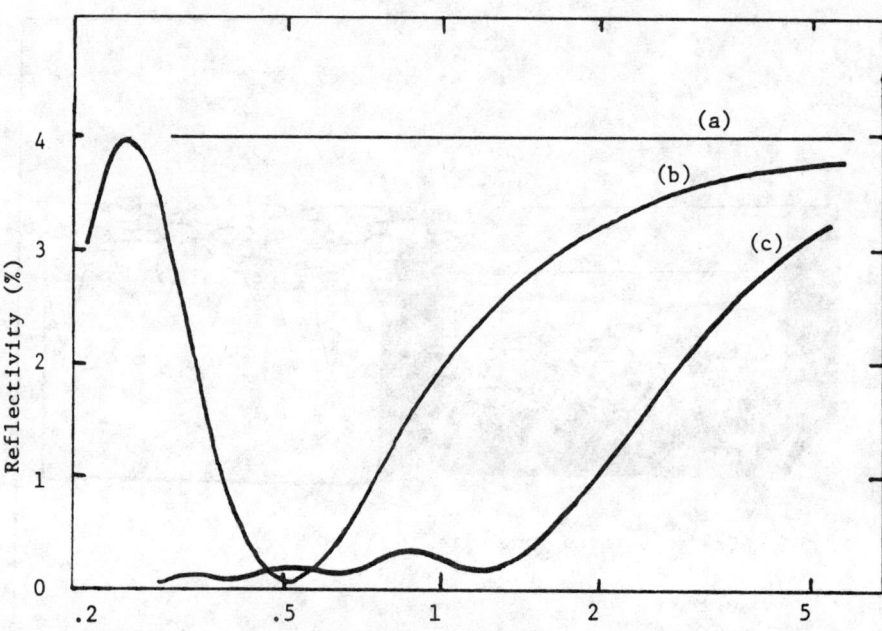

Fig. 4. Reflectivity of glass surfaces which are a) abrupt, b) coated with an antireflection film, and c) graded.

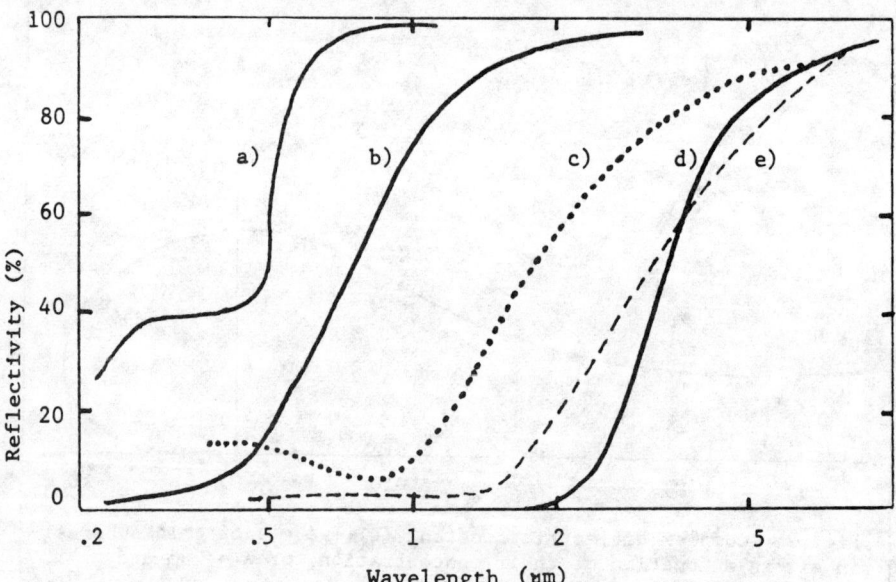

Fig. 5. Reflectivity of graded gold surfaces: a) abrupt, b) 0.1μm grade, d) 1.0μm grade, compared to the experimental reflectivity of c) chrome black and e) nickel black.

The model calculation for Au shown in Fig. 5 illustrates an excellent selective surface. A suitably chosen value for the grade distance D represents an improvement over two popular selective surfaces, nickel blacks and chrome blacks[2]. As shown in Fig. 5, the graded gold has both a lower reflectivity in the visible and as sharp a transition in the infrared to low emissivity. The challenge to the materials scientist is to do as well with less expensive material!

In Table II we present the wavelength λ_c normalized to the grade distance D, where λ_c is defined as the wavelength at which the reflectivity is one-half its "abrupt" value $(R_o/2)$ for a variety of materials with considerably different n and κ.

TABLE II
λ_c/D FOR A VARIETY OF GRADED MATERIALS

Material	n	κ	λ_c/D
ScO_2	1.5	0	6
Si	4	0	6
Au	1.2 → 1.3	1.9 - 13	3.5 → 7
Ti	3	3 - 5	6

From Table II it appears that $kD \lesssim 1$ is indeed a good "rule of thumb" to distinguish an abrupt from a graded transition in dielectric constant.

SCATTERING

For either surface shown in Fig. 2, as the wavelength, λ, is reduced, the reflectivity will first enter a domain where polarization scattering dominates the specular reflectivity, and finally, a domain where geometrical optics and multiple reflections dominates the reflectivity.

There are extensive references in the literature to the optical properties of abrupt but statistically rough surfaces[11]. Instead of utilizing these rigorous theories we have chosen to postulate a surface texture whose scale is considerably less than the wavelength of the incident radiation. Under these circumstances, the reduction in the reflectivity obtained above would not be affected by either scattering or multiple reflections.

It is interesting to note the resemblance between these requirements and the structure of the moth's eye[12] where there is a low reflectivity in the visible. It has corrugations in its surface which produce a graded surface 200Å deep. The solar absorber design of Thornton[13] was motivated by the moth's eye principle of impedance matching which was here approached through optics.

POLARIZATION PHENOMENA

The ordered geometrically graded surface presents additional complications, not only from scattering, but from polarization phenomena due to the shape anisotropy of the surface. Similar effects could occur if compositional variations were accompanied by shape anisotropy in the grains which were correlated at the surface.

We utilize the Maxwell Garnett theory where the shape anisotropy was first calculated by Bragg and Pippard[14] and independently by Cohen et al.[15] On a surface with oriented ellipsoids the theory defines the two independent elements of the tensor dielectric constant as

$$(\tilde{\epsilon}_i - \epsilon_o)/(L_i \tilde{\epsilon}_i + (1 - L_i)\epsilon_o) =$$
$$y(x)\ (\tilde{\epsilon}_1 - \epsilon_o)/(L_i \epsilon_1 + (1 - L_i)\epsilon_o) \quad (7)$$

In Eq. (7), $i = \perp$ or 11, $\tilde{\epsilon}_1$ is the dielectric constant of the ellipsoidal particle, ϵ_o that of the matrix, the L_i are the usual depolarization coefficients ($L_{11} + 2L_\perp = 1$) and $y(x)$ is the spatially dependent volume fraction. Eq. (7) should be distinguished from that of Granquist and Hunderi[16] where a random orientation of ellipsoids within a wavelength of light is assumed.

The graded surface is now a uniaxial surface with an <u>ordinary</u> wave whose E field is polarized perpendicular to the axis of symmetry of the ellipsoid and which has a dielectric constant $\tilde{\epsilon}_{or}$ given by

$$\tilde{\epsilon}_{or} = \tilde{\epsilon}_\perp \quad (8)$$

There is also an extraordinary wave with the E vector confined to the plane defined by the symmetry axis of the ellipsoid and its propagation vector. The appropriate dielectric constant $\tilde{\epsilon}_{EX}$ is given by

$$\tilde{\epsilon}_{EX} = \tilde{\epsilon}_{11} \tilde{\epsilon}_\perp / (\tilde{\epsilon}_\perp \sin^2\theta + \tilde{\epsilon}_{11}\cos^2\theta) \quad (9)$$

where θ is the angle between the propagation vector and the symmetry axis of the ellipsoids. The extraordinary wave is not, in general, polarized perpendicular to its propagation vector.

For such a graded surface, one has, in general, all the complexities of uniaxial optics imposed on that of the grade. However, there are obvious simple geometries. This is the case of normal incidence on "needle" like ellipsoids perpendicular to the surface ($L_{11} = 0$, $L_\perp = 1/2$) or on "razor blade" disks with axes parallel to the surface ($L_{11} = 1$, $L_\perp = 0$).

For "needles" the ordinary and extraordinary waves are degenerate at normal incidence and $\tilde{\varepsilon} = \tilde{\varepsilon}_{or} = \tilde{\varepsilon}_{EX} = \varepsilon$ where

$$(\varepsilon_\perp - \tilde{\varepsilon}_o)/(\varepsilon_\perp + \varepsilon_o) = y(x)(\tilde{\varepsilon}_1 - \varepsilon_o)/(\tilde{\varepsilon}_1 + \varepsilon_o) \quad (10)$$

For "razor blades" with the E field parallel to the "blade" symmetry axis (i.e. normal to the plane of the blades) $\tilde{\varepsilon} = \tilde{\varepsilon}_{or} = \tilde{\varepsilon}_{EX} = \varepsilon$

$$(\tilde{\varepsilon}_{11} - \varepsilon_o)/\tilde{\varepsilon}_{11} = y(x)(\tilde{\varepsilon}_1 - \varepsilon_o)/\tilde{\varepsilon}_1 \quad (11)$$

For "razor blades" with the E field along the "blades" $\tilde{\varepsilon} = \tilde{\varepsilon}_{or} = \tilde{\varepsilon}_\perp$ where

$$\tilde{\varepsilon}_\perp - \varepsilon_o = y(x)(\tilde{\varepsilon}_1 - \varepsilon_o) \quad (12)$$

The differences between Eqs. (10), (11), and (12) are considerable and suggest a variety of interesting experimental possibilities.

EXPERIMENTAL

Of course, one can produce low reflectivities by more complicated schemes. Figure 6 presents experimental reflection data (hemispherical) for an etched ordered eutectic [17] which is clearly not a graded surface. It does, however, make effective use of topology to achieve a very low reflectivity. At very short wavelengths (comparable to the nickel crystallite size of about 100 A, it would have the reflectivity of a rough nickel surface ($R_{eff} = (R_{Ni})^2$). That would be about 20% at .3 microns; the rise in the reflectivity as one goes to shorter wavelengths is the beginning of the transition to that regime. In the visible to infrared, the nickel crystallites are so small compared to the light wavelengths that the rods are better described as an effectively homogeneous medium which has a much lower reflectivity than pure nickel. Since the rods form a rough surface, the reflectivity is still diffuse and even smaller ($R_{eff} = (R_{Ni_{eff}})^2$). At yet longer wavelengths, in which the roughness of the rods is small compared to the wavelength, the reflection should become specular and go to 100%, but that regime (25-40 microns) is beyond our measuring capacity.

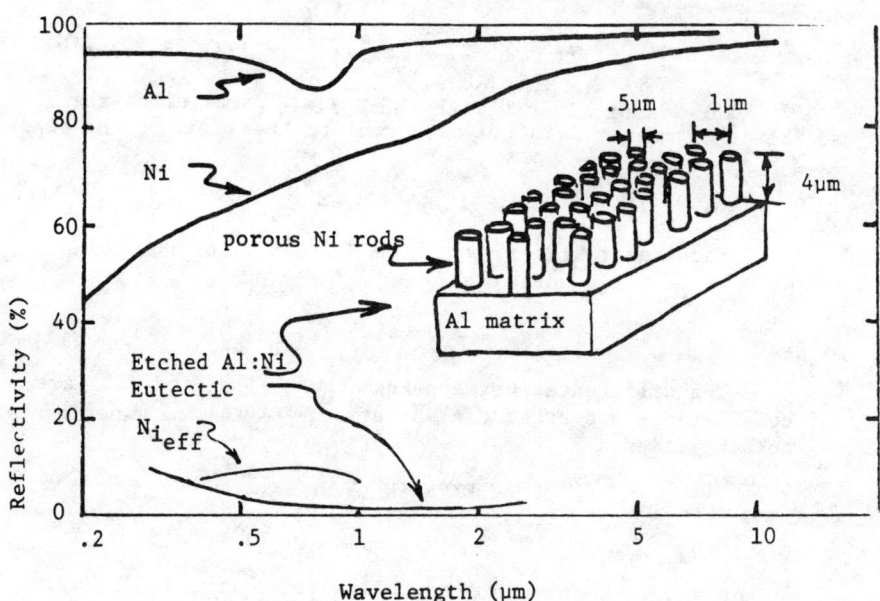

Fig. 6 Reflectivity of an etched Al:Ni eutectic compared to that of its components. The sketch shows the idealized structure on the etched surface. In reality the rods are porous and the structure less regular.

SUMMARY

We have shown in a limited example by an analytic solution, and also by a more general iterative computer program that a graded surface leads to extremely low reflectivities in the domain where we assume the scale of surface texture is considerably less than the wavelength of light. If the material has high reflectivity in the infrared for $\lambda > 6D$ the reflectivity is sufficiently large that such a surface approaches the ideal "selective surface" of Tabor[1]. We have presented model calculations of such surfaces and shown the relative independence of the results over large angles of incidence and wide variation in λ. The advantages over anti-reflection films are apparent. Preliminary experimental data was shown for a non-selective absorber illustrating the variety of approaches possible with textured surfaces. Finally we have pointed out the uniaxial nature of such a surface when it is prepared by geometrical roughness or non-spherical inclusions in a composite medium.

REFERENCES

1. H. Tabor, Low Temperature Engineering Applications of Solar Energy, R. C. Jordan (ed.) ASHRAE, New York, (1976) Ch. IV, and reference therein.

2. L. Melamed and G. M. Kaplin, J. Energy $\underline{1}$ 100 (1977).

3. R. B. Stephens and G. D. Cody, Thin solid Films $\underline{45}$ 19 (1977).

4. I. T. Ritchie and B. Window, Applied Optics $\underline{16}$, 1438 (1977).

5. C. R. Baraona and H. W. Brandhorst, 11th IEEE Photovoltaics Conference (1978) IEEE New York (1975) page 44; J. J. Cuomo, J. F. Ziegler and J. M. Woodal, Applied Physical Letters $\underline{26}$ 557 (1975)

6. D. Stroud, Physical Review B12 3368 (1975); I. Webman, J. Jortner, and M. H. Cohen, Physical Review B $\underline{15}$ 5712 (1977)

7. J. H. Strutt (Lord Rayleigh), Philos. Mag. $\underline{12}$ 51 (1891).

8. F. A. Albini and R. G. Jahn, J. App. Phys. 32, 75 (1961).

9. Handbook of Mathematical Functions. Ed. M. Abramowitz and I. Stegrin, Dover, New York (1965) page 446.

10. M. J. Minot, J. Opt Soc. Am. 66 515 (1976).

11. D. Beaglehole and O. Hunderi, Physical Review B2 309 (1970), C. Acquista, Applied Optics 15 2932 (1976); G. Rasigni, J. P. Palmari, and M. Rasigni, Physical Review B12, 121 (1975)

12. P. B. Clapham and M. C. Hultez, Nature 244 281 (1973)

13. B. S. Thornton, J. Opt. Soc. Am. 65 267 (1975).

14. W. L. Bragg and A. B. Pippard, Acta. Cryst. 6 865 (1953)

15. R. W. Cohen, G. D. Cody, M. D. Coutts and B. Abeles, Physical Review B8 3687 (1973).

16. C. G. Granquist and O. Hunderi, Physical Review B (to be published)

17. R. B. Stephens (to be published).

DISCUSSION

A. J. SIEVERS (Cornell University): Graded films have been used by Perkin-Elmer for a number of years to make absorbers over large wavelength regions. They have been making graded Ti-MgO films for solar energy applications. Etched glass was first made by RCA in the late 1940's. Minneapolis Honeywell picked the idea up for solar energy purposes and developed a double etch which does reduce the reflectivity of glass to less than 1% over the visible spectrum. The whole system is sold commercially by, I believe, Lennox.

CODY: The idea of a graded surface is not novel. I think it was in 1880 that Rayleigh made the comment that it is obvious that something which occurs gradually will have low reflectivity. The interesting questions have to do with the definition of grade, with how low the grade can get, and with how an optimum grade is obtained.

M. H. COHEN (Univ. of Chicago): I gather that you chose the linear variation in dielectric constant because you could solve the problem. A quadratic variation will also give an analytic solution. There are whole families of solutions which you could choose and to some extent it is possible to do functional analysis. One could learn something about optimization.

CODY: I agree with you. We chose the linear variation because we could get an analytic solution. Since then we have gotten away from analytic solutions.

COHEN: The brush surface that you obtained with the Ni-Aℓ eutectic is essentially homogeneous rather than graded - a homogeneous inhomogeniety.

CODY: It turned out to be graded because of the unexpected flopping around of the brush. It is denser at the top than at the bottom but it is quite black.

D. R. McKENZIE (Univ. of Sydney): There has been theoretical work done at Sydney on the inverse problem from yours; holes in a material rather than cylinders or rods. That problem has been solved exactly for the case of a perfect conductor.

CODY: There is interesting work on the moth's eye, which looks very much like the Aℓ-Ni eutectic and which is blacker that it should be. This fact has been alluded to in some of the Australian literature on selective surfaces. Presumeably the reasons are associated with some kinds of grade.

OPTICAL PROPERTIES OF SMALL PARTICLE COMPOSITES: THEORIES AND APPLICATIONS*

W. Lamb, D.M. Wood and N.W. Ashcroft
Laboratory of Atomic and Solid State Physics
and
Materials Science Center
Cornell University, Ithaca, N.Y. 14853

ABSTRACT

The problem of calculating the optical properties of an assembly of small metallic spheres (radius a) dispersed in a dielectric host is approached from the standpoint of multiple scattering theory as formulated for scalar fields and periodic systems by Morse. If k is the largest wave vector of the radiation in either the bulk constituent or in the composite, the leading term for the dielectric function in the limit $ka \to 0$ emerges as the Maxwell-Garnett result. The result is subject to significant modification, however, if structure in the composite large on the scale of a wavelength (e.g. clustering or periodicity in the dilute limit) is present. For both Maxwell-Garnett and effective medium theories the effects of introducing a third component in the form of an insulator coating (spatially correlated, therefore, to the metallic component) are examined, particularly in the frequency dependent reflectivity. Finally, an argument is given for the two component system which leads to the Maxwell-Garnett result from consideration of an effective medium approach for insulator-coated metallic inclusions, these properly reflecting the topology of such composites. For $m(>2)$ component systems an expression for the effective dielectric function is derived by applying a similar argument.

---

*Work supported in part by ERDA, Grant EG-77-5-03-1456 and in part by the NSF through the facilities of the Materials Science Center at Cornell University (Grant DMR-76-81083, Technical Report #2895).

I: INTRODUCTION

By 'small' particles we shall mean those of a size typically found in metal smokes and blacks. Their characteristic dimension, say a, is much less than a wavelength of infra-red and visible light, i.e. $ka \ll 1$ where k is the wave-vector. The particles, though small, are considered large enough that they can be described by bulk physical constants and by the macroscopic equations of electromagnetism. We are concerned here with the response of composite assemblies of those particles to monochromatic radiation with frequency ω. The problem is one of computing the dielectric function $\bar{\epsilon}(\omega, k \to 0)$ of a heterogeneous medium of arbitrary composition, a problem for which no exact solution is likely, though in certain circumstances useful bounds on $\bar{\epsilon}$ can be extracted by exploiting variational principles[1] or convexity arguments[2]. Among approximations to $\bar{\epsilon}$ subject to such bounds are the results of the Maxwell-Garnett (MG) theory[3] and the effective medium theory (EMT)[4,5] which bear close relationships in the electronic theory of alloys to, respectively, the average t-matrix approximation (ATA) and the coherent potential approximation (CPA)[6].

One of the aims of the present paper is to show that the Maxwell-Garnett theory emerges as a limiting case of multiple scattering theor as applied first to a two component periodic array of spherical metal scatterers, and followed by a discussion of the changes required, in order to treat a disordered system. A second aim is to present comparisons of the results of Maxwell-Garnett theory and effective medium theory as applied to model systems of physical interest including, in particular, metallic spheres coated with insulator and embedded in a dielectric host. This example gives an interesting insight into why Maxwell-Garnett and effective-medium-theories disagree and offers a physical basis for establishing an "effective-medium approach" which differs somewhat from the usual one.

II: FORMALISM

Consider a two component composite system in a volume V. It consists of a dielectric host in which is embedded an array of N metallic spheres at a mean density $\rho = N/V$. The electromagnetic properties of the host are characterized by (scalar) dielectric constant $\epsilon_1(\omega)$ and (scalar) permeability $\mu_1(\omega)$: the spheres, on the other hand, are characterized by $\epsilon_2(\omega)$ and $\mu_2(\omega)$. None of the ϵ_i or μ_i are necessarily real. The average dielectric constant of the medium, defined below, will be written

$$\bar{\epsilon} = \bar{\epsilon}(\omega, \eta; ka \to 0) \qquad (1)$$

where η is the volume filling fraction of the metal. In (1) $ka \to 0$ refers to the fact that our analysis is confined in its regime of validity to the case where the smallest of the (generally complex) wavelengths of physical interest, either in a constituent material or in the composite itself, still greatly exceeds a sphere radius a.

To determine $\bar{\epsilon}$ we use an adaptation of Morse's[7] treatment of the band-structure problem of independent electrons moving in periodic fields, which in turn is closely related to the Korringa, Kohn, Rostoker or Green's Function method.[8]

In a heterogeneous medium described by $\epsilon(\underline{r},\omega)$ and $\mu(\underline{r},\omega)$ (where according to position the μ are the ϵ_i and μ_i described above) the equation for magnetic field $\underline{H} = \underline{H}(\underline{r})$ is:

$$\underline{\nabla} \times (\underline{\nabla} \times \underline{H}) - \frac{\omega^2}{c^2} \epsilon(\underline{r}) \mu(\underline{r}) \underline{H} - \epsilon^{-1}(\underline{r}) \underline{\nabla}\epsilon(\underline{r}) \times (\underline{\nabla} \times \underline{H}) = 0, \quad (2)$$

with origin of coordinates taken as the centroid of one of the metallic spheres. Consider first the case in which $\epsilon(\underline{r})$ and $\mu(\underline{r})$ are periodic: then, as is well known, (2) can be rewritten in the interstitial space surrounding the origin as an integral equation of the form:

$$\underline{H}(\underline{r}) = \left(\frac{1}{4\pi}\right) \int_S dS \left\{ G_{\underline{k},\omega_1}(\underline{r},\underline{r}_s) \frac{\partial \underline{H}}{\partial n_s} - \underline{H}(\underline{r}_s) \frac{\partial G_{\underline{k}\omega_1}}{\partial n_s} \right\} \quad (3)$$

where the surface integral is over both sphere and primitive cell boundaries. Here $\omega_1 = \sqrt{\epsilon_1 \mu_1}\, \omega/c$ and n_s denotes an outward normal. The Green's function G appearing in (3) may be written

$$G_{\underline{k}\omega_1}(\underline{r},\underline{r}') = \sum_{\underline{R}} \frac{\exp i\, (\underline{k}\cdot\underline{R} + \omega_1|\underline{r}-\underline{r}'-\underline{R}|)}{|\underline{r}-\underline{r}'-\underline{R}|} \quad (4)$$

where $\{\underline{R}\}$ is the set of N lattice vectors. In a periodic system the integral over the cell boundary vanishes; we are left with an equation involving only an integral over the surface of a single sphere of radius a. This can be reduced further by taking for $\underline{H}(\underline{r})$ the general form:

$$\underline{H}(\underline{r}) = \sum_{\ell m j} A_{\ell m j}\, \mu^{-1}(\underline{r}) f_\ell^{(1)}(\underline{r})\, \underline{L}\, Y_{\ell m j}(\hat{r}) \quad (5a)$$

$$+ \sum_{\ell m j} B_{\ell m j}\, \mu^{-1}(\underline{r})\, \underline{\nabla} \times f_\ell^{(2)}(\underline{r})\, \underline{L}\, Y_{\ell m j}(\hat{r}) \quad (5b)$$

where $\underline{L} = \underline{r} \times \underline{\nabla}$ and the $Y_{\ell m j}$ ($j = 1,2$) are the normalized real and imaginary parts of the usual spherical harmonics. The first set of terms, (5a), correspond to electric multipoles; terms (5b) correspond to magnetic multipoles. Using (5) and performing the standard expansion of $G_{\underline{k}\omega}(\underline{r},\underline{r}')$ in terms of $Y_{\ell m j}(\hat{r})$ and $Y_{\ell' m' j'}(\hat{r}')$ we find that for a given (ℓmj) equation (3) reduces to

$$\sum_{\ell'm'j'}\left\{\left[\left((\Delta g_\ell)^{-1} - i\alpha j_\ell(\alpha)h_\ell^+(\alpha)\right)\delta_{mm'}\delta_{\ell\ell'}\delta_{jj'}\right.\right.$$
$$\left.-4\pi i^{\ell'-\ell}j_\ell(\alpha)j_{\ell'}(\alpha)\sum_{LMJ}C(\ell mj;\ell'm'j';LMJ)\,M_{LMJ}(\underset{\sim}{k\underset{\sim}{w}}_1)\right]A_{\ell'm'j'}$$
$$+\ [\text{corresponding magnetic multipole terms}]\Bigg\} = 0 \quad (6)$$

where $\alpha = w_1 a$ and is thus proportional to ka. In equation (6) the j_ℓ and h_ℓ^+ are spherical Bessel and Hankel functions. The quantities Δg_ℓ (the phase shifts), M (the structure constants) and C (the "selection constants") are defined by

$$\Delta g_\ell = -\frac{a}{f_\ell^{(1)}(a)}\left.\frac{\partial f_\ell^{(1)}}{\partial r}\right|_{a^+} - \alpha j_\ell'(\alpha)/j_\ell(\alpha)\ , \qquad (7a)$$

$$M_{LMJ}(\underset{\sim}{k}\underset{\sim}{w}_1) = i^{1-L}\sum_{\underset{\sim}{R}\neq 0} e^{i\underset{\sim}{k}\cdot\underset{\sim}{R}}\,h_L^+(w_1 R)\,Y_{LMJ}(\hat{R})\ , \qquad (7b)$$

and

$$C(\ell mj;\ell'm'j';LMJ) = \left\{\ell(\ell+1)\ell'(\ell'+1)\right\}^{-\frac{1}{2}} \sum_{\overline{mm'}}\sum_{\overline{jj'}}\sum_{i=xyz}$$
$$\int d\Omega\,Y_{\ell mj}(\Omega)L_i Y_{\overline{\ell mj}}(\Omega)\int d\Omega'\,Y_{\ell'\overline{m}'\overline{j}'}(\Omega')L_i Y_{\ell'm'j'}(\Omega')$$
$$\int d\widetilde{\Omega}\,Y_{\ell'\overline{m}'\overline{j}'}(\widetilde{\Omega})\,Y_{\ell'\overline{m}'\overline{j}'}(\widetilde{\Omega})\,Y_{LMJ}(\widetilde{\Omega})\ . \qquad (7c)$$

We note that the structure constant (7b) can be displayed in reciprocal space as

$$M_{LMJ}(\underset{\sim}{k}\underset{\sim}{w}_1) = -4\pi\int d\underset{\sim}{q}(S_c(\underset{\sim}{q})-1)\frac{Y_{LMJ}(\hat{u}_{\underset{\sim}{q}+\underset{\sim}{k}})\,|\underset{\sim}{q}+\underset{\sim}{k}|^L}{\left((\underset{\sim}{k}+\underset{\sim}{q})^2 - w_1^2\right)w_1^{L+1}} \qquad (8a)$$

where

$$S_c(\underset{\sim}{q}) = \sum_{\underset{\sim}{R}} e^{i\underset{\sim}{q}\cdot\underset{\sim}{R}} \qquad (8b)$$

which for periodic systems reduces (8a) to a sum over reciprocal lattice vectors. To obtain the phase shift (7a) it is only necessary to solve a single sphere scattering problem using the standard boundary conditions on the tangential components of $\underset{\sim}{E}$ and $\underset{\sim}{H}$. Provided, as we assume, ka << 1 we have[9]:

$$\Delta g_\ell = \left(1 - \frac{\epsilon_1}{\epsilon_2}\right)(\ell+1)\ . \qquad (9)$$

For a given frequency w it is clear that the secular problem defined by (6) only has solutions for specific values of k. For these k we <u>define</u> the average dielectric constant $\bar{\epsilon}$ by the relation

$$\bar{\epsilon} = (ck/\omega)^2 . \qquad (10)$$

It is important to note that we are adopting a definition of $\bar{\epsilon}$ appropriate to a quite specific physical context, namely that of transmission of electromagnetic radiation through a composite medium. Other definitions of the average transport coefficients[2,4,10,11] can be given for different physical contexts.

In seeking solutions to (6) the assumptions $ka \ll 1$ and $\mu = 1$ effect enormous simplification; for in this limit the electric and magnetic multipole terms uncouple, and higher order electric multipoles are not induced by <u>radiation</u> effects. This can be seen either directly from (6)[12], or by <u>noting</u> that for scattering from a single sphere of finite conductivity the amplitude of the $(\ell + 1)^{th}$ electric partial wave (or multipole) is of the same order of magnitude as the amplitude of the ℓth magnetic partial wave[13]. Furthermore, it is in general a factor $(ka)^2$ smaller than the ℓth electric partial wave. This is why it was not necessary to display the magnetic multipole terms in (6). However, all orders of electric multipoles can in general be produced by <u>multi-particle</u> interactions, which exist even in the quasi-static limit. While the dipole ($\ell = 1$) term is dominant, higher multipoles can become important[14] at large η.

With ω specified equation (10) gives a solution for k which, with definition (10) for $\bar{\epsilon}$ gives the effective dielectric constant for transmission. We emphasize that this solution for $\bar{\epsilon}$ has been obtained on the assumption that the metallic spheres are arranged periodically. Under these conditions the entire analysis can be carried through by considering a single primitive cell; and we are assured, furthermore, that on this cell the outer surface integral in (3) vanishes.

If we turn now to <u>non-periodic systems</u> the analysis above proceeds via identical assumptions with only slightly more complexity. Equation (3) can still be established except that now the integrals are over (i) the surfaces of N spheres located by their centroids at $\{R\}$, and (ii) the outer bounding surface of the entire assembly. As noted by Morse[7] this last integral will vanish providing the mean density ρ is asymptotically constant. Next, if we assume that H on the surface of a sphere at R differs from $\tilde{H}$ on the surface of $\tilde{a}$ sphere at R' only through a factor $e^{i\tilde{k}\cdot(R - \tilde{R}')}$, then for each sphere in the system we get an equation of the <u>form</u> (6). These equations are not quite identical since the origin of the sum in the definition of M (equation (7b)) is different for each equation. However, they can be made identical if we simply replace M by its configuration average $\langle M \rangle$. With this approximation, only the $\ell = 1$ term survives; retaining only $\ell = 1$ in equation (6) leads to an equation which decouples into a single longitudinal equation and two identical transverse equations. From these we get, after some manipulation:

$$\frac{2(\varepsilon_2-\varepsilon_1)}{(\varepsilon_2+2\varepsilon_1)} \frac{\omega_1^3 \eta}{\rho} \frac{1}{\sqrt{4\pi}} \left\{ M_{002}(\underset{\sim}{k}v_1) + \frac{1}{\sqrt{5}} M_{202}(\underset{\sim}{k}v_1) \right\} = 1 \quad (11)$$

where $\eta = \frac{4\pi}{3}\rho a^3$ is the filling fraction, or packing fraction, of the spheres. A detailed analysis shows that the error consequent on making these assumptions leads to a correction to (11) whose leading order is proportional to $\eta^2 (M - \langle M \rangle)^2$. For small η the approximation is clearly valid. The effects at large η are discussed below.

Finally, in the actual evaluation of the structure constant (7b) we have used the standard expression

$$S(\underset{\sim}{q}) = \langle S_c(\underset{\sim}{q}) \rangle = 1 + \rho\delta(\underset{\sim}{q}) + \rho h(\underset{\sim}{q}) \quad (12)$$

where $h(\underset{\sim}{q})$ is the Fourier transform of $g(r)-1$, the quantity $g(r)$ being the radial distribution function of spheres. A familiar result now emerges: if we assert that there is <u>no</u> correlation between spheres, that is, that they are quite randomly distributed, then $h(\underset{\sim}{q})=0$. Using $S(\underset{\sim}{q})-1 = \rho\delta(\underset{\sim}{q})$ in (8a) and substituting the outcome in (11) we arrive at a simple expression for $\bar{\varepsilon}$, namely

$$\bar{\varepsilon}(\omega, \eta; ka \to 0) = \frac{\varepsilon_1(\varepsilon_2 + \varepsilon_1 + 2\eta(\varepsilon_2-\varepsilon_1))}{\varepsilon_2 + 2\varepsilon_1 - \eta(\varepsilon_2-\varepsilon_1)}, \quad (13)$$

the well known form due originally to Maxwell-Garnett.[3] Notice that, provided we consider only the $\ell = 1$ term, the same result obtains for the periodic case at large η where the $\underset{\sim}{q}$ are restricted to $\underset{\sim}{q} = 0$ and all other reciprocal lattice vectors. The difference between this result and the correct periodic[14] result gives an indication of the importance of angular information, which we have ignored in replacing M by $\langle M \rangle$. In general this only becomes important when ε_1 and ε_2 differ greatly and η is large. The calculation of such effects requires information about the structure of the system <u>beyond</u> the pair distribution function. We are now attempting to treat such effects.

Real composites will lie between the extremes of order discussed above. A system of finite spheres can be at best quasi-random since inter-penetration is forbidden, but the consequent short range correlation effects must become progressively less important as η diminishes from its maximum possible value of 0.74. Thus (13) should certainly be valid for small filling fractions. And if there is little or no clustering (and $M \sim \langle M \rangle$) (13) may actually be valid for large η as well. The reason is that if the disorder is the "equilibrium" type of disorder characteristic of a liquid, then an estimate of $h(\underset{\sim}{q})$ can be obtained from the theory of classical fluids as applied to hard spheres of diameter $2a$ at packing fraction η. In the Percus-Yevick approximation[15], for example, $h(\underset{\sim}{q})$ has the structure

$$h(\underset{\sim}{q}) = a^3 F(qa; \eta) \quad (14)$$

and we see, from (8a) and (11) that corrections arising from such truly equilibrium correlation effects amount to quantities only of order $(\omega_1 a)^2$ which in our assumed limit remain negligible.

For non-equilibrium distributions, such as those prominently exhibiting clustering, we may still conclude that structure small on the scale of a wavelength should produce little change from the Maxwell-Garnett result. On the other hand, structure on the scale of the wavelength, for example periodicity in the dilute limit ($\eta \to 0$), or clustering, may well produce significant deviations from (13) which can be calculated in principle but only if $S(\underset{\sim}{q})$ is known.

III. COATED SPHERES: AN EFFECTIVE MEDIUM

In standard two component effective medium theory, topological symmetry between host and particles is imposed. Thus if the metal is assumed distributed in the form of speheres, so is the interstitial dielectric. Spheres of each component are then taken as embedded in an effective medium in which an effective field $\bar{E}$ is present. Since it is easy to relate fields inside and outside of such inclusions ($E_{in}/E_{out} = 3/(2 + \epsilon_{in}/\epsilon_{out})$) a volume average leads immediately to

$$(1 - \eta)(E_1/\bar{E}) + \eta (E_2/\bar{E}) = 1$$

or

$$\frac{3(1 - \eta)}{2 + \epsilon_1/\bar{\epsilon}} + \frac{3\eta}{2 + \epsilon_2/\bar{\epsilon}} = 1 , \quad (15)$$

as the determining relation for $\bar{\epsilon}$.

The results for $\bar{\epsilon}$ and related functions flowing from (15) and (13) (the Maxwell-Garnett result) are really very different. We shall see in a moment that they should be since the physical statements implied by these expressions are quite distinct. To demonstrate the numerical differences, we show in Figure 1 the reflectivity $R(\omega = \omega_p/5)$ calculated for a composite in which the metal component has a Drude dielectric function (ϵ_2) with $\omega_p \tau = 192$ (ω_p being the plasma frequency). Because of the asserted symmetry between constituents, there is of course, a single curve for effective medium theory. In Maxwell-Garnett theory, on the other hand, there are two curves and this reflects the fact that the specification of fill fraction alone leaves us with a genuine physical ambiguity: i.e. it is clearly necessary to specify whether metal has been added to a dielectric host (up to a maximum packing of 0.74) or vice versa. The curves for these two possibilities are quite different, and this difference is also apparent if we plot $R(\omega)$ at fixed metal fraction, as is done in Figure 2 where η is chosen to be just 1% below the close packing limit for identical metal spheres. It could also be viewed as a 27% fill fraction of dielectric spheres in a metal host, a value of η quite remote from the close packing limit for dielectric spheres. The curves of Figure 2 (note the Mie scattering resonance) therefore correspond to different physical

Figure 1 Reflectivity for a metal-insulator composite at a frequency ω/ω_p = 0.2. Here and in succeeding figures the metallic dielectric function $\varepsilon_2(\omega)$ is evaluated in a Drude model ($\omega_p \tau$ = 192) and the insulating component has dielectric constant ε_1 = 1.0. Curve 1: dielectric taken as host, curve 2: metal as host; both evaluated according to Eqn. (13). The lines at 26% and 74% represent, respectively, the packing fraction limits for dielectric spheres in a metallic matrix and for metal spheres in a dielectric matrix. The standard effective medium result, Eqn. (15), is shown for comparison (EMT).

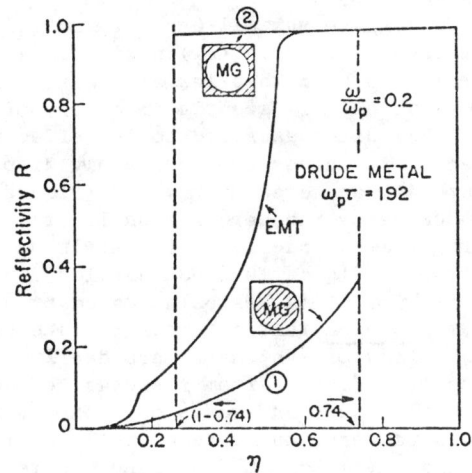

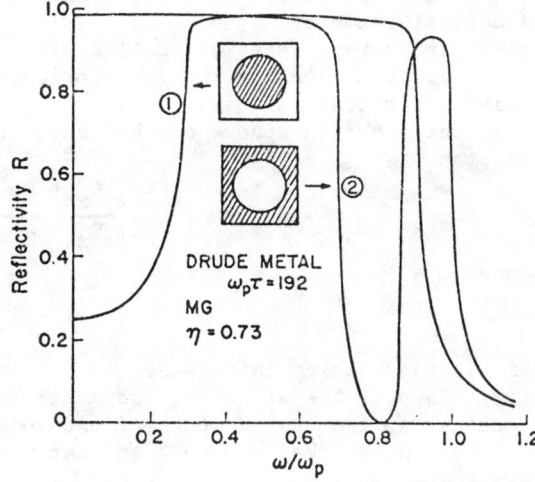

Figure 2 Reflectivity for metal volume fraction 73% according to Eqn. (13); curve 1: dielectric host, curve 2: metal host.

situations.

To compare Maxwell-Garnett and effective medium theories we shall take dielectric (ε_1) always as host. Figure (3) shows $R(\omega)$ according to both theories at a metal filling fraction of $\eta = 0.73$. This value of η is greatly in excess of the percolation limit[16] ($\eta = 0.33$) often ascribed to the effective medium theory. The reflectivity curves are noticeably disparate, and remain so at $\eta=0.4$ (Figure 4), and even at $\eta = 0.2$ (Figure 5), a value of filling fraction now below the percolation limit.

The reason underlying the striking disagreement between these two theories is quite fundamental and has already been pointed out by Smith[17]. It is best elucidated by establishing an <u>effective medium approach</u> that attempts to take proper account of the topology of the kinds of systems we are dealing with here. The argument (which is different from the usual effective medium argument[18]) goes in several steps and rests only on the assumptions that (a) bulk composites possess, on average, full rotational and translational symmetry, and (b) that there is no clustering (as tacitly assumed earlier):

(i) Physical quantities of interest are macroscopic, and the replacement of small sclae details (e.g. particles or inclusions) with an effective medium (i.e. a continuum) is therefore appropriate.

(ii) Prior to averaging, the composite is divided into cells each of which contains a single inclusion surrounded by dielectric. Each cell is then regarded as being embedded in an effective medium provided by the remaining cells. This effective medium is chosen so that the total scattering from all cells vanishes on the average.

(iii) For the average scattering for the system to vanish, the scattering from an average cell must vanish. The average system has rotational symmetry, and the average cell must therefore have spherical symmetry.

(iv) An average spherical cell contains a concentric metal sphere (dielectric constant ε_2) outside of which is a dielectric coating (dielectric constant ε_1).

(v) But a coated sphere can be characterized by a <u>single</u> dielectric constant[19]

$$\varepsilon_{cs} = \varepsilon_1 \left(\frac{2\varepsilon_1+\varepsilon_2 + 2Q^3(\varepsilon_2-\varepsilon_1)}{2\varepsilon_1+\varepsilon_2 - Q^3(\varepsilon_2-\varepsilon_1)} \right) \qquad (16)$$

where

$$Q = 1 - t/a \qquad (17)$$

with t the coating thickness.

(vi) Thus if the effective medium is to have the same dielectric constant as the coated sphere (to ensure, according to (ii), no scattering) we must choose $\bar{\varepsilon}$ so that

$$\bar{\varepsilon} = \varepsilon_{cs} .$$

249

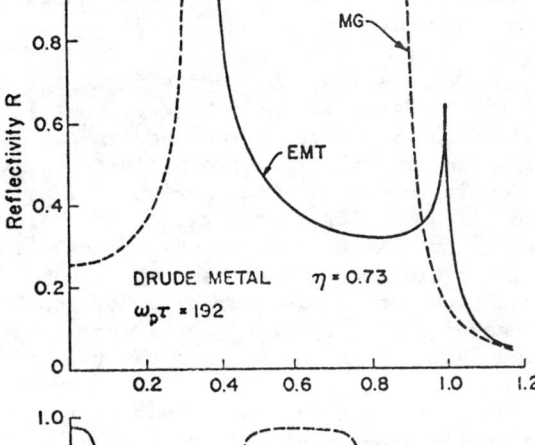

Figure 3 Reflectivity for 73% metal according to Maxwell-Garnett (insulating matrix) and effective medium theories.

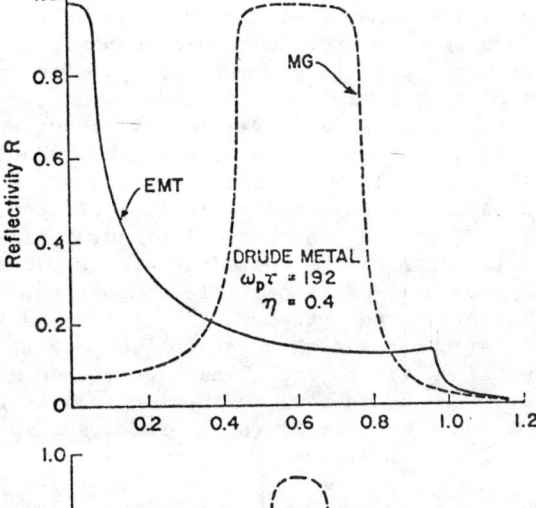

Figure 4 As in Fig. 3, but with metal volume fraction $\eta = 0.4$.

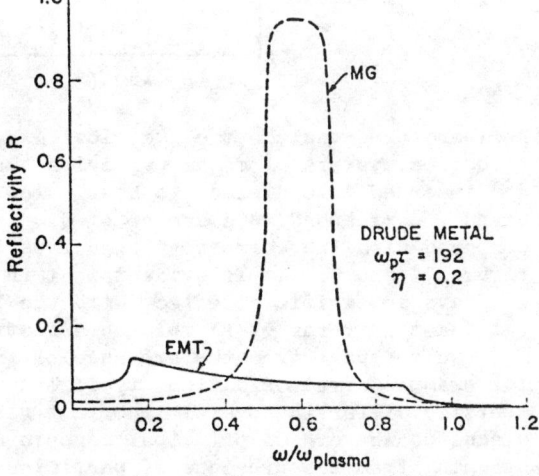

Figure 5 As in Fig. 3, but with metal volume fraction $\eta = 0.2$ (i.e. below "percolation" in the standard effective medium theory).

(vii) But we recognize that the fraction of metal in the coated sphere is Q^3; i.e. $\eta = Q^3$, which shows that (16) is nothing but the Maxwell-Garnett result (13).

Observe that this derivation of $\bar{\epsilon}$ (and that of Smith[17]) is fully self-consistent. The argument given above differs from the standard effective medium argument at step (iii). As noted above, the traditional effective medium approach imposes topological symmetry on the system, for example by taking the average cell to be either a sphere of metal or a sphere of dielectric. In the argument just given, the average cell is a sphere of metal coated with dielectric. For the composites under consideration here, the distribution of dielectric is not topologically equivalent to a sphere[17]. If anything, it is more equivalent to a hollow sphere, and the distinction between the two is electrodynamically important when the behavior of bound surfaces charges is considered.

It follows that an effective medium approach can lead to the Maxwell-Garnett result. Of course, it can also lead to the standard EMT, but should then be considered physically appropriate only when the constituent materials demonstrably enter on an equal footing when fabricating the composite (for example, the case of microcrystallite aggregates as discussed by Landauer[5]). Maxwell-Garnett theory, on the other hand, should be used for situations topologically equivalent to the case where a dielectric is constrained to have a distribution of spherical inclusions.

Provided the physical circumstances warrant it, we may therefore think of the coated sphere as a basis for an effective medium approach to metal particle composites, and the result for $\bar{\epsilon}$ is just Maxwell-Garnett theory. It is interesting to consider what modifications ensue when we consider 3-component systems in which the metal particles are already coated to a thickness t by a dielectric, say ϵ_3, and then randomly dispersed in ϵ_1. The result is that in place of ϵ_2 in (13) (and, of course, in (15) if physically applicable) we must use

$$\epsilon_3 \left(\frac{2\epsilon_3+\epsilon_2 + 2Q^3(\epsilon_2-\epsilon_3)}{2\epsilon_3+\epsilon_2 - Q^3(\epsilon_2-\epsilon_3)} \right) \tag{18}$$

We therefore conclude the numerical examples by showing reflectivities for systems of metal particles clad with insulating coats and immersed in a dielectric host. The curves plotted in figures 6 and 7 correspond to a choice of $Q = 1-t/a = 0 \cdot 8$ and should be compared with the corresponding curves for uncoated particles in figures 4 and 5. It is evident that in both theories the coating can have a significant effect with the influence apparently greater (at least in terms of R) for the effective medium theory.

Quite apart from the presence of insulating coats arising, for example, from oxidation, typical composites encountered experimentally differ in several important respects from the ideal systems discussed above. Of principal concern are corrections possibly stemming from the presence of particles distributed according to

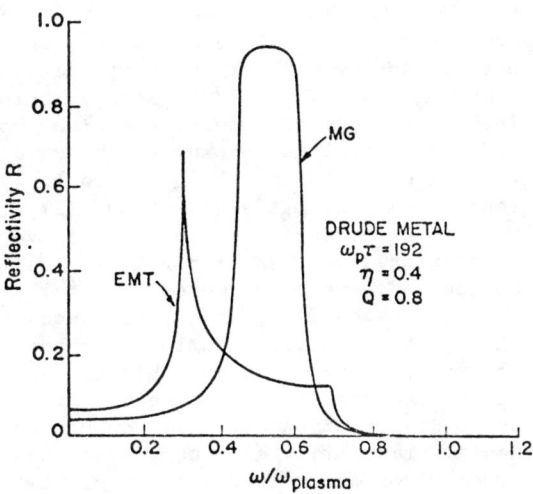

Figure 6 Reflectivities for coated spheres, calculated from Eqn. (18) with $\varepsilon_3 = 2.0$, $Q = .8$, and $\varepsilon_2 =$ Drude result, and from Eqn. (13) (M-G) and Eqn. (15). Here the coated sphere volume fraction is 0.4.

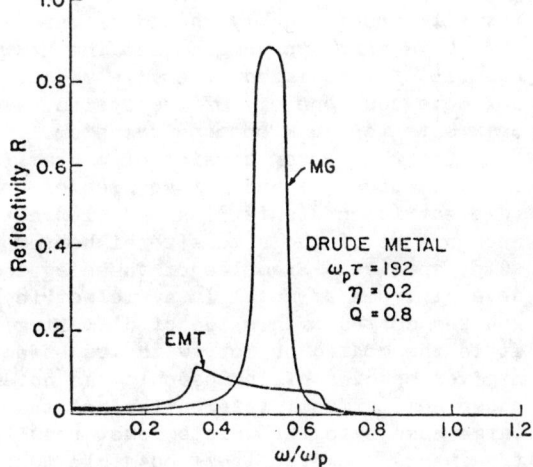

Figure 7 As in Fig. 6, but with coated sphere volume fraction of 0.2.

size, composition and shape. Estimates of the importance of these effects can be made as follows: First, size: a distribution of sphere sizes produces no effect on our result (11) provided (i) $ka_i \ll 1$ for all sphere radii a_i, and (ii) particle size and position are uncorrelated. If such correlations are present then the corrections to (11) are proportional to $\eta \, \langle (\Delta M - \langle \Delta M \rangle)^2 \rangle$ where $\Delta = (a - \langle a \rangle)/a$. Second, composition: a distribution in the composition of spheres (as might result, for example, from different relative thicknesses of oxide coat) leads to a spread in the phase shifts Δg_ℓ. A detailed analysis of this problem shows that to calculate the effect on (11) arising from a spread in Δg_ℓ, we need to evaluate its reciprocal average, and the resultant corrections are then proportional to $(\Delta g_\ell^{-1} - \langle \Delta g_\ell^{-1} \rangle)^2$. These corrections are not necessarily small, but as we shall see below they can be handled, in certain cases, by a different technique. Third, shape; a distribution of shapes (i.e. deviations from sphericity) may occur and this case is difficult to treat rigorously within the framework we have described in Section II. However, shape distribution can be incorporated in our modified effective medium argument, where, because of the necessary rotational symmetry of the <u>average</u> system, it is then quite without effect.

IV: MANY COMPONENT SYSTEMS

Coated-sphere three component systems are highly correlated. There exist, however, other 3 component (or more generally m-component) systems of experimental interest that are far less correlated. For a two component composite we have argued that Maxwell-Garnett theory should be used rather than standard EMT, though we have emphasized that the Maxwell-Garnett theory is an example of a modified effective medium approach which correctly reflects the topology of the system. We now give the corresponding arguments for an m component system.

Let the system consist of a host (dielectric constant ϵ_h and fill fraction η_h) and m-1 species of spherical inclusions (dielectric constants ϵ_i and fill fractions η_i). These inclusions need not be of the same size either among different, or within the same, species. Examples of these systems might be spheres of several types of metal in a dielectric matrix, or identical metal spheres coated with oxide of differing thickness (in which case it is the coated object, with its effective ϵ, that constitutes a given species of inclusion). As noted above, the spread in phase shifts from distributions of these inclusions can become large enough to throw into doubt results based on a two component treatment. However it is possible to treat cases of this kind by a further modification of our effective medium argument. That argument followed 7 steps, the first two which remain unchanged. Proceeding to the third and fourth we now have, instead:

(iii)' Each species of inclusion has an associated average cell and for the average scattering to vanish, the weighted average scattering from these cells must vanish, the weights being determined by the average cell volume fractions. The average system has rotational symmetry (since the inclusions are taken as uncorrelated) and each average cell is spherical.

(iv)' Each average cell contains an inclusion (dielectric constant ϵ_i) of a given species, and coated by the host (dielectric constant ϵ_h). The coating thickness is simply proportional to the average sphere radius for that species[20].

We need not progress to step (v) because at this point we recognize that the system of coated spheres just constructed has precisely the topology that is assumed in the derivation of the standard EMT. All coated spheres genuinely enter on an equal footing, and we therefore deduce the following: the result for $\bar{\epsilon}$ for an m-component small particle composite can be obtained from the usual EMT approach as applied to an m-1 component composite consisting of host-coated-spheres. The result for $\bar{\epsilon}$ reduces, after some simplification, to:

$$3 \sum_{i=1}^{m-1} \eta_i \left[2 + \frac{\epsilon_h}{\bar{\epsilon}} \left\{ \frac{3\epsilon_i + 2\eta_h(\epsilon_h - \epsilon_i)}{3\epsilon_h - \eta_h(\epsilon_h - \epsilon_i)} \right\} \right]^{-1} = 1 - \eta_h . \qquad (19)$$

This result is neither the standard m-component EMT nor a straightforward average of Maxwell-Garnett theory. (For a discussion of the shortcomings of the usual many-component Maxwell-Garnett expression, see the review article by Landauer[21].) It does, however, accurately reflect the topology of the system. Each inclusion is surrounded by (and therefore correlated with) host, but inclusions are uncorrelated with each other. We end by noting that the Maxwell-Garnett result is recovered from (19) when either m = 2 or all ϵ_i are identical. An investigation of the practical implications of this expression is in progress.

Acknowledgements:

We wish to thank Dr. J. Dobson, Prof. R. Buhrman and Prof. A.J. Sievers for helpful discussions during the course of this work.

REFERENCES AND FOOTNOTES

*Work supported in part by the Energy Research and Development Agency, Grant EG-77-5-03-1456, and in part by the National Science Foundation through the facilities of the Materials Science Center, Grant DMR 76-81083 (Technical Report No. 2895)

1. Z. Hashin and S. Shtrikman, J. Appl. Phys. $\underline{33}$, 3125 (1962).
2. D.J. Bergman, Phys. Rev. $\underline{B14}$, 1531 (1976); see also p. 46 of these proceedings.
3. J.C. Maxwell-Garnett, Philos, Trans. R. Soc. Lond., $\underline{203}$, 385 (1904); $\underline{205}$, 237 (1906) (apparently a close relative of J.C.M. Garnett, see Ref. 21).
4. D.A.G. Bruggeman, Ann. Phys. (Leipz.) $\underline{24}$, 636 (1935).
5. R. Landauer, J. Appl. Phys. $\underline{23}$, 779 (1952).
6. See, for example, R.J. Elliot, J.A. Krumhansl, and P.L. Leath, Rev. Mod. Phys. $\underline{46}$, 465 (1974), and J.E. Gubernatis and J.A. Krumhansl, J. Appl. Phys. $\underline{46}$, 1875 (1975).
7. P.M. Morse, Proc. Nat. Acad. Sci. $\underline{42}$, 276 (1956).
8. J. Korringa, Physica 13, 392 (1947); W. Kohn and N. Rostoker, Phys. Rev. $\underline{94}$, 1111 (1954).
9. Notice that unlike the quantum mechanical analog $\lim_{\ell \to \infty} \Delta g_\ell \neq 0$.
10. I. Webman, J. Jortner, and M. H. Cohen, Phys. Rev. $\underline{B15}$, 5712 (1977).
11. D. Stroud, Phys. Rev. $\underline{B12}$, 3368 (1975).
12. The behavior of the $j_\ell(\alpha)$ for small ka guarantees this.
13. M. Born and E. Wolf, Principles of Optics (5th Edition) (Pergamon, 1975) pp 650-651.
14. D.R. McKenzie, p. 295 of these proceedings, and W.T. Doyle, p. 301 of these proceedings.
15. N.W. Ashcroft and J. Lekner, Phys. Rev. $\underline{145}$, 83 (1966).
16. D.M. Wood and N.W. Ashcroft, Phil. Mag. $\underline{35}$, 269 (1977).
17. G.B. Smith, J. Phys. $\underline{D10}$, L39 (1977).
18. J.A. Krumhansl, Amorphous Magnetism (Proceedings of the International Symposium on Amorphous Magnetism, 1972) (Plenum, N.Y. 1975) pp 15-25.
19. H.C. Van de Hulst, Light Scattering by Small Particles (Wiley, 1957).
20. This choice insures that if a single species of inclusion is formally treated as several species the results are unchanged.
21. R. Landauer, p. 2 of these proceedings, Section 3.

DISCUSSION

R. LANDAUER (IBM, Yorktown Heights): What is the difference between your result and a multiple component Maxwell-Garnett theory which has been written down a number of times in dielectric theory?

ASHCROFT: One of the curves I showed was for a three component version of the theories, and there is clearly a difference. Characteristics of the effective medium theory appear in the denominator in the standard way.

J. KORRINGA (Ohio State University): I would like to make a remark on the history of the subject. It started with Poisson, perhaps, and in the beginning many papers were related to optical or electromagnetic properties. Some of the theories were developed for acoustic properties. In particular, the KKR method to which you referred was developed in 1896 by a Russian, Kasterin. Ewald and others developed and used the optical analogues to this theory. Secondly, the theory which you have derived (that the Maxwell-Garnett approximation may be considered as a self-consistent theory) was known in acoustics for zero frequency in 1950. The theory was due to Mackenzie, who showed that films or surface coatings give the Maxwell-Garnett equation and that in the static case this equation is an extreme value for the elastic properties of the medium, considering all possible arrangements of the constituents. I have always considered the static version of the Maxwell-Garnett approximation as an extreme or bound. Of course, it is well known that if this approximation is inverted, one goes from an upper bound to a lower bound, or vice versa, so in applying the theory one must be very sure which is the embedding medium and which is the embedded. In cases where the concentrations are nearly equal, as is often true in heterogeneous materials, that choice may be a dilemma.

ASHCROFT: In our case, it is quite clear which is to be the host and which is to be the inclusion. That choice is set up by the geometry in the beginning. For example if a face centered cubic lattice is being used, there is no possibility of having a filling factor greater than 74%. At 74% the spheres will touch and the percolation regions begin. The result, of course, is very different if one starts with metal and adds inclusions of insulator. The asymmetry is quite real and it is necessary at the beginning to specify exactly what the connectedness of the system is.

S. NAM (University of Dayton): Your treatment of scattering is done for long wavelengths. Have you tried to determine where the long wavelength approximation breaks down?

ASHCROFT: We have no quantitative results on that problem as yet. As Stroud has proposed, a regime occurs where the magnetic multipole terms cannot be neglected any longer. A similar result will surely come out of this as well.

CHAPTER IV: CONTRIBUTED PAPERS:
OPTICAL PROPERTIES

ON THE ANOMALOUS ABSORPTION OF ULTRAFINE PARTICLES IN THE FAR INFRARED*

D. Pramanik, R. A. Buhrman, and A. J. Sievers
Dept. of Physics, Cornell University, Ithaca, N.Y. 14850

ABSTRACT

The far-infrared absorption coefficients α of 100 Å particles of a number of metals and also of amorphous and crystalline Al_2O_3 have been measured at 4.2K and are orders of magnitude larger than effective medium calculations of α which rely on bulk dielectric constants. These results complement earlier measurements on Al particles[1]. We find that: 1. α varies linearly with filling fraction f for $f \leq 15\%$ 2. The magnitude of α does not depend to any great extent on whether the particles are free standing or are embedded in a KBr matrix, and 3. Particles of Au and Pd which do not form oxide coatings have larger absorption than do particles of Al or Cu with oxide coatings. A comparison of the resonance absorption in the visible part of the spectrum for Au and oxide-coated Cu particles dispersed in KBr reveals that the key role of the oxide coating is to transfer oscillator strength from the far-infrared to the small particle resonance in the visible.

[1] C. G. Granqvist, R. A. Buhrman, J. Wyns, and A. J. Sievers, Phys. Rev. Lett., <u>37</u>, 625 (1976).
* This work is supported by the ERDA contract #EG-77-5-03-1456.

OPTICAL AND INFRARED REFLECTANCE OF METAL-INSULATOR COMPOSITES*

N. E. Russell, E. M. Yam and D. B. Tanner
The Ohio State University, Columbus, Ohio 43210

ABSTRACT

The reflectance in the infrared and visible (0.08 eV to 3.2 eV) has been measured for aluminum small particles compacted into potassium chloride. The aluminum particles were prepared by evaporation in inert gas atmosphere and had mean diameters of 200Å and 600Å. The volume fraction, f, of metal in the samples studied was between f = 0.03 and f ~ 1. For f ~ 0.15 the specimens exhibited finite dc conductivity. The main feature of the data is a decreasing reflectance with increasing frequency. The magnitude of the reflectance increases with metal concentration. Comparison of experiment with simple theories shows that the self consistent theory predicts the general form of the reflectance if the low concentration for first achieving conduction is taken into consideration. Kramers-Kronig inversion of the reflectances yield frequency dependent conductivities which are not like those of a simple metal.

INTRODUCTION

There has been some disagreement in the literature recently over the dielectric function appropriate to describe the optical properties of composite systems. In particular Stroud[1] has proposed that the effective medium approximation[2] (EMA) should be used while Gittleman and Abeles[3] have advocated use of the Maxwell-Garnett theory[4] (MGT).

EXPERIMENT

We have measured the reflectance in the visible and infrared of composite systems formed by compacting together small particles of aluminum and KCl. The particles were made by the smoke method: evaporating aluminum in an argon atmosphere. Argon pressure was about 0.2 Torr to produce d ~ 200Å particles and 0.5 Torr giving d ~ 600Å particles.

The particles were thoroughly mixed with KCl powder, 40 μM < d < 100μM, pressed into a disk shaped pellet and mounted into a grating spectrometer for reflectance measurements. Four gratings were used to cover 600 cm^{-1} to 26000 cm^{-1} (0.075 eV to 3.25 eV) with thermocouple, lead sulfide and photomultiplier detectors as appropriate.

Figure 1 shows the reflectance of d ~ 200Å aluminum particles, of volume fraction f = 0.03, f = 0.18 and f = 0.30 metal. The low concentration shows a rather flat reflectance. The two higher concentrations show a high low frequency reflectance which falls rapidly as the frequency is increased. The general shape of the reflectance is characteristic of a poor conductor. This is consistent with the observation that both of these samples were conducting, with $\sigma \sim 1\Omega^{-1}$ cm^{-1}

Figure 2 shows the reflectance of composites made from d ~ 600Å Al particles in KCl with f - 0.03, f = 0.50, f = 0.75 and f ≈ 1. The f ~ 1 sample was prepared in the same way as the others except that no KCl was used. The concentration is not known very accurately. The sample looks like rather old aluminum foil and is quite brittle. The shapes of the reflectances are similar to those in Figure 1 while the magnitudes are somewhat lower. Because the pellets used in these measurements were reground and repressed five to ten times, these samples were more thoroughly mixed than those whose reflectance is shown in Figure 1.

THEORY

The EMA dielectric function ε_{EMA}, arises

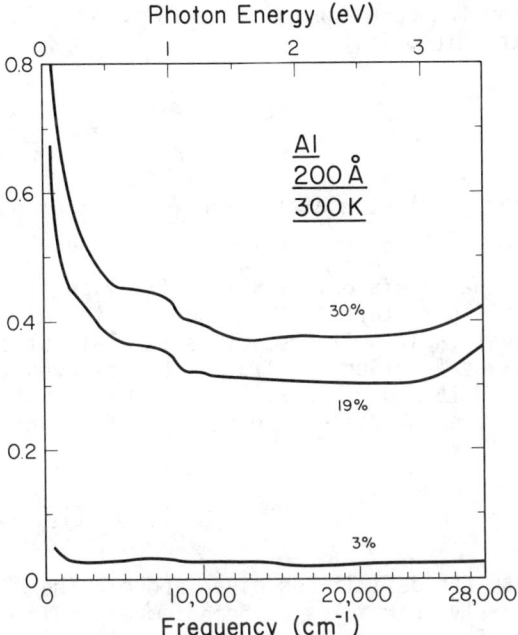

Fig. 1. Reflectance of three concentrations of aluminum particles in KCl.

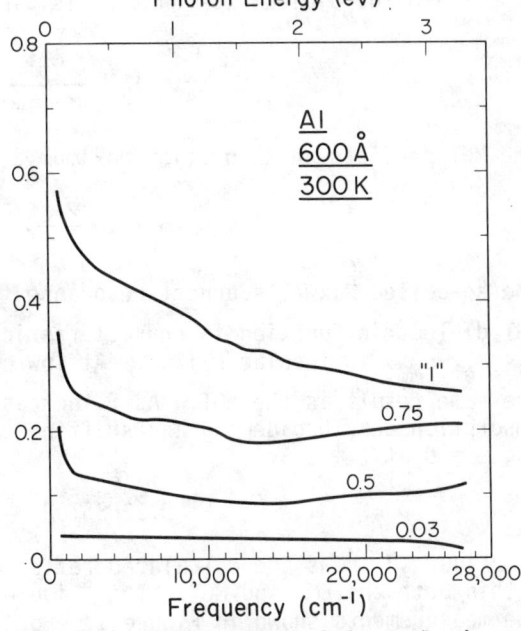

Fig. 2. Reflectance of four aluminum small particle composites.

from the solution of a quadratic equation which, assuming spherical particles, is:

$$f\frac{\varepsilon_{EMA} - \varepsilon_A}{2\varepsilon_{EMA} + \varepsilon_A} + (1-f)\frac{\varepsilon_{EMA} - \varepsilon_B}{2\varepsilon_{EMA} + \varepsilon_B} = 0 \qquad (1)$$

where ε_A is the dielectric function of one of the components, say the metal, of volume fraction f, and ε_B is the dielectric function of the insulator, of volume fraction (1 - f). The EMA has the appealing feature of treating both the A and B particles on an equivalent basis. It predicts a metal-insulator transition at metal volume fraction f = 1/3. (The observed transitions occur between f = 0.15 and f = 0.6)

The MGT dielectric function for spherical particles is

$$\varepsilon_{MGT} = \varepsilon_B + \varepsilon_B \frac{3f(\varepsilon_A - \varepsilon_B)}{(1-f)\varepsilon_A + (2+f)\varepsilon_B} \qquad (2)$$

with the definitions given above. The MGT treats the type A (metal, usually) particles as being embedded in a uniform background of type B. There is no metal-insulator transition below f = 1, although some authors interchange the A and B particles as the metal concentration increases.

If a metal dielectric function is used for type A:

$$\varepsilon_A = 1 - \frac{\omega_p^2}{\omega^2 + i\omega/\tau} \qquad (3)$$

the MGT predicts an absorption maximum at

$$\omega_o = \omega_p / (1 + \frac{2+f}{1-f} \varepsilon_B)^{\frac{1}{2}} \qquad (4)$$

the so-called Maxwell-Garnett resonance frequency. For $\omega > \omega_o$ the MGT dielectric function is characteristic of a metal while for $\omega < \omega_o$, ε_{MGT} is insulator-like. At low concentrations the EMA gives the same result as the MGT. As f increases the EMA predicts the absorption peak broadening and shifting to lower frequencies, reaching ω = o at f = 1/3.

DISCUSSION

Figure 3 shows the calculated reflectance of composite systems within both the EMA and MGT. This figure should be compared with the measurements shown in Figure 1. For the metal we have used a

Drude dielectric function with parameters appropriate to aluminum at 300K. For the insulator we have used a frequency independent dielectric constant ε_i = 2.1. The dielectric is taken to be the host in the MGT. For f = 0.03 the two theories produce essentially the same result at these frequencies: a constant low reflectance governed by the insulator dielectric constant. The resonance frequency of equation (4) is ω_o = 50000 cm^{-1} (6.3 eV). For the higher concentrations the two theories produced qualitatively different results.

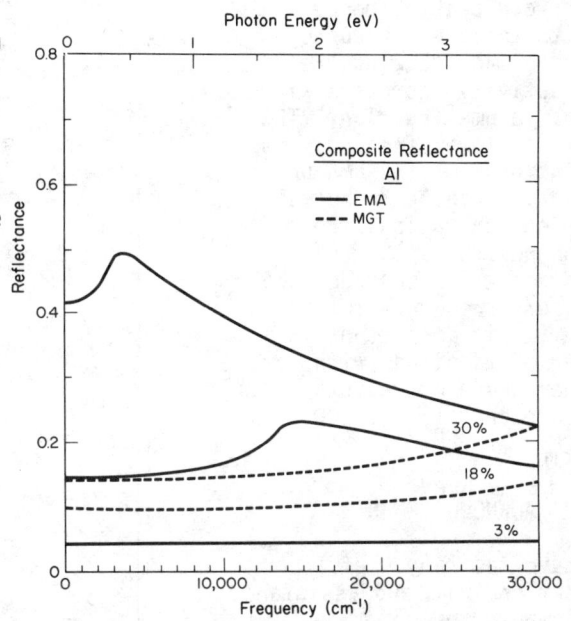

Fig. 3. Calculated reflectance.

The MGT (dashed line) is small at low frequency and rises at high frequencies. The EMA is higher than the MGT at low frequency, produces a weak maximum in the infrared, and then decreases with increasing frequency. The maximum shifts to lower frequency and higher reflectance as the metal concentration increases, reaching zero frequency and unity at f = 1/3 (the percolation transition). Since the samples shown in Figure 1 are conducting, the absence of a peak in the measured reflectance should not be surprising. Before any detailed comparison of experiment with theory can be attempted, a theory is needed which either (1) predicts the percolation transition correctly from the structure of the samples or (2) uses experimentally determined percolation parameters as input.

No attempt has been made to fit the reflectance measurements shown in Figure 2. Instead a Kramers-Kronig[5] analysis of the reflectance was performed. In calculating the integral the measured reflectance was extrapolated to zero frequency using a Hagen-Reubens form: $R \sim 1 - A\omega^{1/2}$ with A chosen such that a smooth connection to experiment was made. At high frequencies a $1/\omega^2$ followed by a $1/\omega^4$ extrapolation was used. Figure 4 shows the frequency dependent conductivity $\sigma_1(\omega) = \omega\varepsilon_2(\omega)/4\pi$, where ε_2 is the imaginary part of the dielectric function, as determined by the Kramers-Kronig analysis. The important feature of $\sigma_1(\omega)$, which is independent of the extrapolation procedure, is the broad peak. The peak shifts to lower frequency and the magnitude of the conductivity increases with metal volume fraction. The shift to lower frequency is reminiscent of the EMA behavior in Figure 3, but at high concentrations the MGT resonance, equation (4), also goes down in frequency. The zero frequency intercept of $\sigma_1(\omega)$ is consistent in order of magnitude with rather crude dc resistance measurements ($\sigma \sim 1\text{-}100$ $\Omega^{-1}\text{cm}^{-1}$).

ACKNOWLEDGMENTS

We have had many useful conversations with and assistance in computer programming from Prof. David Stroud. We thank Prof. F. P. Dickey for the kind loan of the spectrometer.

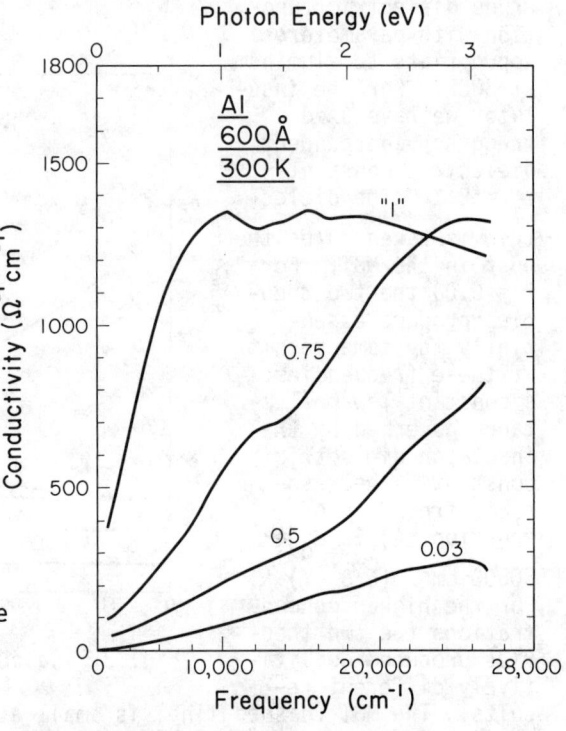

Fig. 4. Kramer-Kronig derived conductivities.

REFERENCES

*Research supported by The Ohio State Program for Energy Research, Education and Public Service.
1. D. Stroud, Phys. Rev. B12, 3368 (1975).
2. D. A. G. Bruggeman, Ann. Phys. (Leipz.) 24, 636 (1935).
3. J. I. Gittleman and B. Abeles, Phys. Rev. B15, 3272 (1977).
4. J. C. Maxwell-Garnett, Philos. Trans. Roy. Soc. A203, 385 (1904).
5. Frederick Wooten, Optical Properties of Solids (Academic Press, 1972), Appendix G.

FAR INFRARED ABSORPTION IN METAL-INSULATOR COMPOSITES*

N. E. Russell, G. L. Carr and D. B. Tanner
The Ohio State University, Columbus, Oh. 43210

ABSTRACT

The absorption coefficient of composite systems, formed by compacting together small metal particles and finely ground KCl powder, has been measured at far infrared frequencies (4-400 cm^{-1}). The measurements were made on particles of aluminum, prepared by inert gas evaporation, having mean diameters in the range 200 Å to 1000 Å and on particles of palladium, with mean diameter 2 µm. Measurements were made for metal filling factors, f, between 0.001 and 0.10. The absorption coefficient is nearly linear in concentration for small concentrations but rises more rapidly than linearly at larger values of f, indicating the onset of a metal-insulator transition. The dependence of the absorption coefficient on particle size is very weak. A theory which includes eddy current (magnetic dipole) absorption gives a fair description of the data for the largest particles but falls several orders of magnitude below the data for the smallest ones.

INTRODUCTION

In this paper we present results of measurements of the far infrared transmission in composite systems. The specimens studied were made of small metal particles in an insulating matrix. The absorption in such small particle samples has been observed[1,2] to be many orders of magnitude larger than would be predicted by simple theory, for example that of Garnett.[3]

EXPERIMENTAL DETAILS

Aluminum small particles were made by the smoke evaporation method described by Granqvist and Buhrman.[4] Aluminum foil is evaporated from a tungsten filament in the presence of a fraction of a Torr of argon gas. The smoke particles collect on the sides and top of a glass cylinder which surrounds the filament. By varying the argon pressure we have been able to make particle batches with average diameters of 200 Å to 1000 Å, as determined by scanning electron microscopy, and a fairly narrow range of diameters. In addition palladium small particles with average diameter of 2 µm were obtained from a commercial source.[5]

The metal particles were thoroughly mixed with finely ground (40 µm < d < 120 µm) KCl powder and pressed, under vacuum, into a solid disc. The samples were reground and repressed three to five times in an attempt to achieve uniformity. X-ray emission measurements in the electron microscope indicated a fairly uniform spatial distribution of aluminum particles.

The far infrared measurements were made on samples at 4.2 K using lamellar grating[6] (4-40 cm^{-1}) and Michelson (20-200 cm^{-1})

interferometers in conjunction with a germanium bolometer-detector operating at 1.2 K. From the measured transmission T we calculate the absorption coefficient from

$$\alpha(\omega) = \alpha_0 - \left[\ln T(\omega)\right]/x \qquad (1)$$

where x is the thickness of the specimen and ω is the far infrared frequency. To account approximately for the reflectance of the sample and to compensate for detector nonlinearities α_0 is chosen so that $\alpha(0) \equiv 0$, as it must.

EXPERIMENTAL RESULTS

Figure 1 shows the absorption coefficient for aluminum small particles, d~ 600 Å, between 8 cm^{-1} and 70 cm^{-1} for Al concentrations of 0.003, 0.01 and 0.03 by volume. The data are shown at low resolution. In higher resolution measurements an interference pattern appears at low frequencies due to multiple internal reflections between the plane faces of the sample. From the fringe spacing the index of refraction of the composite system can be obtained. The index so found is larger than that of the pressed KCl by about 10% in the f = 0.03 sample.

The size dependence of the absorption coefficient in these samples is unusual. Both d = 200 Å and d = 800 Å particles have <u>smaller</u> values of the absorption coefficient at a given frequency and concentration than do the d = 600 Å particles. We find $\alpha(200\ \text{Å}) \sim (1/6)\alpha(600\ \text{Å})$ and $\alpha(800\ \text{Å}) \sim (\frac{1}{3})\alpha(600\ \text{Å})$. The <u>shapes</u> of the curves are quite similar.

Figure 2 shows the absorption coefficient of palladium small particles, d = 2 μm, between 5 and 70 cm^{-1}, for Pd concentrations of 0.001, 0.0029, 0.01, 0.029 and 0.096 by volume, at low resolution.

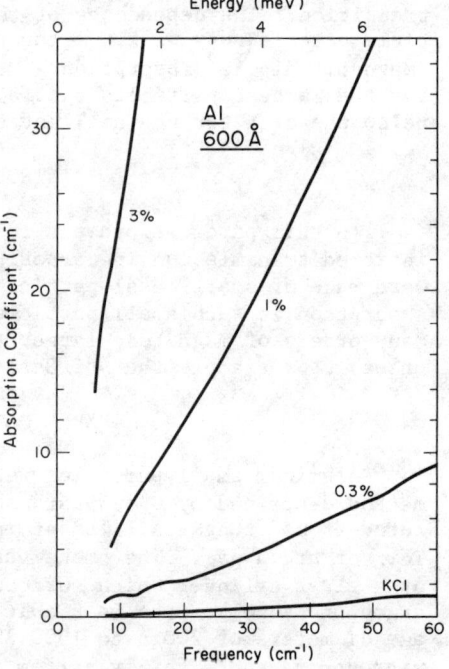

Fig. 1. Absorption coefficient of Al small particles.

The magnitude of the absorption coefficient is of the same order as the <u>much</u> smaller aluminum particles. However the shape is quite different: there is a tendency towards saturation in α at frequencies above 50 cm^{-1}.

THEORY

At metal volume fraction $f \leq 0.03$ the Maxwell-Garnett[3] theory (MGT) and effective medium approximation[8] (EMA) for the composite dielectric function give identical values for the absorption coefficient in the far-infrared. The MGT dielectric function is

$$\varepsilon_{MGT} = \varepsilon_i + \varepsilon_i \frac{3f(\varepsilon_m - \varepsilon_i)}{(1+f)\varepsilon_m + (2+f)\varepsilon_i} \quad (2)$$

where ε_m is the dielectric function of the metal and ε_i is that of the insulator. The absorption coefficient is given by $\alpha_{MGT} = \omega\varepsilon_2/nc$ where c is the speed of light, $\varepsilon_2 = \text{Im}(\varepsilon_{MGT})$ and $n = \text{Re}(\varepsilon_{MGT})^{\frac{1}{2}} \simeq (\varepsilon_i)^{\frac{1}{2}}$. If ε_i is constant and ε_m has the Drude form, the calculated absorption coefficient is quadratic in frequency (observed in low frequency experiments) but too small by up to five orders of magnitude. The values are independent of size (on $\alpha \sim 1/d$ at the lowest frequencies if one puts $\tau \simeq d/v_f$, where d is the particle diameter and v_f, the fermi velocity, into the Drude experession.) As an example, using Drude parameters appropriate to aluminum at 70 cm^{-1} $\alpha_{MGT} \sim 0.03$ f cm^{-1} whereas experiment gives $\alpha \sim (1000-6000)$f cm^{-1}.

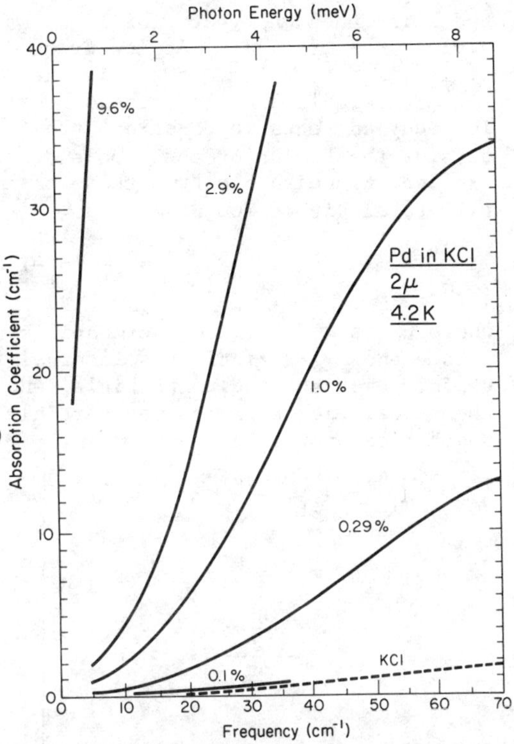

Fig. 2. Absorption coefficient of Pd small particles.

As has been pointed out previously,[1] magnetic dipole (eddy) current losses are larger than the electric dipole losses considered in the simple MGT for metallic particle sizes greater than about 50 Å diameter. A complete theory of this effect has recently been given by Stroud.[9] In an ac field a single spherical small particle has a definite magnetic dipole moment given by[10]

$$\vec{m} = \Omega\gamma\vec{H}_{app} \quad (3)$$

where $\vec{H}_{app}$ is the applied field, Ω is the particle volume and

$$\gamma = -\frac{3}{8\pi}\left[1 - \frac{3}{(ak)^2} + \frac{3}{ak}\cot(ak)\right] \quad (4)$$

is the magnetic polarizibility in which a is the particle radius and k is the wave vector of light waves of frequency ω in the particle: $k = \omega(\varepsilon_m)^{1/2}/c$. At low frequencies the absorption is governed by $\text{Im}(\gamma) \simeq a^2\omega\sigma/10c^2$ where σ is the dc conductivity.

To calculate the absorption coefficient of a composite including eddy currents we construct the following argument: from the outside the fields of the dipole described by Equations (2) and (3) are indistinguishable from those of a uniformly magnetized sphere with total dipole moment

$$\vec{m} = \frac{\Omega}{4\pi}(\mu_m - 1)\vec{H}_{in} = \Omega\vec{M} \tag{5}$$

where μ_m is a fictitious permeability, M the dipole moment/unit volume and H_{in} a fictitious internal field. Note that H_{in} would be uniform over the entire particle, whereas because of the skin depth the actual fields inside the particle are not uniform at all. For a spherical particle

$$\vec{H}_{in} = \vec{H}_{app} - \frac{4\pi}{3}\vec{M}$$

and

$$\mu_m = \frac{1 + \frac{8\pi}{3}\gamma}{1 - \frac{4\pi}{3}\gamma}$$

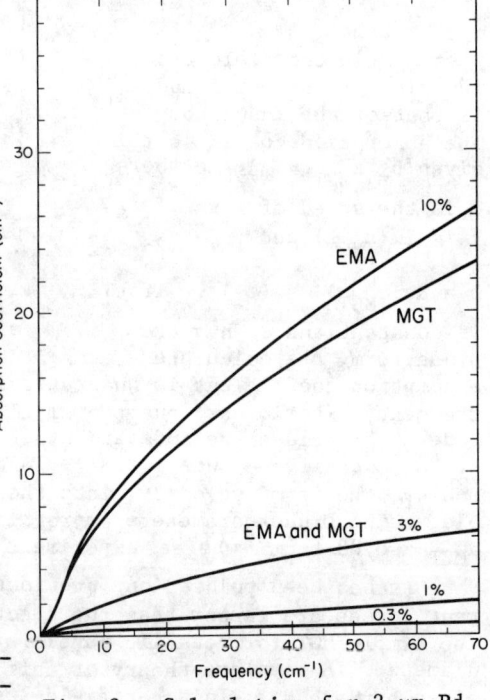

Fig. 3. Calculation for 2 μm Pd particles including eddy currents.

The MGT expression for the composite permeability is of the same form of as Equation (1) with ε replaced by μ. The absorption coefficient is given by $\alpha = 2\omega\kappa/c$ where $\kappa = \text{Im}(\varepsilon_{MGT}\mu_{MGT})^{1/2}$. Figure 3 shows the calculated absorption coefficient for the 2 μm Pd particles. At 70 cm^{-1} the calculation is smaller than experiment by about an order of magnitude. The qualitative shape of the calculation is similar to experiment. The absorption coefficients start out quadratic but the calculations bend over much sooner than do the experimental data. In calculating the curves in Figure 3 we have used Drude parameters appropriate for Pd.[11] Curves closer to experiment can be obtained by treating the Drude parameters as free variables.

The size dependence of the absorption due to the magnetic dipole effect varies initially as d^2. For the smaller aluminum particles the calculated absorption coefficients are smaller than experiment by two orders of magnitude. For example, for d = 500 Å aluminum particles we obtain $\alpha \simeq 70$ f at $\omega = 70$ cm^{-1}.

CONCENTRATION DEPENDENCE

For any batch of particles we observe an increase in absorption coefficient with increasing particle concentration. Figure 4 shows the values of the measured absorption coefficient in the Pd particles at several frequencies, normalized by the value of the absorption coefficient at that frequency for f = 0.01. The absorption coefficient is linear in concentration for $f \lesssim 0.03$ but rises rapidly for $f \simeq 0.1$. The predictions of the MGT and EMA are also shown. The MGT predicts a linear concentration dependence for $f \ll 1$ (no metal-insulator transition) while the EMA gives a metal-insulator transition at f = 1/3 where α would be very large indeed. The increasing absorption at $f \simeq 0.1$ is due to the onset of the metal-insulator transition which in these metal-KCl samples occurs in the range 0.15 < f < 0.25.

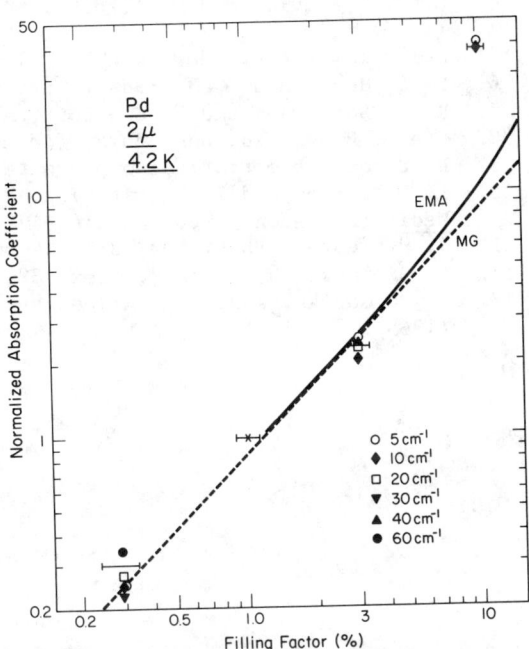

Fig. 4. Normalized absorption coefficient versus concentration.

CONCLUSIONS

We conclude that the magnetic dipole absorption is sufficient to explain the far infrared absorption in relatively large (d $\simeq$ 2 µm) small particles but is unable to account for absorption in smaller particles. Possible explanations include surface layers such as oxide coats,[12] quantum effects[13] or accumulation of many small particles into a macroparticle.

REFERENCES

*Research supported by the Ohio State Program for Energy Research, Education and Public Service.

1. D. B. Tanner, A. J. Sievers and R. A. Buhrman, Phys. Rev. B<u>11</u>, 1330 (1975).
2. C. G. Granqvist, R. A. Buhrman, J. Wyns and A. J. Sievers, Phys. Rev. Letters <u>37</u>, 625 (1976).

3. J. C. Maxwell Garnett, Philos. Trans. Roy. Soc. London 203, 385 (1904); 205, 237 (1906).
4. C. G. Granqvist and R. A. Buhrman, J. Appl. Phys. 47, 2200 (1976).
5. Leico Industries, New York, New York.
6. R. L. Henry and D. B. Tanner, preprint.
7. R. B. Sanderson and H. E. Scott, Appl. Opt. 10, 1097 (1971).
8. D.A.G. Bruggeman, Ann. Phys. (Leipz.) 24, 636 (1935).
9. D. Stroud, preprint; also presented at the ETOPIM conference.
10. L. D. Landau and E. M. Lifshitz, Electrodynamics of Continuous Media (Pergamon, 1960) Sections 72 and 73.
11. J. H. Weaver, Phys. Rev. B11, 1416 (1975).
12. E. Šimánek, Phys. Rev. Letters 38, 1161 (1977).
13. C. P. Gor'kov and G. M. Éliashberg, Sov. Phys. JETP 21, 940 (1965).

MICROWAVE PROPAGATION IN POWDERED SEMICONDUCTORS[*]

J. E. Sansonetti[†] and J. K. Furdyna
Physics Department, Purdue University,
West Lafayette, Indiana 47907

ABSTRACT

Microwave propagation in powdered semiconductors is a potentially important technique for characterization of semiconducting materials. The problem also contains valuable insights into the interaction of electromagnetic waves with inhomogeneous media in general. We have chosen powdered silicon to study the behavior of such an inhomogeneous system because of the wide range of material parameters available and the ability to vary the permittivity of the grains in situ by varying the temperature and/or the applied dc magnetic field. A series of samples, with grain size ca. 100 μm and a wide range of conductivities, was investigated at temperatures from 77°K to 300°K and magnetic fields from 0 to 20 kG. We show that the interaction can be satisfactorily formulated in terms of a theoretical model involving an effective depolarizing factor for the powder medium as a whole.

INTRODUCTION

The investigation of microwave propagation in powdered semiconductors can yield considerable information about inhomogeneous media. This is possible primarily because the electrical properties of the semiconductor can be varied in situ while the geometry of the powder remains unchanged. For example, by changing the temperature of a sample or placing it in a magnetic field, the contribution of the free carriers to the dielectric constant can be varied by orders of magnitude. This flexibility permits the separation of geometrical effects from those due to material characteristics. In investigating semiconductor powders, the range of constitutive parameters is further extended by the fact that semiconductors are available with a wide range of electrical properties.

Microwave propagation in the powder can be described by an effective dielectric constant, which is a function of the dielectric constants of the constituent materials. The model we will employ to calculate the effective dielectric constant is a modification of the approach used to calculate the Clausius-Mossotti equation, and yields the Maxwell-Garnett[1] and "Effective Medium" theory[2] results in the appropriate limits. The expressions obtained are very similar in form to the Maxwell-Garnett relations, except that they have been generalized to include the effects of non-spherical grains, of grain-grain interactions, and of the anisotropy imposed by the presence of the external dc magnetic field.

---

[*] Supported by NSF/MRL Program 76-00889

[†] NSF Energy Trainee

THE MODEL

Let us consider a powder in vacuum. We denote all quantities pertaining to the powder as a whole by the subscript p, and the quantities related to the grain material by the subscript g. Taking volume averages, we can write

$$V_p \langle \vec{P}_p \rangle = V_g \langle \vec{P}_g \rangle , \qquad (1)$$

where V denotes volume and $\vec{P}$ denotes polarization. This can be rewritten in terms of the electric fields as

$$(\kappa_p - 1) \langle \vec{E}_p \rangle = C_g (\kappa_g - 1) \langle \vec{E}_g \rangle , \qquad (2)$$

where κ is the relative dielectric constant and C_g is the volume packing fraction, V_g/V_p.

The average electric field <u>within</u> the grain is

$$\langle \vec{E}_g \rangle = \langle \vec{E}_{Loc} \rangle - L_g \langle \vec{P}_g \rangle , \qquad (3)$$

where $\langle \vec{E}_{Loc} \rangle$ is the average local field and L_g is the depolarizing factor for a single grain ($L_g = 1/3$ for a sphere). Since, in a powder, there are a collection of particles with different shapes, L_g is actually an average value for a statistical distribution. Equation (3) can be rewritten as

$$\langle \vec{E}_{Loc} \rangle = [1 + L_g(\kappa_g - 1)] \langle \vec{E}_g \rangle . \qquad (4)$$

The local field is equal to the average electric field in the powder medium plus a contribution due to the grain-grain interaction, the latter being proportional to the polarization of each grain, $\langle P_g \rangle$, and thus also to $\langle P_p \rangle$. This can be expressed as

$$\langle \vec{E}_{Loc} \rangle = \langle \vec{E}_p \rangle + S \langle \vec{P}_p \rangle = [1 + S(\kappa_p - 1)] \langle \vec{E}_p \rangle , \qquad (5)$$

where S is analogous to the Lorentz factor in the Clausius-Mossotti (CM) equation. For a system of point dipoles, for which the CM equation is usually derived, the factor S can be shown to be 1/3 for several cases. However, since we do not have such a system in a powder, we will keep S as an empirical factor.

Using Eqs. (2), (4), and (5), we can eliminate the electric fields and obtain

$$\kappa_p = 1 + \frac{C_g}{L_p} \frac{\kappa_g - 1}{(\kappa_g - 1) + 1/L_p} , \qquad (6)$$

where $L_p = L_g - SC_g$ is the effective depolarization factor for the powder[3]. Equation (6) can also be written in terms of the electric susceptibilities as

$$\chi_p = C_g \frac{\chi_g}{1 + L_p \chi_g} \quad . \tag{7}$$

In cases where the powder is not in a vacuum but embedded in a medium with a dielectric constant K_e, we can show that Maxwell's equations and the boundary conditions can be reduced to the same mathematical form as for a powder in a vacuum by replacing K_g and K_p with K_g/K_e and K_p/K_e, respectively. Then Eq. (6) takes the form

$$K_p = K_e \left\{ 1 + \frac{C_g}{L_p} \frac{K_g - K_e}{(K_g - K_e) + K_e/L_p} \right\} \quad , \tag{8}$$

which reduces to the Maxwell-Garnett relation if we let $L_p = (1-C_g)/3$. Setting L_p to this is equivalent to assuming spherical grains ($L_g = 1/3$) and a small packing fraction, $C_g \ll 1$. Under these conditions the "effective medium" theory and Eq. (8) also become equivalent.

A semiconductor in a dc magnetic field $\vec{B}$ can be described by three response functions corresponding to three principal polarizations (normal modes). These are: a longitudinal polarization, parallel to $\vec{B}$; and two transverse circular polarizations, one polarized in the same sense as the cyclotron motion of the charge carriers (thus called the "Cyclotron Resonance Active," or CRA mode), and the other polarized in the opposite sense to the cyclotron motion ("Cyclotron Resonance Inactive," or CRI). In this paper we shall only be interested in the transverse response functions. Since the cyclotron motion of the charge carriers reverses when the magnetic field changes sign, we can write the two transverse dielectric constants (CRA and CRI) of the semiconductor as a single function

$$K_g = K_\ell + \frac{i\sigma}{\omega \epsilon_o} \frac{1}{1-i(\omega-\omega_c)\tau} \quad , \tag{9}$$

where K_ℓ is the lattice contribution to the dielectric constant, σ is the dc conductivity, ω is the angular frequency, τ is the carrier relaxation time, ϵ_o is the free space permittivity, $\omega_c = eB/m^*$ is the cyclotron frequency, which can be either positive or negative depending on the sign of the magnetic field, and m^* is the electron effective mass. Thus Eq. (9) represents the CRA response for $\omega_c > 0$; and the CRI response for $\omega_c < 0$. To be rigorously correct, statistical averaging of the free carrier term in Eq. (9) must be done to allow for the possible dependence of τ on the energy. In the theoretical calculations presented here this correction will be neglected.

It has been shown by Galeener[4] that anisotropic spheroids excited by one of the principal polarizations will respond as if they were isotropic, with the dielectric constant corresponding to that polarization. We will assume that this is also true for the grains of the powder, and thus that normal modes propagate without mixing.

Putting the above dielectric constant into our theoretical model, the dielectric constant of the multiparticle medium consisting of semiconductor powder in a vacuum can be expressed as follows:

$$\kappa_p = \kappa_{\ell p} + \frac{i\sigma_p}{\omega\epsilon_o} \frac{1}{1-i(\omega-\omega_c-\omega_p^2/\omega)\tau} \quad , \tag{10}$$

where
$$\kappa_{\ell p} = 1 + \frac{C_g}{L_p} \frac{\kappa_\ell - 1}{(\kappa_\ell - 1) + 1/L_p} \quad ,$$

$$\sigma_p = \sigma_g \left[\frac{C_g}{L_p^2} \frac{1}{(\kappa_\ell - 1) + 1/L_p} \right] \quad ,$$

$$\omega_p^2 = \frac{\sigma_g}{\epsilon_o \tau} \frac{1}{(\kappa_\ell - 1) + 1/L_p} \quad .$$

By arranging the expression for the dielectric constant in this fashion, we can immediately see one of the major consequences of the powder geometry: the resonance in the free carrier term, instead of occurring at the cyclotron frequency, i.e., $\omega = \omega_c$, will now occur when

$$\omega = \omega_c + \omega_p^2/\omega \quad .$$

This is analogous to the "plasma shift"[5] of the cyclotron resonance in an isolated particle, but depends directly on the grain-grain interaction through the powder depolarization factor L_p.

It has been shown by Ford, Furdyna, and Werner[6] that small semiconductor particles can also have a magnetic polarizability, which is proportional to $(a/\lambda_o)^2$, where a is the radius of the grain and λ_o is the free space wavelength. Since, for the particles we have studied, this factor is about 7×10^{-4}, any contributions to the relative magnetic permeability will be small and shall be neglected.

Having formulated the effective dielectric constant for the powder, we can write the complex propagation constant k for a circularly polarized wave propagating in a powder medium parallel to $\vec{B}$ (the Faraday geometry):

$$k \equiv \alpha + i\beta = \frac{\omega}{c} \sqrt{\kappa_p}$$

where α and β are the phase and attenuation coefficients, respectively. It is these quantities, α and β, which we measure in microwave transmission experiments on powders.

EXPERIMENT AND RESULTS

To examine the validity of the theoretical model, we have studied microwave magneto-transmission in a series of powdered n-type Si samples with room temperature resistivities ranging from 0.1 Ω-cm to 270 Ω-cm. Each sample was powdered with mortar and pestle and sieved to obtain the same size distribution for all samples (44 μ < a < 52 μ).

It was then loosely packed into a cylindrical sample holder and the volume packing fraction, C_g, was measured. The packing fraction in the samples so prepared was between 0.47 and 0.51. Each sample was then inserted into the dewar in the 35 GHz microwave bridge shown in Fig. 1, and the phase and amplitude of the transmitted beam were measured as the magnetic field was swept from 0 to 20 kG. This was repeated at seven temperatures between 77 and 300 K.

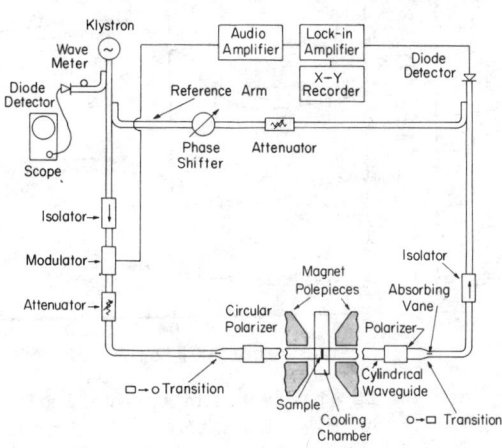

Fig. 1. Schematic diagram of the 35 GHz microwave bridge.

The electrical parameters of the samples were estimated by making similar microwave transmission measurements on solid slabs cut from the same ingots as the samples. The values of the lattice dielectric constant and of the longitudinal and transverse electron effective masses for Si are known. Using these parameters, we estimated the carrier concentration and relaxation time by fitting the phase and attenuation data with a computer program written by J. R. Dixon, Jr.[7] to simulate microwave transmission in slabs. These fits indicated that the energy dependence of τ plays a significant role in determining the dielectric constant of the silicon; therefore the values of n and τ were allowed to vary slightly from the estimates to obtain an optimal fit to the powder data. Since, in a powder, the crystallites are randomly oriented, we account approximately for the anisotropy of the band structure of silicon by using an average effective mass equivalent to the cyclotron resonance mass obtained in the $\langle 111 \rangle$ direction, $m^* = 0.27\ m_0$, in calculating K_g for the average grain.

With K_g determined by the slab transmission data, and C_g measured, we have reduced the number of unknown quantities in the expression for K_p to one, viz. the effective depolarizing factor for the powder L_p. A value of L_p was chosen for each sample to give the best agreement with the experimental data. Examples of the experimental data and theoretical fits are given in Figs. 2 and 3.

Figure 2 demonstrates the temperature and magnetic field dependences of the phase shift attenuation of microwaves transmitted through the sample in the Faraday geometry. The particular sample illustrated has a packing fraction $C_g = 0.47$ and $L_p = 0.11$. Within this one set of data, which are all fit with the same value of L_p, the real part of the dielectric constant K_g of the silicon ranges from 4.0 to 20.0 and the imaginary part from 0.8 to 16.0. The dashed lines show the theoretical fits and for comparison the dotted lines show theoretical curves obtained with $L_p = 1/3(1-C_g) = 0.17$.

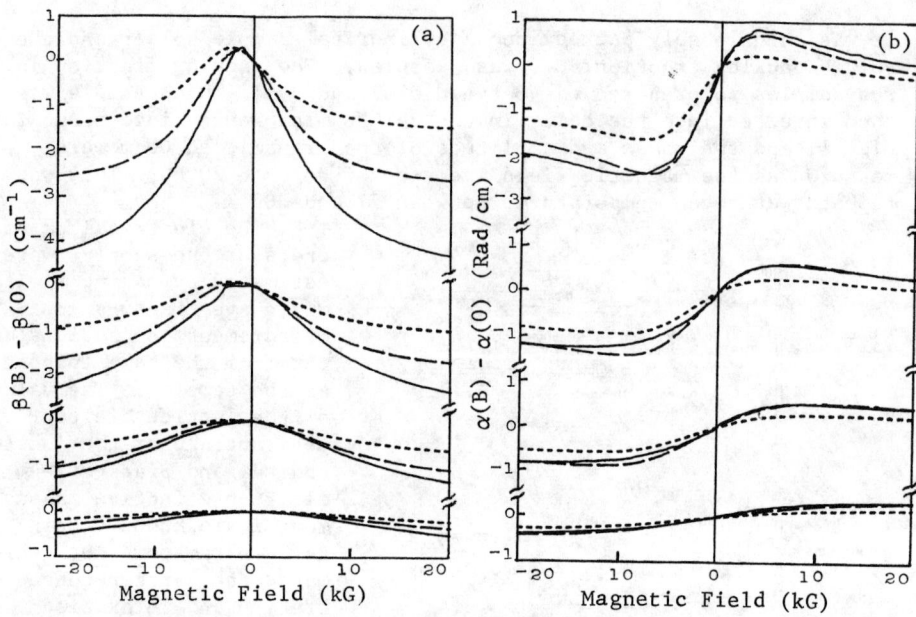

Fig. 2. Attenuation (a) and phase shift (b) data and theoretical fits for n-type Si powder (30 Ω-cm at 300 K). From top to bottom the data was taken at 77, 107, 134, and 182 K, respectively.

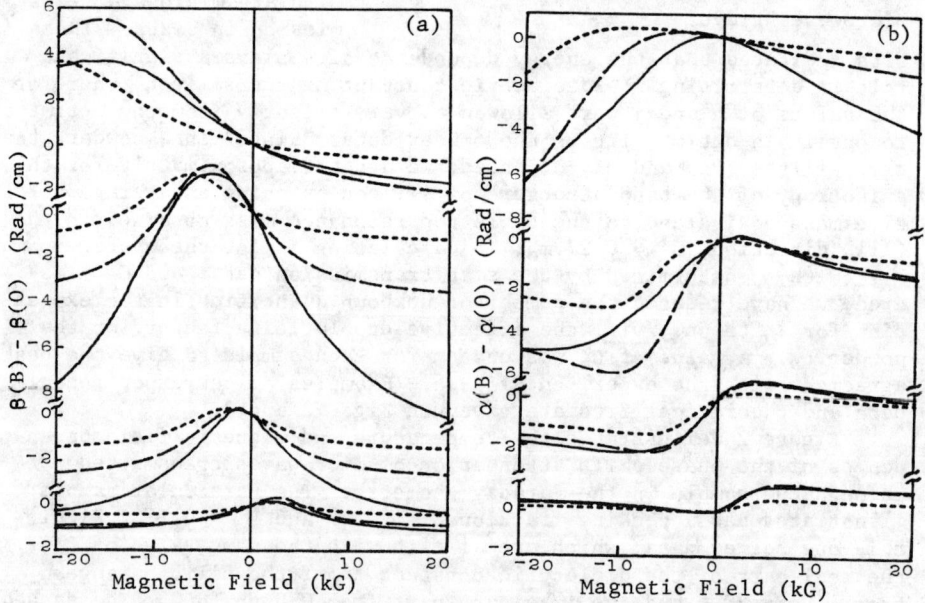

Fig. 3. Attenuation (a) and phase shift (b) data and theoretical fits for four samples at 77 K. From top to bottom the samples had room temperature resistivities of 2, 14, 30, and 100 Ω-cm.

Figure 3 shows phase and attenuation data for four different samples, all measured at 77 K. It illustrates the development of the "plasma shift" of cyclotron resonance as the number of carriers increases. For all samples used in this graph the packing fractions were about 0.50 and the value of L_p used was between 0.06 and 0.11. The range of the real part of K_g for these samples was -5.50 to 30.0 and that of the imaginary part was 0.1 to 35.0.

CONCLUSION

We have presented a model which characterizes the powder by an effective dielectric constant determined solely by the dielectric constant of the constituents, their relative concentrations, and an effective depolarization factor. The latter accounts for the scattering due to the grains and the grain-grain interaction and hence it is a geometrical factor which depends upon the size and shape of the grains and upon the inter-grain distance in a particular powder sample. This makes the model much more appropriate for higher packing fractions than similar models which assume $L_p = (1-C_g)/3$. It also avoids the assumption of a homogeneous medium outside the grain of interest, which the "effective medium" theory entails. We have compared theoretical calculations based on our model with experimental data for a wide range of permittivities. The agreement of the model with the data is qualitatively excellent, and quantitatively quite encouraging. The quantitative discrepancies are due, at least in part, to the inaccuracy of K_g used: in formulating K_g we have, e.g., neglected the possible dependence of τ on energy. One of the drawbacks of the model itself is its assumption of complete mode separation, e.g., that K_p for the CRA polarization is determined totally by K_g for the CRA waves. In spite of these shortcomings, the overall agreement of data and theory indicates the viability of the concepts employed, suggesting further quantitative investigation.

REFERENCES

1. J. C. Maxwell-Garnett, Philos. Trans. R. Soc. Lond. 203, 385 (1904).
2. I. Webman, J. Jortner, and M. Cohen, Phys. Rev. B15, 5712 (1977), and references therein.
3. The concept of the powder depolarizing factor was briefly considered by F. L. Galeener, K. K. Chen, and J. K. Furdyna, Appl. Phys. Letters 16, 387 (1970); also by F. L. Galeener, Ph.D. Thesis, Purdue University, 1970 (unpublished); and by K. K. Chen, Ph.D. Thesis, Purdue University, 1972 (unpublished).
4. F. L. Galeener, Phys. Rev. Lett. 22, 1292 (1969).
5. G. Dresselhaus, A. Kip, and C. Kittel, Phys. Rev. 98, 368 (1955) and 100, 618 (1955).
6. G. W. Ford, J. K. Furdyna, and S. A. Werner, Phys. Rev. B12, 1452 (1975).
7. J. R. Dixon, Jr., Purdue University (unpublished).

OPTICAL PROPERTIES OF SMALL-PARTICLE COMPOSITES

Ronald Fuchs
Ames Lab-USERDA, Iowa State Univ., Ames, Iowa 50011

ABSTRACT

A new theory for the effective dielectric constant of a small-particle composite system includes broadening of single-particle resonances by dipolar interactions and percolation at high particle density. Comparisons are made with effective medium theory, Maxwell-Garnett theory, resistor network models, and with optical measurements on metallic and dielectric composites.

INTRODUCTION

Both the Maxwell-Garnett theory (MGT)[1] and the effective medium theory (EMT)[2] have been used to calculate the effective dielectric constant and optical absorption of small-particle composite systems. However, neither theory is adequate. The infrared absorption spectrum of powders of ionic crystal cubes has a broad peak between the TO and LO frequencies.[3] The MGT (for cubes) agrees roughly with experiment, but the experimental peak width is much greater than the theoretical width, and the predicted absorption peaks associated with the individual cube modes are not observed. An absorption peak often appears at the TO frequency; this is a mode with depolarization factor n=0, and it is not contained in the MGT. A system of metallic spheres (Ag in a SiO_2 matrix) has an absorption peak much broader than predicted by the MGT.[4]

The EMT theory for spheres has a mode with depolarization factor n=0 if the particle filling fraction f > 1/3; however, the sphere mode with n ~ 1/3 is absent for these f values. This theory therefore does not explain how absorption peaks corresponding to n=0 and n ~ 1/3 can appear simultaneously in ionic-crystal powders, and it also does not give the observed absorption peak for Ag spheres when f=0.39.

The new theory of the effective dielectric constant is an extension of the MGT which can treat particles of arbitrary shape. It includes both broadening effects of dipolar interactions between particles and a n=0 mode arising from percolation or clustering.

GENERAL THEORY

The electric dipole moment of a system of small particles in vacuum, with a uniform applied field E_o, can be written[5]

$$M = v\langle\chi\rangle E_o, \qquad (1)$$

where

$$\langle\chi\rangle = \sum_m \frac{C_m}{\chi^{-1} + 4\pi n_m}. \qquad (2)$$

Here v is the volume of the particles, which are composed of a material with dielectric susceptibility χ, while C_m and n_m are the strengths and depolarization factors of the dipole modes labeled by the index m. Imagine that the particle system forms a needle-shaped sample with the long axis in the direction of E_o. If the sample has a homogenous effective susceptibility χ_{eff}, then one also has

$$M = V \chi_{eff} E_o, \qquad (3)$$

where V is the volume of the needle. From Eqs. (1)-(3) one finds

$$\chi_{eff} = f \sum_m \frac{C_m}{\chi^{-1}+4\pi n_m}, \qquad (4)$$

where $f=v/V$ is the particle filling fraction.

The depolarization factors n_m lie in the range $0 \leq n_m \leq 1$ and are densely distributed for a many-particle system, so it is convenient to replace the sum over the discrete mode index m by an integral over the continuous variable n: $\sum_m C_m \to \int C(n)D(n)dn = \int g(n)dn$, where $D(n)$ is the density of modes and $g(n) = C(n)D(n)$. Eq. (4) then becomes

$$\chi_{eff} = f[C_o \chi + \int_0^1 \frac{g(n)dn}{\chi^{-1}+4\pi n}]. \qquad (5)$$

The separately written term $C_o\chi$ comes from the mode with $n_m=0$; all other modes, with $n_m>0$, are included in the integral. The function $g(n)$ contains the dipole strength and density of the particle resonance modes, whereas C_o is the strength of the percolation mode. The quantities $g(n)$ and C_o depend only on the structure of the composite system, including the filling fraction f (they are independent of the material χ), and they obey the sum rule

$$C_o + \int_0^1 g(n)dn = 1. \qquad (6)$$

GAUSSIAN BROADENING MODEL

Every valid theory of the effective susceptibility can be cast into the form of Eq. (5), and different theories will yield different results for C_o and $g(n)$. An extension of the MGT will be described; it includes random dipolar interactions which cause a Gaussian broadening function to appear in the theory.

First consider a randomly oriented single particle of volume v_p, in a field E. Its dipole moment is

$$\mu = v_p \langle \chi_p \rangle E, \qquad (7)$$

where $\langle \chi_p \rangle$ is the single particle susceptibility, expressible in the form of Eq. (2). For a sphere, $\langle \chi_p \rangle = (\chi^{-1}+4\pi/3)^{-1}$; for a cube, see references 5 and 6. Imagine applying an electric field to a powder consisting of identical particles, and draw a large sphere with one of the particles at its center. The dipole moment of the central particle is

$$\mu_i = v_p \langle \chi_p \rangle (E_o + h_i \mu_i / v_p), \tag{8}$$

where E_o is the field at the central particle arising from the system outside the sphere, and $h_i \mu_i / v_p$ is the sum of the dipolar fields of the particles inside the sphere. The index i labels a given particle configuration, and h_i is a geometrical factor containing the dipolar sum. One can solve Eq. (8) for μ_i and take a configuration average by introducing the probability P_i of a configuration i. The average susceptibility of an interacting particle can then be written

$$\langle \chi_I \rangle = (v_p E_o)^{-1} \sum_i P_i \mu_i \tag{9}$$

The configuration average can be found by assuming that the particles other than the central particle are independently and randomly distributed in the large sphere, excluding a small spherical volume of radius r_{min} containing the central particle. That is, correlation between the "other" particles is neglected, and this allows the central limit theorem to be used, with the result

$$\langle \chi_I \rangle = \int_{-\infty}^{\infty} \frac{P(h) dh}{\langle \chi_p \rangle^{-1} - h} \tag{10}$$

where $P(h)$ is a Gaussian broadening function,

$$P(h) = (w\sqrt{2\pi})^{-1} \exp(-\tfrac{1}{2} h^2/w^2), \tag{11}$$

with

$$w^2 = \frac{16}{15} \pi f v_p (r_{min})^{-3}. \tag{12}$$

Finally the Maxwell-Garnett theory is applied to these interacting particles with susceptibility $\langle \chi_I \rangle$, and one finds that the effective susceptibility of the composite system is

$$\chi_{eff} = [\frac{1}{\chi} + \frac{1-f}{f \langle \chi_I \rangle}]^{-1}. \tag{13}$$

If χ_{eff} is written in the form of Eq. (5), the percolation mode C_o does not appear explicitly, and a tail of the broadened single-particle resonance function $g(n)$ appears in the unphysical region $n < 0$. The percolation mode is introduced by taking its strength to be equal to the total strength of the unphysical tail,

$$C_o = \int_{-\infty}^{0} g(n)dn, \qquad (14)$$

and thereafter setting $g(n) = 0$ for $n<0$.

If f is very small or if interactions between particles are neglected, $P(h) \to \delta(h)$ and $\langle \chi_I \rangle \to \langle \chi_p \rangle$. Eq. (13) then reduces to the usual Maxwell-Garnett theory for particles of arbitrary shape.[5]

RESULTS

The particle resonance function $g(n)$ has been calculated for interacting spheres using the Gaussian broadening model, and it is shown in Fig. 1 using four values of f. The percolation mode C_o is simulated by a peak at $n=0$. As f increases, the sphere mode broadens and its center shifts from $n=1/3$ to smaller values of n, while the percolation mode grows stronger. Fig. 1 also shows the EMT theory for spheres, written in the form of Eq. (5). The function $g(n)$ broadens asymmetrically as f increases, and for $f>1/3$, the critical concentration for the onset of percolation, there is no remnant of a particle resonance peak.

A random resistor network also has been used to simulate a composite system. An electric field is applied in the z direction to a cubic resistor network with host resistors of unit conductivity containing randomly located defect resistors with conductivity $1+4\pi\chi$. Each defect resistor in the z direction contributes $\chi \cdot \Delta V$ to the dipole moment, ΔV being the potential difference across the resistor. The dipole moment of a single defect resistor in the z direction becomes infinite if $\chi^{-1} = -4\pi/3$ and thus the defect simulates a sphere with depolarization factor $n=1/3$. Several defects simulate a group of interacting spheres. Defects in the x and y directions do not contribute to the dipole moment in the z

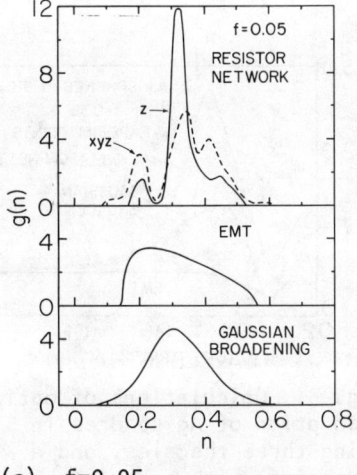

(a) f=0.05

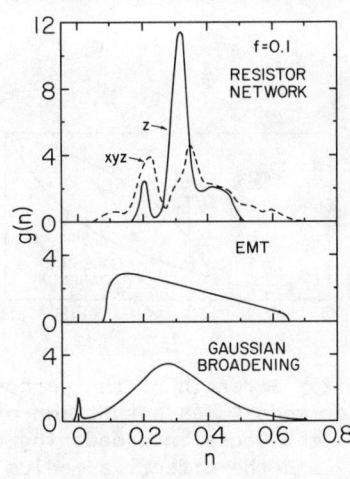

(b) f=0.1

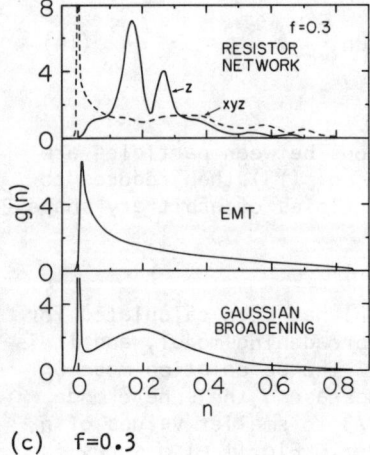
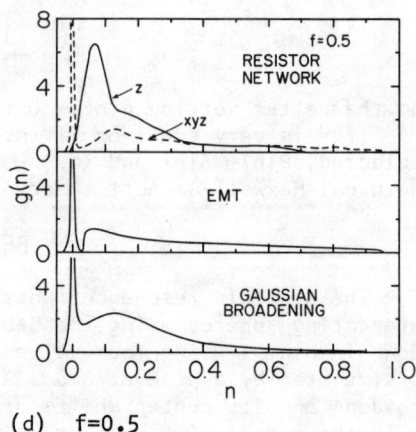

(c) f=0.3   (d) f=0.5

Fig. 1. Mode strength $g(n)$ as a function of depolarization factor n, with four values of the filling fraction f. Calculations using the Gaussian broadening model, the effective medium theory, and resistor network models are compared.

direction, but they broaden resonance structure arising from nearby z-directed defects. A calculation of the total dipole moment of the network gives an effective susceptibility in the form of Eq. (5). Fig. (1) shows the function $g(n)$ for networks with defects in only the z direction (solid curve) and with equal defect concentrations in the x, y, and z directions (dashed curve). Percolation occurs only in the xyz model.

Figure 2 contrasts the gradual increase of the percolation constant C_0 in the Gaussian broadening theory as f increases, with the sudden increase of C_0 in the EMT at the critical concentration f=1/3.

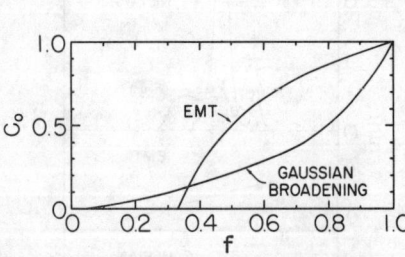

Fig. 2. Strength of the percolation mode C_0 as a function of f for the Gaussian broadening model and the effective medium theory.

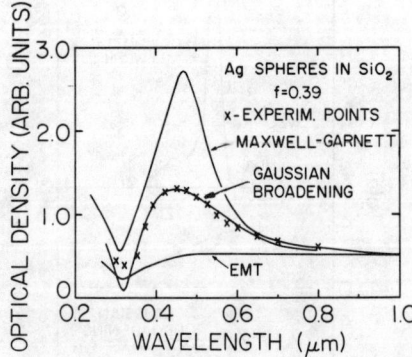

Fig. 3. Calculations of optical absorption of Ag spheres in SiO_2 using three theories, and a comparison with experiment.

Eq. (5) can be adapted easily to find the effective dielectric constant ϵ_{eff} of a system of particles with dielectric constant ϵ in a host with dielectric constant ϵ_h by writing $\chi = (\epsilon/\epsilon_h-1)/4\pi$ and $\chi_{eff} = (\epsilon_{eff}/\epsilon_h-1)/4\pi$. The Gaussian broadening theory has been used to calculate the optical absorption of Ag spheres in a SiO_2 host[4], with a filling fraction f=0.39. It is evident from Fig. 3 that the new theory is in better agreement with experiment than either the MGT or the EMT. The theory has also been used to calculate the infrared absorption spectrum of ionic crystal powders. It reproduces qualitatively both the broad peak between ω_{TO} and ω_{LO} arising from particle modes and the narrow peak at ω_{TO} arising from clustering.

CONCLUSION

The Gaussian broadening model gives robust particle modes and describes powders in which the particles retain their identity for relatively large f values, but it becomes incorrect for f close to 1. It is qualitatively similar to the z-directed resistor network. In the EMT, just as in the xyz resistor network, the particles lose their identity as f increases, a behavior consistent with the particle-host symmetry.

The absence of a sharp percolation threshold in the Gaussian broadening model may be physically reasonable for samples consisting of a thin powder layer, in which particle clustering and deviations from randomness occur. It is a phenomenological model, in the sense that the cutoff radius r_{min} in Eq. (12) was somewhat arbitrarily chosen as $r_{min} = \frac{1}{2}(v_p)^{\frac{1}{3}}$, a value for which the mode broadening agrees qualitatively with the resistor network calculations. The percolation strength C_o arises in an unnatural manner, and it is inaccurate to neglect higher multipole interactions between particles. A combination of Gaussian broadening with the EMT might give a better theory for the effective dielectric constant.

ACKNOWLEDGEMENT

This work was supported by the U.S. Energy Research and Development Administration, Division of Physical Research.

REFERENCES

1. J. C. Maxwell-Garnett, Philosoph. Trans. R. Soc. Lond., 203, 385 (1904).
2. R. Landauer, J. Appl. Phys. 23, 779 (1952).
3. L. Genzel and T. P. Martin, Phys. Status Solidi B51, 91 (1972).
4. J. I. Gittleman and B. Abeles, Phys. Rev. B15, 3273 (1977).
5. R. Fuchs, Phys. Rev. B11, 1732 (1975).
6. D. Langbein, J. Phys. A4, 627 (1976).

SELF-CONSISTENT THEORY OF ELECTROMAGNETIC WAVE PROPAGATION IN COMPOSITE MEDIA

D. Stroud and F. P. Pan
Dept. of Physics, The Ohio State Univ., Columbus, Ohio 43210

ABSTRACT

A self-consistent theory is developed to treat the propagation of electromagnetic waves in composite media. The theory reduces to the effective-medium approximation in the static limit, but unlike the latter is not necessarily limited to composites in which the particles are small compared to the wavelength of the electromagnetic field. The self-consistency condition on the effective dielectric functions ε_{eff} reduces to the requirement that ε_{eff} be chosen so that the forward-scattering amplitude of particles embedded in this medium should vanish on the average. The approximation is applied to far infrared absorption in a model granular metal. The absorption due to induced eddy currents, properly included in the new theory, but neglected in the existing quasistatic theories, is shown to dominate the classical absorpiton coefficient of such composites below the insulator-metal transition even for very small particles.

This paper will appear in Physical Review B.

ISSN: 0094-243X/78/282/$1.50 Copyright 1978 American Institute of Physics

OPTICAL CONSTANTS OF CERMET MATERIALS
INCLUDING PROXIMITY EFFECTS

D.R. McKenzie
R.C. McPhedran
School of Physics, University of Sydney, 2006. N.S.W.
Australia.

ABSTRACT

An extended Rayleigh formulation for the solution of transport properties of arrays of spheres of one material embedded in another is applied to the case of the optical constants of an array of spherical particles of gold in a dielectric. The formulation includes, unlike the Maxwell-Garnett theory, the effects of proximity. The optical constants of the metal are assumed to be the bulk values. A model is presented for applying the results based on the simple cubic lattice to the case of gold cermets. New experimental results are presented for $Au-Al_2O_3$ cermets. The new theory explains effects present in these and previously available data.

INTRODUCTION

It is the aim of this paper to find the implications of the techniques described in our earlier paper when applied to the problem of determining the optical constants n and k of a cermet alloy. The model we will use here will assume the metal particles to be spherical in shape and to be arranged in the non-metallic matrix in a disordered manner rather similar to the positions of molecules in a liquid at a given point in time. The model will further assume that the optical constants of the metal spheres are those of the bulk metal. We have chosen to assume spherical particles and bulk values of the optical constants to avoid the introduction of undetermined parameters describing ellipticity of the particles and variation of the optical constants with particle size. Previous authors have introduced such parameters in conjunction with the Maxwell-Garnett theory. It is our belief that proximity effects should be taken into account in a treatment of the optical constants of a cermet before any effect of ellipticity or particle size dependence of the metal optical constants is introduced. Gold cermets have been studied more extensively than those of other metals so we shall concentrate our attention on those.

EXPERIMENTAL

There are a number of determinations in the literature[1] of the optical constants of gold cermets. Lissberger and Nelson found the optical constant of $Au-MgF_2$ cermets with volume fraction f = 0.2 and f = 0.3. Fan and Zavracky[2] have presented

ISSN: 0094-243X/78/283/$1.50 Copyright 1978 American Institute of Physics

results for Au-MgO cermets prepared by r.f. sputtering from a
composite target with f = .25. We present in figure 1 results
for Au-Al$_2$O$_3$ cermets with volume fractions ranging from f = .096
to f = 0.307 together with previous measurements. These cermets
were prepared by vacuum evaporation from dual sources, the Al$_2$O$_3$
from an electron beam gun and the Au from a resistively heated
tungs ten helix. The composition of the film was varied by
altering the relative evaporation rates from the two sources.
Films of graded composition could also be prepared by changing
the relative evaporation rates during coating. The optical
constants of each film were determined as a function of wavelength
by measuring four wavelength dependent properties. These were
the reflectance of the film when evaporated onto a glass slide
from both the air and the glass side; the transmittance of the
film on glass and the reflectance of the film on a copper substrate.
The optical constants n and k were determined by a least squares
fitting procedure at each wavelength. The thickness of the film
was not allowed to vary in the fitting procedure but was determined using a "Talysurf" mechanical step-height instrument to a
precision of 50 Å.

THEORETICAL

Although the experimental measurements vary from sample to
sample, certain features are evident. Firstly, the resonance is
much broader than predicted by the Maxwell-Garnett theory, which
predicts for volume fractions up to the touching limit of 63%
volume fraction of spheres^3, a sharp resonance centred at 0.6μm
approximately. Secondly, the region over which k is measured to
be large extends to wavelengths longer than 1.0μm, again in con-
tradiction of the Maxwell-Garnett theory. These features of the
experimental results give rise to the good solar absorbing
properties of gold cermets which are not predicted by the
Maxwell-Garnett theory.

The consequences of including the effects of interaction
between particles for simple cubic lattices of various volume
fractions are shown in figure 2, calculated using the Extended
Rayleigh Formulism[4,5] (E.R.F.). A cermet, of course, is a disordered structure, but many of the features will be reproduced
if we postulate a layer model which consists of layers of
regular lattice of various volume fractions. If the layers are
parallel to the applied field, then the resultant dielectric
constant will be

$$\varepsilon = \sum_i w_i \, \varepsilon_i \, (f_i) \qquad (1)$$

where the weight w_i of the i th volume fraction is chosen
according to a radial distribution function. This model will
then have the same most probable nearest neighbour distance and

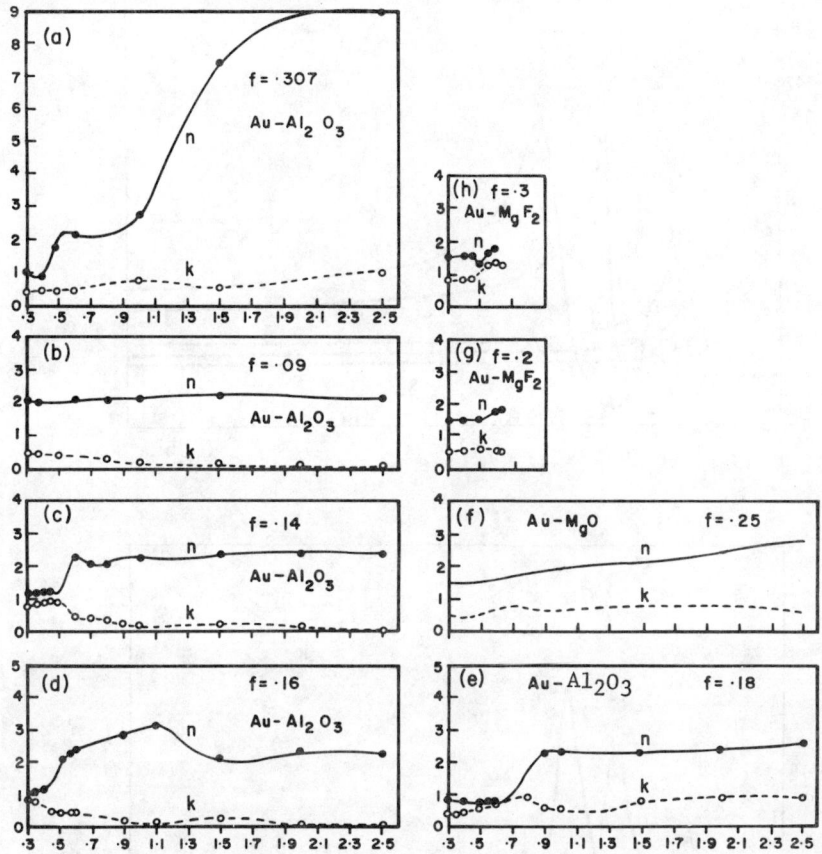

Figure 1.

Measured optical constants of gold cermets. (a) to (e) are our results for Au-Al_2O_3 cermets. (f) is the result obtained by Fan and Zavracky for an Au-MgO cermet of nominal volume fraction f = 0.25 (g) and (h) are the results of Lissberger and Nelson, for Au-MgF_2 cermets.

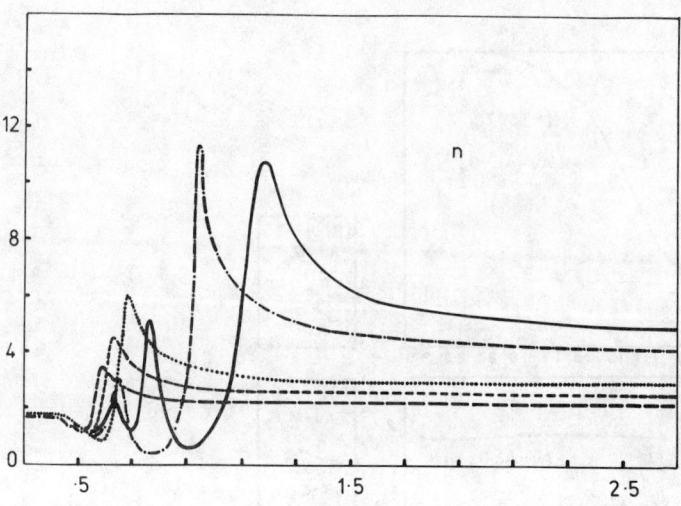

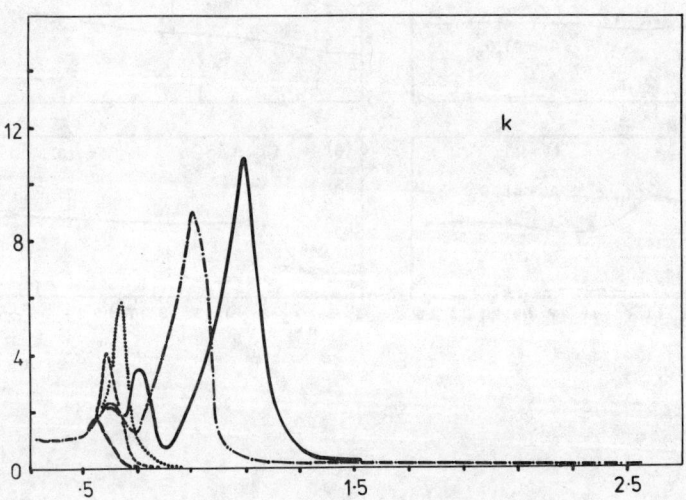

Figure 2.

The values of n and k predicted by the E.R.F. for the following volume fractions of the simple cubic lattice:

——— f = 0.515,	━━━━ f = 0.217,
—·— f = 0.50,	━━ ━━ f = 0.171.
······ f = 0.282,	

the same distribution about that distance as the cermet. It is found that by appropriate weighting of the six volume fractions of figure 2, the broadening of the resonance and the extension of significant values of k to long wavelengths are qualitatively reproduced.

CONCLUSIONS

Features of the experimental results for the optical constants of cermets are reproduced using our model. These features are incapable of explanation in terms of the Maxwell-Garnett theory. Further comparison with experiment will necessitate the measurement of radial distribution functions for cermets by such techniques as electron microscopy. Only when cermets have been accurately characterized in this way will it be possible to fully explain their optical properties, using models capable of taking proximity effects into account.

The model of a cermet described here is based on the simple cubic lattice, which has only 6 nearest neighbours for each sphere. A more elaborate treatment would include admixtures of the other two cubic lattices so that as well as achieving the same radial distribution function as the cermet, the model would also have the same average spatial distribution of near neighbour spheres.

REFERENCES

1. P.H. Lissberger and R.G. Nelson, Thin Solid Films, 21, 159 (1974).
2. J.C.C. Fan and P.M. Zavracky, Applied Physics Letters, 29, 478 (1976).
3. G.D. Scott, Nature, 188, 908 (1960).
4. R.C. McPhedran and D.R. McKenzie, Proc. Roy. Soc. Lond. (1977) (to appear).
5. D.R. McKenzie, R.C. McPhedran and G.H. Derrick, paper submitted to Proc. Roy. Soc. Lond.
6. J.C. Maxwell-Garnett, Phil. Trans. Roy. Soc. (Lond.) 203, 385 (1904).

THE STRUCTURAL COMPOSITION AND ITS INFLUENCE ON THE OPTICAL PROPERTIES OF GOLD BLACK*

P. O'Neill,[+] A. Ignatiev and C. Doland
Department of Physics, Houston, Texas 77004

ABSTRACT

A model is developed for the optical properties of gold black within the realm of effective medium theory. This model assumes that the strands of gold black can be approximated by spheroids for which the depolarization factor is well known from electrostatic theory. This model is applied to several gold blacks produced by inert-gas evaporation. It is found that excellent agreement between experimental and theoretical transmittance can be achieved in the solar spectrum using a distribution of spheroids (gold black strands) that is closely approximated by a log-normal function. Experimental results for packing factor and sample thickness were used without modification. Experimental data for the electron collision frequency was modified to account for particle size effects.

INTRODUCTION

Recent interest in the utilization of solar energy has prompted this investigation into an understanding of the basic physical concepts underlying the absorption of solar radiation by a material. Materials known as "solar black" (gold black, carbon black, and many others) have been classically used as good solar absorbers[1]. These materials, generally stable only below ~300°C, are noted for their extremely low reflectivity throughout the solar spectrum, $R < 1\%$ and yet they are absorptive enough to provide complete extinction within a few microns of material. The basis for such ideal optical properties lies in the particulate, low density nature of the films[2]. However, the precise dependence of absorptance on microscopic structure is unknown.

The analysis of the dependence of solar absorptance on absorber microscopic structure has been undertaken for a test solar black: gold black. The information obtained will be useful in the development of a high-temperature stable solar absorber through the concept of modifying the surface morphology of a material so as to obtain the optimal structural properties for maximum solar radiation absorption.

EXPERIMENTAL

Gold black films were deposited by inert-gas evaporation onto glass and metallic substrates and electron microscope grids as described previously[2,3]. The evaporations were done at Helium pressures of 1 to 20 torr and resulted in films consisting of loosely packed gold particles whose size increased from 40Å to 95Å with increasing helium pressure. The structural detail of the films was monitored by transmission electron microscopy (T.E.M.), and it was found that the films consist of crosslinked chain-of-spheres[3]. Median particle size was determined from the T.E.M. micrographs[3] and from X-ray particle size broadening[4]. The film thickness t was determined using scanning electron microscopy (S.E.M.) as described previously[3].

ISSN: 0094-243X/78/288/$1.50 Copyright 1978 American Institute of Physics

Spectrophotometric techniques[5] were used to measure the transmittance and reflectance of the gold blacks in the wavelength range of 0.35μ to 2.40μ. The measurements were made on a Beckman DK-2A spectrometer and were corrected for the substrate glass cover slips. These measurements have shown that the reflectance of the gold blacks is less than 1% at normal incidence over the noted wavelength range. Thus, the absorption coefficient $\alpha(\lambda)$ was determined by: $T(\lambda) = \exp(\alpha(\lambda)t)$ where $T(\lambda)$ is the measured transmittance.

THEORETICAL MODEL

Examination of a gold black T.E.M. micrograph readily establishes the fact that gold black consists of chains-of-spheres that interconnect so as to form completed (conducting) and interrupted (capacitive) strands. This structure led Zaeschmar and Nedoluha[6] to develop the capacitor theory of the optical properties of gold black. Using this theory, good agreement between theory and experiment is achieved for wavelengths between 3μ and 100μ by assuming that a single capacitor gap approximates the average gap between interrupted strands. However when the capacitor theory is applied in the solar spectrum (.3μ to 2.5μ), the theory does not agree well with experiment[7].

The basic idea of a model based on gold black strands seems reasonable but requires modification. The model suggested in the following paragraphs achieves this modification by assuming that each gold black strand (chain-of-spheres) can be approximated by a spheroid (ellipsoid of revolution) with a semimajor to semiminor axis ratio m, and that a distribution of various spheroid sizes (spheroids with various values of m) exists for the complete gold black film.

The theoretical model to be applied to the optical properties of gold black is a generalization of the theory presented by Genzel and Martin[8] and is referred to as continuum theory. When continuum theory is applied to the case of spheroids (with various semimajor to semiminor axis ratios m) imbedded at random orientations in air, it can be shown that[9]:

$$<\varepsilon(\lambda)> = \frac{1-f + \varepsilon(\lambda)f/3 \sum_{m=1}^{\infty} \tilde{g}(m,\lambda)\rho(m)\Delta m}{1-f + f/3 \sum_{m=1}^{\infty} \tilde{g}(m,\lambda)\rho(m)\Delta m} \quad (1)$$

where: $<\varepsilon>$ = effective dielectric function composite ($<\varepsilon> = <\varepsilon_1> + i<\varepsilon_2>$)

ε = dielectric function of the gold strands ($\varepsilon = \varepsilon_1 + i\varepsilon_2$)

f = packing factor = density of gold black/density of bulk gold

$P(m) = \rho(m)\Delta m$ = fraction of spheroid distribution occupied by spheroids with axis ratio m within the interval Δm.

$$\tilde{g}(m,\lambda) = \sum_{i=1}^{3} \frac{1}{1 + (\varepsilon(\lambda)-1)L_i(m)}$$

$$L_1(m) = \frac{1}{m^2-1} \left\{ \frac{m}{(m^2-1)^{1/2}} \text{Ln} \left[m + (m^2-1)^{1/2}\right] - 1 \right\} \text{ for } m \neq 1$$

$$= 1/3 \text{ for } m = 1$$

$$L_2(m) = L_3(m) = [1-L_1(m)]/2$$

In Equation (1), $L_i(m)$ (i=1,2,3) are depolarization factors[10] associated with a spheroid in a uniform electric field parallel (i=1) and perpendicular (i=2,3) to it's semimajor axis.

Applying Equation (1) to gold black by associating the gold black strands (or chains-of-spheres) with a distribution of spheroids and using the fact that $f \ll 1$, Equation (1) reduces to:

$$<\varepsilon_1(\lambda)> \simeq 1.0 \tag{2a}$$

$$<\varepsilon_2(\lambda)>/\lambda \simeq \sum_{j=1}^{N_e} g(m_j, \lambda) \, P(m_j) \tag{2b}$$

where: $g(m_j, \lambda) = \sum_{i=1}^{3} \dfrac{f/3 \; \varepsilon_2(\lambda)/\lambda}{\delta_i^2(m_j) + 2L_i(m_j)\delta_i(m_j)\varepsilon_1(\lambda) + L_i^2(m_j)|\varepsilon(\lambda)|^2}$

$$\delta_i(m_j) = 1 - L_i(m_j)$$

and the variable m has been replaced by m_j which spans the range m=1 to $m \sim \infty$ in N_e distinct steps.

In order to apply Equation (2) to gold black, a microscopic model of the gold strands is required to provide $\varepsilon(\omega)$. For this, the dielectric function of bulk gold[11] was used with the exception that γ(electron collision frequency) was calculated such that the small size of the gold particles could be accounted for as suggested by Kriebeg and Fragstein[12]:

$$\gamma = 1.7 \, \gamma_{Au}(1 + 2\ell/\bar{a}) \tag{3}$$

Where:

γ_{Au} = d.c. collision frequency for bulk gold = 4.16×10^{13} sec^{-1}

ℓ = mean free path in bulk gold = 340 Å

$\bar{a}$ = particle size in Å (from X-ray broadening or T.E.M.)

The factor 1.7 arises due to the fact that γ is an optical rather than a direct current collision frequency[13]. Thus, the dielectric function consists of Drude and interband terms:

$$\varepsilon_1(\omega) = P(\omega) + 1 - \frac{\omega_p^2}{\omega^2 + \gamma^2} \tag{4a}$$

$$\varepsilon_2(\omega) = \frac{\omega_p^2 \gamma}{\omega(\omega^2 + \gamma^2)} + \varepsilon_2^I(\omega) \tag{4b}$$

Where for bulk gold:[11]

  $P(\omega) = 6$ for $\lambda \gtrsim 1.5\mu$ and determined from Kramers-Kronig contribution from $\varepsilon_2^I(\omega)$ otherwise

  ω_p = plasma frequency = $1.4 \cdot 10^{16}$ sec^{-1}

  ω = frequency (sec^{-1}) = $2\pi c/\lambda$

  c = speed of light in vacuum

  λ = wavelength

  $\varepsilon_2^I(\omega)$ = Interband contribution to the dielectric function for gold crystallites

Equation (2) will now be examined quantitatively. An absorption coefficient can be associated with each strand of given m_j and Equation (2b) predicts that it will be proportional to $g(m_j, \lambda)$. The wavelength and spheroid size (m) dependence of this absorption coefficient is shown in Figure 1.

In Figure 1 each curve has been normalized to unity at its peak and it may be noted that each spheroid (each value of m) produces a different absorption peak. Figure 1 also indicates that a distribution of spheroids is capable of producing absorption over the entire solar spectrum. The longer spheroids (larger m's) produce broader absorption, peak at a longer wavelength, and have higher absorption magnitudes.

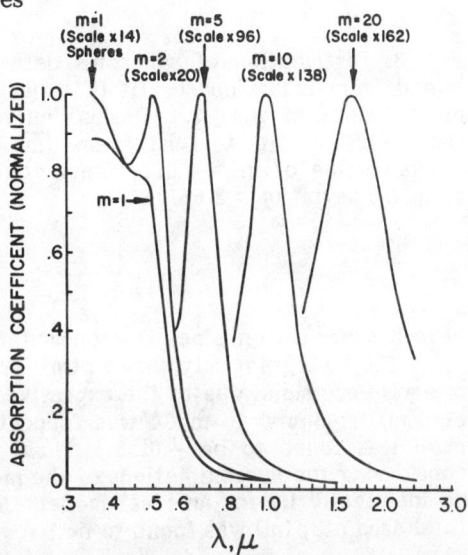

Fig. 1. Absorption coefficient associated with an array consisting of spheroids of given m as a function of wavelength λ.

RESULTS OF APPLICATION TO GOLD BLACK

The distribution of spheroids (gold black strand lengths) in each gold black sample was found by determining the distribution $\rho(m)$ such that the experimental absorption coefficient $\alpha(\lambda)$ was fit by that predicted by the model Equation (2). A least squares technique using a preselected set of discrete values of m_j was chosen such that each $g(m_j, \lambda)$ was linearly independent over the range $\lambda = 0.35$ to 2.40μ was utilized. The results of this fitting procedure for a typical gold black sample are shown in Figure 2 where theoretical and experimental transmittances are given.

The distribution of strand lengths $\rho(m)$ that was determined to produce the theoretical data of Figure 2 is shown in Figure 3.

The histogram represents the $\rho(m)$ determined from the least squares fitting procedure and the dashed curve is a log-normal distribution function (LNDF)[14] that has been fit to the histogram (assuming an independent variable

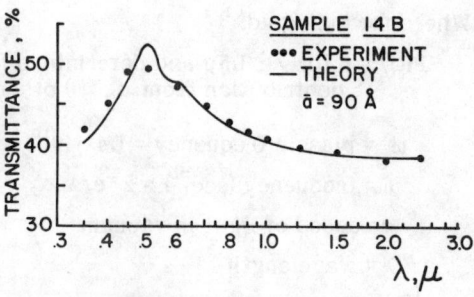

Fig. 2. Theoretical fit to experimental transmittance (corrected for glass substrate) as a function of wavelength λ for a typical gold black sample.

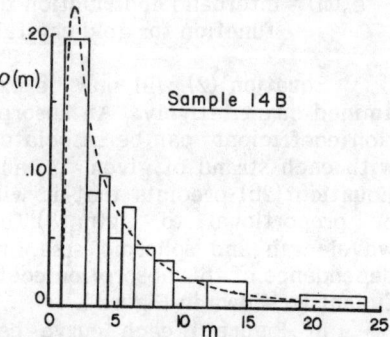

Fig. 3. Distribution of spheroids determined from least squares fit of experimental data of Fig. 2. Dashed curve shows LNDF fit to histogram ($\overline{m}$ = median value of m = 4.0, σ = geometric standard deviation = 3.66).

of m-1 rather than m since the sphere (m = 1) is the smallest allowed entity).

The histogram only shows $\rho(m)$ for m=1 to m=24 since $\rho(m)$ for m=24 to m ~∞ is very small due to the excessive length Δm of this segment. The area of $\rho(m)$ from m=1 to m=24 was found to be ~ 0.95 and that from m=24 to m ~∞ was found to be ~ 0.35. Thus, the total area (~1.3) is reasonable considering the approximations of the model and the fact that this area would be identically 1.0 for an ideal model. For the ten samples investigated, the total area of $\rho(m)$ was found to lie between ~ 0.9 and 1.3.

PHYSICAL SIGNIFICANCE OF LNDF

As noted previously, the LNDF provides a good approximation to the distribution $\rho(m)$ (for m=1 to m=24). The physical significance of this result can be established by assuming a growth model similar to that used to explain the log-normal distribution of particle sizes for particles grown by inert-gas evaporation[14]. The strand growth, however, is assumed to be the result of binary collisions that occur well away from the melt between the spherical particles produced just above the melt. In the region near the substrate the particles are cooler and larger and therefore do not coalesce like a liquid into a single particle after the collision. The result is a strand or chain-of-spheres (sintered[15] at the point of contact) with log-normal distribution in the strand volume. Since this volume is roughly $2\pi b^3 m$ (b=spheroid semiminor axis ~ constant, m = semimajor/minor axis ratio), the reproductive nature of the LNDF yields a log-normal distribution in strand length m.

This model accounts for the production of strand lengths m up to a certain size (say m~12). The production of the longer strands is assumed to simply be the result of overlapping shorter strands that eventually interconnect to form large scale conducting paths.

CONCLUSION

The spheroid model of gold black presented here appears to provide more detailed knowledge of the effect of microscopic structure on gold black optical properties than was previously possible. Specifically, it indicates that absorption throughout much of the solar spectrum is due to the presence of a distribution of small 50Å to 100 Å diameter gold strands or chains-of-spheres that are ~12 or less spheres long and a physical basis for this distribution is proposed.

ACKNOWLEDGEMENTS

The authors thank Dr. A. F. Hildebrandt for contributing to many discussions concerning the microscopic properties of good solar energy absorbers. They also thank Dr. W. W. Wendlandt for the use of the spectrophotometric instruments in his laboratory.

*Supported in part by ERDA and The University of Houston Solar Energy Laboratory.
+National Science Foundation Trainee.

REFERENCES

1. A. H. Pfund, J. Opt. Soc. Am. 23, 375 (1933).
2. L. Harris, The Optical Properties of Metal Blacks and Carbon Blacks (The Eppley Foundation for Research, Newport, R. I., Monograph Series No. 1, Dec. 1967).
3. C. Doland, P. O'Neill and A. Ignatiev, J. Vac. Sci. Technol. 14, 259 (1977).
4. B. E. Warren, X-ray Diffraction; (Addison-Wesley, Mass., 1967).
5. W. W. Wendlandt and A. G. Hecht, Reflectance Spectroscopy; (J. Wiley & Sons, New York, 1966).
6. G. Zaeschmar and A. Nedoluha, J. Opt. Soc. Am. 62, 348 (1972).
7. P. O'Neill, C. Doland and A. Ignatiev, Applied Optics, Nov. (1977).
8. L. Genzel and T. P. Martin, Surface Science 34, 33 (1973).
9. P. O'Neill, "The Structural Composition and It's Influence on the Optical Properties of Gold Blacks and Gold Smokes," Dissertation, The University of Houston, 1977.
10. J. A. Stratton, Electromagnetic Theory; (McGraw Hill, New York, 1941).
11. M. L. Thèye, Phys. Rev. B2, 3060 (1970).
12. U. Kreibig and C. v. Fragstein, Z. Phys. 224, 307 (1969).
13. F. Abelès, in Optical Properties of Solids, edited by F. Abelès (North-Holland, Amsterdam, 1972). p. 93.
14. C. G. Granqvist and R. A. Buhrman, J. Appl. Phys. 47, 2200 (1975).
15. G. C. Kuczynski, Trans. A.I.M.E., 169 (1949).

EXACT SOLUTIONS FOR TRANSPORT PROPERTIES OF ARRAYS OF SPHERES

R. C. McPhedran and D. R. McKenzie
School of Physics, University of Sydney,
Sydney, N.S.W., 2006, Australia

ABSTRACT

The problems which involve calculation of transport properties (for example: dielectric constant, optical constants, thermal and electrical conductivity, resistance to fluid flow and Lamé constants) of an array of spheres of one material embedded in another are mathematically analogous. The first order solution was given by J. C. Maxwell (1873) and the second order solution for spheres arranged in the simple cubic lattice was given by Lord Rayleigh (1892). The method of Lord Rayleigh is shown to be mathematically rigorous and capable of extension to give exact values of the transport property for each of the three cubic lattices. When applied to the conductivity problem, the solutions have the correct behaviour in that, for perfectly conducting spheres, the array conductivity becomes infinite when the spheres touch. The solutions have been verified by comparison with experiment for the simple cubic and body centred cubic lattices. The exact solutions enable the limits of validity of the widely used Maxwell solution, also known as the Maxwell-Garnett solution, to be obtained. We have used the exact solution for the regular lattice to establish useful approximations for the transport properties of a disordered lattice.

INTRODUCTION

There have been three distinct methods of approaching the problem of the transport properties of inhomogeneous materials. The first of these is based on the Maxwell[1] approximation in which each particle of the dispersed phase is regarded as an isolated dipole in a uniform field. This approximation gives rise to the widely used formula known as the Maxwell-Garnett[2] formula. In an attempt to get agreement between this formula and experimental results on the optical constants of cermets, for example, many authors have invoked such effects as distribution of particle shapes, a distribution in orientation of elliptical particles, and a variation of the optical constants of the metallic phase with particle size.

The second method is based on the use of approximations which are asymptotically true as the particles, assumed spherical, approach touching. The first treatment of this type is that of Keller[3] and recently, notable advances have been made by Batchelor and O'Brien.[4]

The third approach has its origins in a classic paper of Lord Rayleigh[5] in which he considered a simple cubic lattice of conducting spheres and derived an expression for the conductivity taking into account contributions from both the octupoles and dipoles induced on each sphere. Rayleigh's method has received sporadic attention from later authors[6-9].

Of these approaches we consider the third to be the most powerful, since it is not restricted to small volume fractions of the dispersed phase, as is the first, nor to high volume fractions as is the second. It can be used to define the limits of validity of various approximations such as those of Maxwell-Garnett or Keller when applied to the regular lattices. It is a rigorous treatment in that full field expansions are used and it is capable of taking into account the strong proximity effects which arise when spheres come close together and which make the dipole treatments highly inaccurate. These proximity effects must be taken into account before it can be decided whether other effects such as distribution of ellipticity or orientation of the particles need be invoked. All the analysis to follow will be for the case of spherical particles. The language of electrical conductivity will be used.

THEORETICAL

Consider a lattice of identical spheres, radius a, arranged in cubic array in a medium. We expand the potential in the medium around a particle at the origin in terms of spherical harmonics:

$$V_c(r,\theta,\phi) = A_o + \sum_{\ell=1}^{\infty} \sum_{m=-\ell}^{\ell} (A_{\ell m} r^{\ell} + B_{\ell m} r^{-\ell-1}) Y_{\ell m}(\theta,\phi) \qquad (1)$$

From the boundary conditions we find

$$A_{\ell m} = \frac{B_{\ell m}}{T_\ell a^{2\ell+1}} \qquad (2)$$

with

$$T_\ell = \frac{(1-\sigma)}{(\sigma + (\ell+1)/\ell)} \qquad (3)$$

where σ is the ratio of the sphere conductivity to the matrix conductivity. The conventional cubic unit cell is chosen to have side unity for each of the simple cubic, body centred cubic and face centred cubic lattices. Rayleigh devised an identity, reasoning that the part of the potential expansion not singular at the origin is due to terms originating at infinity and at the other lattice sites. The identity, which has been shown to be rigorously true,[10,11] has the form

$$A_o + \sum_{\ell=1}^{\infty} \sum_{m=-L}^{L} A_{2\ell-1,m} \; r^{2\ell-1} \; Y_{2\ell-1,m}(\theta,\phi)$$

$$\equiv \sum_{\ell=1}^{\infty} \sum_{m=-L}^{L} \sum_{i=1}^{\infty} B_{2\ell-1,m} \; \rho_i^{-2\ell} \; Y_{2\ell-1,m}(\theta_i,\phi_i) + E_o x. \tag{4}$$

Coordinates (ρ_i,θ_i,ϕ_i) are measured relative to the centre of the ith sphere. (Here $L = 2\ell-1$). For each lattice the x axis is chosen to lie along a direction in which spheres first come into contact, and the macroscopic electric field E is taken parallel to the x axis. By application of Green's theorem to an appropriate unit cell, it is found that the coefficient $B_{1,0}$ alone is related to the conductivity ε of the composite, thus:

$$\varepsilon = 1 - \frac{4\pi B_1}{E} \qquad \text{(SC)} \tag{5}$$

$$\varepsilon = 1 - \frac{8\pi B_1}{E} \qquad \text{(BCC)} \tag{6}$$

$$\varepsilon = 1 - \frac{16\pi B_1}{E} \qquad \text{(FCC)}. \tag{7}$$

The equations (4) are solved for the $B_{\ell,m}$ by repeated differentiation. For the simple cubic lattice coefficients $B_{2\ell-1,m}$ for $m \geq 0$ were included in the formulation and an additional set of equations found by applying Rayleigh's identity at the point $(\frac{1}{4\sqrt{3}}, \frac{1}{4\sqrt{3}}, \frac{1}{4\sqrt{3}})$ as well as at the origin. The effect of neglecting terms having $m \neq 0$ was found to be negligible and consequently they were not included in the formulations for the BCC and FCC lattices.

For the SC and BCC lattices, the calculations of conductivity made using the Extended Rayleigh Formulation (E.R.F.) have been shown to be accurate by comparing the numerical results with experimental measurements[10,11]. These comparisons of theory and experiment are shown in fig. 1.

Another test on the validity of the E.R.F. is that the calculated curves of conductivity (ε) against volume fraction (f) should diverge for arrays of perfectly conducting spheres at the critical volume fraction (f_c) for which the spheres first touch. In fig. 2 we show curves for the three cubic lattices, which demonstrate the appropriate divergence at the critical volume fractions. It is to be noted that as $f \to f_c$, sphere proximity effects become more and more important, and multipoles of ever higher order have to be incorporated into the calculations if accuracy is to be preserved. In fact, we use multipoles of order up to 2^{199} in our computer program implementation of the E.R.F.

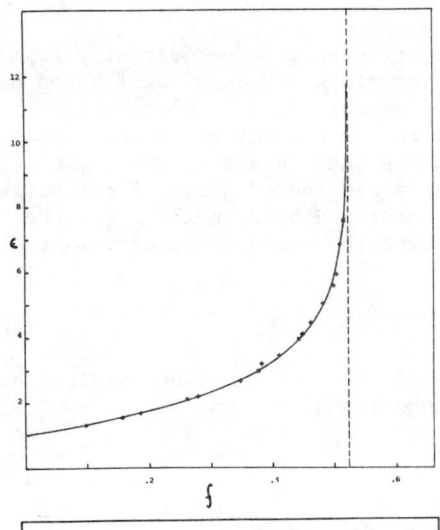

Figure 1

(a) The experimental measurements of Kharadly and Jackson (+) and of Meredith and Tobias (x) are compared with the exact theoretical curve for the SC lattice.

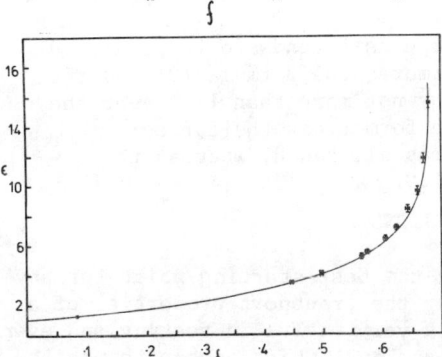

(b) Experimental measurements for the BCC lattice are compared with the exact theoretical curve. Error bars on the experimental points are shown.

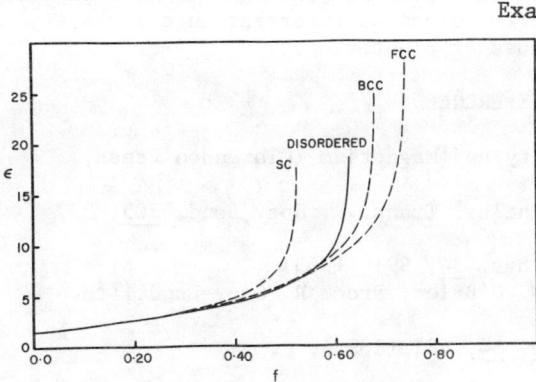

Figure 2

Exact theoretical curves for the SC, BCC and FCC lattices of perfectly conducting spheres are shown. The fourth curve is obtained from (9) and refers to a disordered lattice.

The fourth curve of fig. 2 has not been obtained directly from the E.R.F. Rather, we have used a technique of rescaling[11] based on the idea that as spheres come close together in a regular lattice ($f \to f_c$), the dominant contribution to the current comes from the nearest neighbour interaction. Knowing that interaction for the SCL, we rescale to find its magnitude for a disordered array of perfectly conducting spheres of the type characterized by Bernal[12]. For the SCL, Keller's formula[3] is accurate over the whole range of volume fractions:

$$\varepsilon = -\frac{\pi}{2} \ln(f_c - f). \tag{8}$$

Our re-scaled formula for a disordered array of the type obtained by gravitational packing of ball-bearings is:

$$\varepsilon = 0.05398 - 2.0675 \ln(0.63 - f). \tag{9}$$

This formula has the asymptotic slope as f tends to f_c predicted by Batchelor and O'Brien.[4] Furthermore, the formula (9) and the Maxwell-Garnett equation disagree by not more than 1.5% over the range of f from 0.0 to 0.5, but the former has the correct divergence at $f = f_c$ when the spheres all touch, whereas the latter does not diverge until $f = 1.0$.

CONCLUSIONS

We believe the E.R.F. provides the best starting point for any attack on the problem of calculating the transport properties of a regular lattice. Furthermore, it is very useful in testing and even in suggesting techniques such as rescaling, which enable the establishment of formulae for disordered lattices. Our investigations using the E.R.F. have shown the importance of proximity effects, which arise whenever spheres in a lattice approach each other, and which rule out the application of dipole models. In our next paper, we will show that proximity effects can be important in cermets even at low volume fractions, because of clustering.

REFERENCES

1. J. C. Maxwell, Electricity and Magnetism (Clarendon Press, Oxford, 1873), p. 365.
2. J. C. Maxwell-Garnett, Philos. Trans. R. Soc. Lond. **205**, 237 (1906).
3. J. B. Keller, J. Appl. Phys. **34**, 991 (1963)
4. G. K. Batchelor and R. W. O'Brien, Proc. R. Soc. Lond., to appear.
5. Lord Rayleigh, Phil. Mag. **34**, 481 (1892).

6. R. E. Meredith and C. W. Tobias, J. Appl. Phys. $\underline{31}$, 1270 (1960).
7. M. M. Z. Kharadly and W. Jackson, Proc. Elect. Engrs. $\underline{100}$, 199 (1952).
8. M. G. Bertaux, G. Bienfait and J. Jolivet, Ann. Geophys. $\underline{31}$, 191 (1975).
9. D. A. de Vries, Mededelingen Landbouwhogeschool Wageningen 52, 1 (1952).
10. R. C. McPhedran and D. R. McKenzie, Proc. R. Soc. Lond., to appear.
11. D. R. McKenzie, R. C. McPhedran and G. H. Derrick, paper submitted to Proc. R. Soc. Lond.
12. J. D. Bernal, Proc. R. Soc. Lond., $\underline{280A}$, 299 (1962).

THE PERMITTIVITY OF CUBIC ARRAYS OF SPHERES

W. T. Doyle
Dartmouth College, Hanover, N.H.

ABSTRACT

The mean permittivity of an heterogeneous medium consisting of a cubical array of spheres of one permittivity embedded in a continuous medium of a different permittivity has been calculated for arbitrary volume ratios up to closest packing for s.c., b.c.c., and f.c.c. lattices. The results are compared with approximate formulas and with experiment. For conducting spheres the calculated permittivities exhibit an anomaly at the critical volume ratio where the spheres just touch, in contrast with the predictions of the Clausius-Mossotti relation. The dielectric criterion for the occurrence of a metal-insulator transition is thus lattice dependent, coinciding in all lattices with the onset of macroscopic conductivity. The exact calculations are applied to the Maxwell-Garnett problem and large departures from the usual dipole approximation are found.

INTRODUCTION

The first calculation of the linear response of an heterogeneous medium composed of spheres was made by Poisson in his theory of magnetism, the acknowledged model for the equivalent expressions for electrical permittivity (Clausius-Mossotti) and electrical conductivity (Maxwell). Even the analogous expressions for optical refractivity (Lorentz-Lorenz) and metal optics (Maxwell Garnett), as well as similar results for thermal and particle diffusivity are implicit in Poisson's dipole approximation.

The general form of the result is expressed in Maxwell's relation[1] for the mean conductivity K_m of an array of spheres of conductivity K_1 embedded in a continuous medium of conductivity K_2,

$$K_m = K_2 \frac{[(K_1+2K_2)/(K_2-K_1)] - 2p}{[(K_1+2K_2)/(K_2-K_1)] + p} \quad (1)$$

where p is the total volume fraction occupied by the spheres. If a is the sphere radius and n is the number of spheres per unit volume, we have

$$p = \frac{4\pi}{3} a^3 n . \quad (2)$$

Identical relations hold for the other response functions or generalized "permittivities". Thus, if we interpret the K's as electrical permittivities and let K_1 go to infinity we get the Clausius-

Mossotti equation. If we then put $K = n^2$ throughout, the result is essentially the Lorentz-Lorenz relation. Finally, if we insert the frequency dependent optical constants of the sphere material into n_1^2 we get the Maxwell Garnett expression for the optical constants of the composite medium $N_m^2 = K_m$. In the optical applications it is assumed that the wavelength of the light is large compared with the radius of the spheres. In all applications the results hold only for moderate values of p. This restriction has often been ignored. In the Maxwell Garnett problem it is usually ignored. As we shall see, large departures from the simple dipole formulas occur when p approaches closest packing.

CALCULATED PERMITTIVITY

In the dipole approximation (1) only the particle density enters, not the spatial distribution of the spheres. In higher approximations a definite distribution must be assumed. For cubic arrays of spheres, Rayleigh[2] developed a general method that may be carried to any degree of approximation. In brief, Rayleigh's method involves expanding the potential near one of the spheres due to all of the others in two different series of regular and irregular solid harmonics, with the expansion functions centered on the sphere of interest in one of the series and on each of the neighboring spheres in the other series. Including non-axial terms, we may write

$$A_0 - rY_1^o + \sum_{n,m} A_n^m r^n Y_n^m = \sum_{n,m,i} B_n^m Y_n^m / r_i'^{(n+1)} \qquad (3)$$

where n (odd) ranges from 1 to the highest multipole order included, m ranges from 0 to n in multiples of 4, and i ranges over all of the other spheres in the lattice. E is the external field strength, later set to unity, and r and r_i' are the coordinates of the field point relative to the central sphere and to the neighboring spheres, respectively. Boundary conditions at the surface of the sphere lead to the relations

$$B_n^m = (1 - K_1/K_2)/(K_1/K_2 + (n+1)/n) \, a^{2n+1} \, A_n^m . \qquad (4)$$

Moreover, applying Green's theorem to the unit cell, Rayleigh showed that

$$K_m/K_2 = 1 - 4\pi B_1^o/\alpha^3 \qquad (5)$$

where α is the lattice constant. Depending upon the number of multipoles to be included, as many more equations as are needed may be obtained from (3) by differentiating both sides with respect to the poles of the successive spherical harmonics of the central sphere and then setting the coordinates relative to the central sphere equal to zero. The sums over i, for fixed n and m, expressed in dimensionless form, are the usual multipole lattice sums S_n^m. We are thus led to a system of equations which must be solved simultaneously for B_1^0, and by (5), for K_m/K_2.

We have solved these equations numerically for a wide range of the parameters K_1/K_2 and p in s.c., b.c.c., and f.c.c. lattices. For large values of K_1/K_2 near $p = p_c$, the critical volume ratio where the spheres just touch, high order multipoles must be included for accurate results. Large departures from the dipole approximation occur, as may be seen in Figure 1 where the exact calculations are compared with the Clausius-Mossotti relation (1), labelled CM. The Clausius-Mossotti relation is equivalent to truncating series (3) at the dipole term. Rayleigh's approximation, R, is obtained by truncating series (3) at the quadrupole term and solving the two equations by elimination, ignoring the last term of the second equation. The points shown are experimental values.[3,4] All of the curves shown are for K_1/K_2 infinite, corresponding to perfectly conducting spheres in an insulating medium.

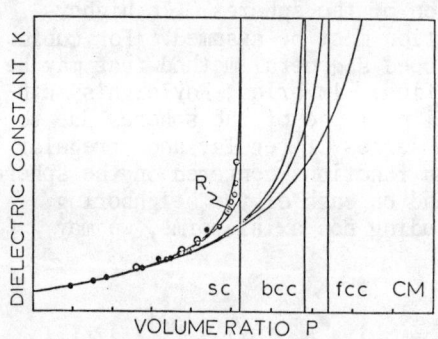

Fig. 1. Dielectric Constant $K = K_m/K_2$ in cubic lattices as a function of volume ratio p.

In Figure 2 K_m/K_2 is shown as a function of the dipole polarizability of the spheres, for the important special case of closest packing. The dotted lines show the predictions of the Clausius-Mossotti relation. Again, large departures from the dipole approximation occur for large K_1/K_2 near $p = p_c$. Because of the different ways K_1/K_2 and p enter the equations, there is no direct scaling relation between the functions in Figures 1 and 2.

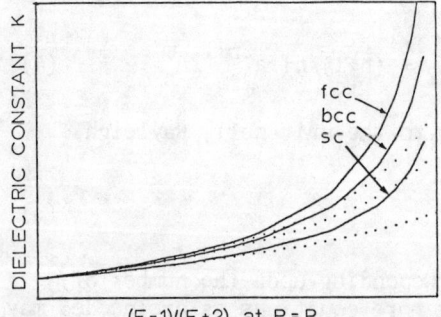

Fig. 2. Dielectric Constant $K = K_m/K_2$ in cubic lattices as a function of dipolar polarizability at closest packing.

While no simple approximation to the multipole expansion serves to represent the results over the entire range of p and K_1/K_2, in the special case of perfectly conducting spheres in a s.c. lattice, the simple interpolation function

$$K_m = K_2 [1 - (\pi/2) \ln(1 - 6p/\pi)] \qquad (6)$$

gives results indistinguishable from the exact calculations on the scale of Figure 1. This function was chosen to have a logarithmic singularity at $p = p_c = \pi/6$ with $K_m/K_2 = 1$ and $dK_m/dp = 3$ at $p = 0$.

METAL-INSULATOR TRANSITION

An interesting feature of Figure 1 is the singularity exhibited by the permittivity in each of the lattices as the spheres just touch. This is a dielectric manifestation of the impending metal-insulator transition. We note that when higher multipoles are included the transition is lattice dependent. In the dipole approximation the permittivity anomaly occurs at $p = 1$, giving rise to the well-known "$4\pi/3$ polarization catastrophe", invoked in ferroelectric transitions. It has also been used as a criterion for the occurrence of the metal-insulator transition, where it is known as the Herzfeld's metallization criterion.[5] It does not seem to be generally known that this criterion for distinguishing insulators from metals is very much older, going back to Goldhammer,[6] who first clearly stated this metallization criterion in 1911. It is clear from Figure 1 that the dipole criterion is inadequate and must be replaced by a condition involving all of the multipoles. A more general criterion may be stated using the system of linear equations in B_n^m obtained from equations (3) and (4). Metallization occurs when the determinant of the matrix multiplying the B_n^m equals zero. In general, all multipoles are involved in the transition. If we truncate the system at the dipole term we recover the usual dipole criterion. In the case of conducting spheres, the metal-insulator transition occurs when the spheres just touch and not at $p = 1$.

MAXWELL GARNETT THEORY

The same equations that govern the static permittivity may be used to examine the effects of higher multipoles on the optical properties of granular materials. The result is effectively the multipole generalization of the Maxwell Garnett theory.[7] In applying the Maxwell Garnett theory it is assumed that if the particles are much smaller than the wavelength of the light only dipoles need be considered. This is wrong. While it is true that with isolated particles only dipole terms need be considered in the long wavelength limit, it is not permissible to neglect higher order multipoles when neighboring particles approach within distances comparable with their diameter, regardless of particle size. At high particle densities the non-uniform fields in the neighborhood of each particle excite all the multipoles in its neighbors.

This is shown in Figure 3 where we plot the real and imaginary

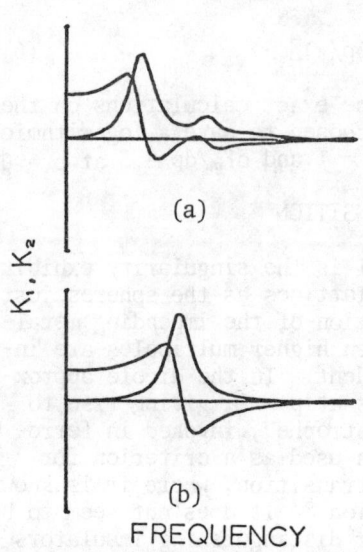

Fig. 3. Complex dielectric construct of a s.c. array of Drude metal spheres, (a) Nine multipole terms, (b) Maxwell Garnett.

parts of the dielectric constant of a medium of Drude metal spheres ($\gamma = 1/\tau = \omega_p/10$). The lower curves show the Maxwell Garnett approximations (dipoles only) and the upper curves show an approximation involving nine multipole terms. The case shown is for a s.c. lattice with p approaching contact with p = .52. Other lattices show similar effects. The higher multipoles manifest themselves both in additional resonances and in producing a shift in the peak position of the dipole resonance. These effects enter at volume ratios that are comparable with those encountered in applications of Maxwell Garnett equations. In our problem the critical volume ratio p_c is also the highest possible ratio. Although higher volume ratios are possible with spheres of differing sizes, and with particles of differing shapes, in these cases too we may expect that higher order multipole effects will enter. Higher multipoles will be excited in even the smallest particles at high density. Moreover, a distribution of particle shapes provides an additional source of multipole excitation. For a particle of irregular shape, even uniform fields excite higher multipoles. Higher multipoles may be expected to be important in determining the optical properties of dense granular media, particularly in regularly textured structures.

REFERENCES

1. J.C. Maxwell, A Treatise on Electricity and Magnetism, (Clarendon Press, Oxford, 1891), 3rd. ed., Vol. 2, p. 57.
2. J.W.S. Rayleigh, Phil. Mag. 34, 481 (1892).
3. R.E. Meredith, and C.W. Tobias, J. Appl. Phys. 31, 1270 (1960).
4. M.M.Z. Kharadly, and W. Jackson, Proc. Inst. Elect. Engrs. 100, Part III, 199 (1953).
5. K.F. Herzfeld, Phys. Rev. 29, 701 (1927).
6. D.A. Goldhammer, Dispersion und Absorption des Lichtes, (Teubner, Leipzig, 1913), p. 29, 130.
7. J.C. Maxwell Garnett, Philos. Trans. R. Soc. Lond. 203, 385 (1904); A205, 237 (1906).

ON THE CORRELATION BETWEEN OPTICAL PROPERTIES
AND CHEMICAL/METALLURGICAL CONSTITUTION OF
Cr_2O_3/Cr THIN FILMS*

R. Chang and W. F. Hall
Science Center, Rockwell International Corporation
Thousand Oaks, Calif. 91360

ABSTRACT

A theoretical model correlating the optical properties and chemical/metallurgical constitution of Cr_2O_3/Cr thin films for solar thermal applications is outlined. Preliminary results of comparison of calculated and experimental optical properties via the proposed approach are encouraging. It should be possible to synthesize in the computer dielectric/metal composite films of optimum thickness, composition, and composition gradient for the best selectivity in solar thermal applications and to confirm the theoretical prediction by means of directed experimental film processing and property evaluations.

INTRODUCTION

The search for efficient selective absorbers with well-defined spectral response for solar thermal energy conversion has focused attention on the optical properties of inhomogeneous media. In a previous communication,[1] we reported a close correlation between the reflectivity of $A\ell_2O_3/A\ell$ thin films and its chemical/metallurgical constitution assuming that metallic aluminum is present in the film as finely dispersed spherical particles with diameters small compared to the wave length of light. This assumption permits the calculation of an effective complex refractive index for the film from the known optical constants of $A\ell$ and $A\ell_2O_3$. For the small volume fractions ($\lesssim 0.01$) of metal encountered in these films one obtains essentially the same effective refractive index for the film whether the Mie scattering theory[2], the Maxwell-Garnett formula[3], or the mean field approach[4] is used for its derivation. The good agreement between theory (involving no adjustable parameters) and experiment for the $A\ell_2O_3$ thin films encourages us to speculate that the dark appearance of other films formed on metallic substrates may be understood within the same framework. This paper reports our investigation of Cr_2O_3/Cr thin films.

APPROACH

Two recent experimental studies appearing in the literature[5,6] on Cr_2O_3/Cr thin films contain sufficient information on chemical/

---

*For presentation at the Conference on Electrical Transport and Optical Properties of Inhomogeneous Media of the American Physical Society, September 7-9, 1977, Columbus, Ohio.

metallurgical consititution for our theoretical investigation. As the volume fraction of metallic chromium reported in these studies varies from 0.01 to as large as 0.66, the Maxwell-Garnett formula[3] for estimating the effective refractive index of a film from the known bulk optical constants of the components is no longer valid. The mean field approach,[4] although by no means exact, appears to be the best choice. The approach assumes not only that the metal component is present as spheres small in comparison with the wave length of light but also that the dielectric component which fills all the volume not occupied by metal is in the form of dielectric spheres. This assumption ensures no predisposition to any particular metal/dielectric composition and is in accord with the experimentally observed metallurgical structures of the Cr_2O_3/Cr films discussed above. The small particle assumption enables the electric field to be calculable for a uniform dielectric sphere immersed in a static external electric field which is constant at infinity and reduces the situation to one of dipole dominance and negligible spatial electric field variation within the small spheres.

The effective complex dielectric constant $\tilde{\varepsilon}_{eff}$ for a composite film having n components of volume fraction f_i and dielectric constant $\tilde{\varepsilon}_i$ according to the mean field approximation is,

$$\sum_{i=1}^{n} \frac{3f_i}{2 + \tilde{\varepsilon}_i/\tilde{\varepsilon}_{eff}} = 1 \tag{1a}$$

$$\sum_{i=1}^{n} f_i = 1 \tag{1b}$$

Next the Rouard's method[7] is used to calculate its optical properties. For a multi-film composed of k layers of thicknesses d_k, d_{k-1}, ..., complex refractive indices $\tilde{n}_k$, $\tilde{n}_{k-1}$, ..., and bounded by a substrate having complex refractive index $\tilde{n}_{k+1}$ at the kth layer and air (or vacuum) at the first layer, one starts calculation with the kth layer next to the supporting substrate. Since a single film bounded by two surfaces possesses an effective reflection coefficient and accompaning phase change, the film may be replaced by a single surface with these properties. The effective Fresnel coefficient $\tilde{F}_k$ for the kth layer is, for normal incidence, given by,

$$\tilde{F}_k = \frac{\tilde{R}_k + \tilde{R}_{k+1} \cdot \exp(-2i\delta_k)}{1 + \tilde{R}_k \cdot \tilde{R}_{k+1} \cdot \exp(-2i\delta_k)} \tag{2a}$$

$$\tilde{R}_k = \frac{\tilde{n}_{k-1} - \tilde{n}_k}{\tilde{n}_{k-1} + \tilde{n}_k} \tag{2b}$$

$$\tilde{R}_{k+1} = \frac{\tilde{n}_k - \tilde{n}_{k+1}}{\tilde{n}_k + \tilde{n}_{k+1}} \tag{2c}$$

$$\delta_k = 2\pi d_k \tilde{n}_k / \lambda \tag{2d}$$

where λ is the wave length of incident light. The effective Fresnel coefficient $\tilde{F}_{k-1}$ for the $(k-1)^{th}$ layer is in turn given by,

$$\tilde{F}_{k-1} = \frac{\tilde{R}_{k-1} + \tilde{F}_k \cdot \exp(-2i\delta_{k-1})}{1 + \tilde{R}_{k-1} \cdot \tilde{F}_k \cdot \exp(-2i\delta_{k-1})} \tag{3}$$

The process is repeated until one reaches the first layer. The reflectivity of the multi-film is then given by the square of the modulus of $\tilde{F}_1$,

$$R = (\tilde{F}_1 \cdot \tilde{F}_1^*) \tag{4}$$

Equations (2) to (4) are simplified, for a single layer film of thickness d and effective refractive index $\tilde{n}_{eff}$ bounded by a metallic substrate of refractive index $\tilde{n}_m$ and air (or vacuum), to the following,

$$\tilde{F} = \frac{\tilde{R} + \tilde{R}_m \cdot \exp(-2i\delta)}{1 + \tilde{R} \cdot \tilde{R}_m \cdot \exp(-2i\delta)} \tag{5a}$$

$$\tilde{R} = \frac{1 - \tilde{n}_{eff}}{1 + \tilde{n}_{eff}} \tag{5b}$$

$$\tilde{R}_m = \frac{\tilde{n}_{eff} - \tilde{n}_m}{\tilde{n}_{eff} + \tilde{n}_m} \tag{5c}$$

$$\delta = 2\pi d\, \tilde{n}_{eff} / \lambda \tag{5d}$$

and

$$R = (\tilde{F} \cdot \tilde{F}^*) \tag{6}$$

COMPARISON WITH EXPERIMENTAL DATA OF FAN AND SPURA[5]

Fan and Spura made reflectivity and transmission measurements of Cr_2O_3/Cr films of various compositions formed on single crystal BaF_2 substrates and obtained the complex refractive indices by fitting the experimental data to expressions relating the reflectivity and transmittance to the complex dielectric constants. They have not given the error limits of their measurements and standard deviations of the complex refractive indices. We calculate independently the effective refractive indices of Cr_2O_3/Cr films (of same chemical compositions as those of Fan and Spura) according to equations (1a) and (1b) employing literature values of the bulk optical constants of Cr[8] and Cr_2O_3.[9] The results are compared in Table I at $\lambda = 0.6$ micron.

Since the films were made by means of rf sputtering, we suspect that their zero volume percent Cr film might actually contain a small amount (say up to one volume percent) of metallic chromium. Also the absorption coefficient α quoted by them for the 11 volume percent Cr film appears to be too low (see Fig. 2 of Ref. 5). A general inspection of the results shown in Table I suggests that our calculated optical constants are about 10 to 20 percent larger than the experimentally evaluated optical constants of Fan and Spura. The agreement is considered good. With more accurate bulk optical constants of Cr and Cr_2O_3 than presently available and with further improvement in the theory of obtaining effective refractive indices of a composite film from the known bulk optical constants of the components, better agreement between theory and experiment than that shown in Table I is expected.

Table I. Comparison of Experimental (Fan and Spura[5]) and Calculated (this paper) Optical Constants ($\lambda = 0.6$ micron) of Cr_2O_3/Cr Films of Various Compositions

Film Composition Volume Percent Cr	n		k	
	Experimental	Calculated	Experimental	Calculated
1	2.25	2.36	0.07	0.06
11	2.25	2.65	0.14	0.34
21	2.48	2.84	0.69	0.74
29	2.75	2.92	0.84	1.07
35	2.89	2.96	1.10	1.33

DISCUSSION OF GRADED Cr_2O_3/Cr FILMS OF SOWELL AND MATTOX[6]

To reduce the number of steps in our calculation of the reflectivity of Cr_2O_3/Cr films having identical graded composition as that of Sowell and Mattox's, we arbitrarily divide their 1800Å film into six layers of 300Å thick each having a uniform composition corresponding to the average value of the respective layer. The effective refractive indices of the six layers are evaluated from the bulk optical constants of Cr[8] and Cr_2O_3[9] at three wave lengths ($\lambda = 0.4$, 0.5, 0.6 micron) according to equation (1). The results are summarized in Table II. The reflectivities of the composite film on Ni substrate are then calculated according to equations (2) and (4) and compared with the experimentally observed values by Sowell and Mattox in Table III.

Table II. Calculated Effective Refractive Indices of Six Cr_2O_3/Cr Layers of 300Å Each Having Chemical Composition Equivalent to Sowell and Mattox's Graded Film

Layer No.	Average Composition* Volume Percent Cr	$\lambda = 0.4\mu$ n	k	$\lambda = 0.5\mu$ n	k	$\lambda = 0.6\mu$ n	k
	Bulk Cr, see Ref. 8	1.74	3.66	2.92	4.55	3.55	4.48
	Bulk Cr_2O_3, see Ref. 9	2.60	0.036	2.38	0.077	2.33	0.042
1	24.0	2.634	1.010	2.825	0.994	2.883	0.866
2	46.9	2.422	1.693	2.787	1.950	2.993	1.873
3	56.7	2.292	1.979	2.725	2.394	3.007	2.353
4	61.1	2.226	2.112	2.698	2.608	3.020	2.581
5	63.8	2.184	2.195	2.685	2.744	3.033	2.722
6	65.6	2.156	2.251	2.677	2.833	3.043	2.815
Ni Substrate	–	1.36	2.45	1.54	2.98	1.80	3.41

*The apparent presence of Ni near the substrate layers, see Fig. 5 of reference 6 on composition profile as determined by sputter-Auger, is ignored. Please see text for further discussions.

Table III. Comparison of Experimental (Sowell and Mattox[6]) and Calculated (this paper) Reflectivities of Graded Cr_2O_3/Cr Films

	Reflectivity at		
	$\lambda = 0.4\mu$	$\lambda = 0.5\mu$	$\lambda = 0.6\mu$
Experimental	0.028	0.028	0.028
Calculated (six layers)	0.225	0.223	0.214
Calculated (first of the six layers replaced with 300Å pure Cr_2O_3)	0.097	0.032	0.116

The disagreement between the experimental and calculated reflectivities shown in Table III is not surprising. We suspect that the composition profile reported by Sowell and Mattox (see Fig. 5 of reference 6) could be implicated by ion-beam-induced mixing and diffusion. The apparent presence of metallic nickel near the substrate layers is suggestive of such implications. Ion-beam-induced mixing and diffusion might result in an apparent composition profile richer in the first layer and poorer in the subsequent layers (towards substrate) of the more volatile component. The surface layer might actually contain less metallic chromium than that shown in Table II. For example, upon replacing the first layer containing originally 24.0 volume percent Cr by a pure Cr_2O_3 layer of same thickness, the reflectivity at $\lambda = 0.5$ micron dropped from 0.223 to 0.032 (see Table III) but the optical interference effect returned. Our calculation supports Sowell and Mattox's claim that a continuously variable Cr/O composition would not exhibit optical interference effects. Impedance matching of an absorbing surface by grading the refractive index can, in principle, improve the absorbance of selective surfaces.[10]

SUMMARY

The theoretical treatment presented above on the correlation between optical properties and chemical/metallurgical constitution of Cr_2O_3/Cr thin films yields encouraging results. Further improvement of the method of calculating effective refractive indices of a composite film from bulk optical properties of the components over that of the mean field approach is desirable. It should be possible to synthesize selective Cr_2O_3/Cr and other dielectric/metal absorber films of optimum thickness, composition, and composition gradient for best optical properties in solar thermal applications and subsequently to confirm these theoretical predictions by means of directed experimental film processing and property evaluations.

REFERENCES

1. R. Chang and W. F. Hall, accepted for publication, Thin Solid Films.
2. G. Mie, Ann. d. Physik 25, 377 (1909).
3. J. C. Maxwell-Garnett, Phil. Trans. Roy. Soc. 203, 385 (1904).
4. D. M. Wood and N. W. Ashcroft, Phil. Mag. 35, 269 (1977).
5. J.C.C. Fan and S. A. Spura, Appl. Phys. Letters 30, 511 (1977).
6. R. R. Sowell and D. M. Mattox, to appear in Plating.
7. O. S. Heavens, <u>Optical Properties of Thin Solid Films</u> (Dover, N.Y., 1965), pp. 63-66.
8. Landolt-Bornstein, Band II, Teil 8, <u>Optical Constants</u> (Springer-Verlag, Berlin, 1962), pp. 1-8.
9. R. I. Frank and W. L. Moberg, J. Vac. Sci. Tech. 4, 133 (1967).
10. I. T. Ritchie and B. Window, Applied Optics 16, 1438 (1977).

NEAR INFRARED REFLECTIVITY AT HIGH TEMPERATURE OF LAYERS OF SILICON CONTAINING OXYGEN

S. O. Sari, K. D. Masterson, H. S. Gurev
Optical Sciences Center, University of Arizona, Tucson, Arizona 85721

ABSTRACT

Optical reflectivity measurements in the wavelength range of 1 μm to 15 μm are discussed for layers of silane-vapor-deposited silicon on metallic silver. These spectra are dominated by absorption due to both high oxygen concentration in silicon and enhanced intrinsic absorption at the fundamental energy threshold. Some evidence for residual absorption due to localized states below this energy gap also seems to be implied experimentally. Temperature-dependent reflectivity measurements between room temperature and 980°K have been obtained. Interpretation is given of small observed lineshifts, reversible shifts of the silicon edge, and increased film absorption as a function of temperature. X-ray diffraction measurements related to these optical experiments are compared to existing results for sputtered silicon.

The near-infrared optical spectrum of films of silane-vapor-deposited silicon is dominated by a number of features: resonance absorption due to a relatively high percentage of oxygen in the semiconductor lattice, the onset of fundamental intrinsic absorption near 1 eV, and the possible occurrence of residual losses below this edge energy due to localized states. It is our purpose here to discuss these and related effects which have been observed in optical measurements on silicon layers under 5 μm in thickness deposited on optically opaque highly reflecting silver substrates. Such heterogeneous structures have a potential utility as selective absorbers as has been reviewed previously.[1] A preliminary discussion of the role of oxygen in near-infrared silicon film properties has been given elsewhere.[2,3]

Most silicon layers which we examined were prepared by vapor deposition of silane in a helium or argon carrier gas (1:99 volume ratio) at approximately 650°C at a rate near 20 Å/sec on opaque evaporated silver on pyrex substrates. Optical spectra were taken with Perkin-Elmer 137 and 450 infrared and visible spectrometers, and temperature-dependent reflectivity measurements were made between 300 K and 980 K with an evacuable high-temperature salt-prism reflectance spectrometer. Oxygen and other trace element content was determined using K_α x-ray fluorescence analysis with both proton and electron excitation using energies such that the particle range was only a fraction of the semiconductor film thickness. Since x-ray self-absorption can lead to error in interpreting concentration even in very thin layers, this combination of methods[3] gave confidence in our results. Concentration measurements for oxygen gave values of 2 to 3 × 10^{20} cm^3 for the films examined. In

ISSN: 0094-243X/78/311/$1.50 Copyright 1978 American Institute of Physics

addition, x-ray diffraction analysis of several of these layer combinations indicated a general absence of sharp Bragg structure in the silicon except for coherent scattering at near forward angles, although the (111) silicon plane could be identified. Several of the prominent crystal planes for the silver underlayer could be observed, yielding peak locations and structure factors approximating those expected for metallic silver.

The optical spectrum obtained for a silicon-on-silver sample about 3.4 μm in thickness is shown in Fig. 1. The large peak near 10 μm is due to absorption by oxygen and is large in this instance

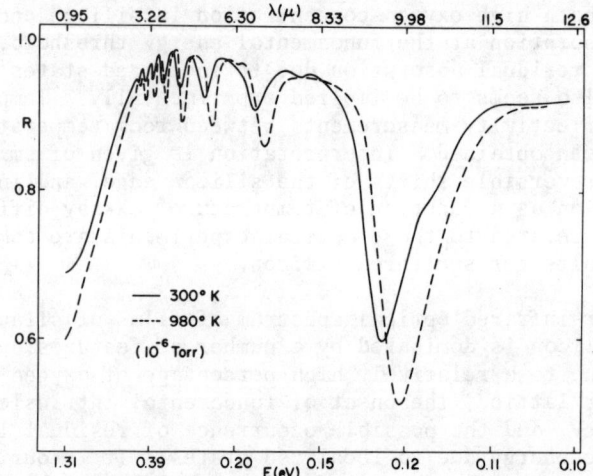

Fig. 1. Near-infrared reflectivity spectrum of silicon-on silver at 300 K and 900 K showing oxygen absorption near 10 μm and onset of intrinsic absorption for energies E ≳ 0.3 eV. Shown is a silicon film of thickness such that a longitudinal mode of the film is in near resonance with an oxygen resonance transition near 10 μm.

because a cavity mode of the film, as determined by its thickness ℓ, is in near resonance with the prominent oxygen transition near 10 μm. For oxygen in crystalline silicon the transition is believed to correspond to the state near 1100 cm^{-1}. By choosing the thickness of the film so that a given fringe mode is detuned from the impurity resonance energy, the observed reflectivity drop can be separated into components which are primarily cavity-like and impurity-like. This latter case is shown in Fig. 2. The reflectivity of a nearly transparent silicon layer which takes these properties into account may be calculated in the limit in which the absorption length of radiation is many times the film thickness. This holds in the case of weak absorption well below the semiconductor absorption edge and yields

$$1 - R = \frac{2n\Lambda}{1 + [\varepsilon(\infty) - 1]\cos^2 k\ell} \qquad (1)$$

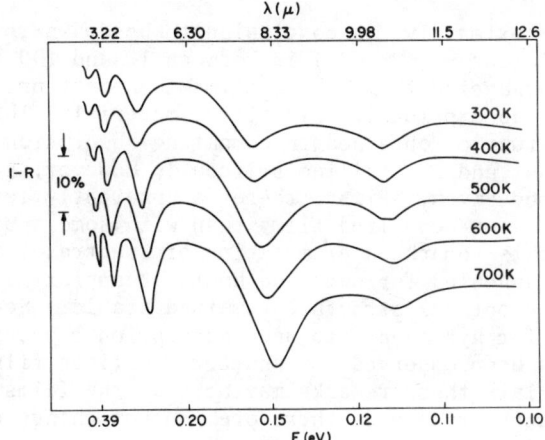

Fig. 2. Reflectivity at increasing temperatures for a silicon film of mode frequency detuned from the oxygen resonance transition frequency. The oxygen line appears near 10 μm.

where $c^2 k^2/\omega^2 = \text{Re } \varepsilon \simeq \varepsilon(\infty) = n^2$ and where

$$\Lambda = \frac{1-R_m}{2} + \frac{4\pi\ell}{nc}\frac{\sigma_0}{1+\omega^2\tau^2} + 4\pi\frac{N\ell}{V}\frac{e^{*2}\xi}{Mcn}\frac{\omega^2\gamma}{(\omega_0^2-\omega^2)^2+\gamma^2\omega^2}. \quad (2)$$

Equation (1) ignores a small phase shift at the outermost silicon interface which modifies this reflectivity slightly as reported in Ref. 2. In Eq. (2) are shown the contributions, respectively, of the metal substrate with reflectivity R_m, of free-carrier conductivity having a relaxation time τ, and of the oxygen transition in question. This absorption has a Lorentzian resonance shape multiplied by an oscillator strength factor depending on the effective charge e* and on the reduced mass M of the excitation. The factor ξ is a correction factor to the impurity line strength due to local lattice polarization or line multiplicity.[3] Numerical fits to obtain impurity parameters have yielded line strengths $(N/V)(e^{*2}\xi/M)$ of 7 to 15 × 10^{25} sec^{-2} giving a value for ξ of 10 to 20 under the assumption that e* = e. We have typically found the effect of free carriers to be small at the wavelengths considered.

The reflectivity at shorter wavelengths, as again shown in Fig. 1, is characterized by residual absorption below the onset of intrinsic absorption as is evident from some numerical fits to data extended from the 10 μm oxygen region. This is also consistent with the tendency for interference fringe structure to be averaged for $\lambda \lesssim 4$ μm until front surface reflectivity begins to dominate the spectrum for energies greater than 0.8 eV. The optical constants of silicon near its intrinsic edge may be determined from a full analysis of its reflectivity, but we have presently carried this out only approximately. A conclusion to be drawn is that the absorption coefficient α (=$2K\omega/c$) is between 10 and 100 times larger at a given wavelength in these films than that predicted on the basis of bulk absorption in crystalline material. This seems similar to the situation obtained in comparing absorption in rf sputtered amorphous and crystalline silicon.[4] However, some differences need to be given. First, there is apparently less thermal annealing in our vapor-deposited films than with some sputtered films. We base this remark on examination of spectra of films which were temperature annealed for over 100 hours at approximately 900 K (Ref. 5) and whose optical structure remained stable. Second, little evidence is seen for hydrogen resonance absorption expected near 4.8 and 5.4 μm as has been observed for sputtered silicon films.[6] A hypothesis to explain these remarks may be that the films are self-annealed during deposition and, therefore, do not change under further treatment. Insofar as oxygen is concerned, the annealing rate $1/\tau_A$ of oxygen in polycrystalline silicon has been examined in experiments on dichroic relaxation[7] and shows that

$$\frac{1}{\tau_A} = \Omega_0 \, e^{-E/kT} \qquad (3)$$

where the activation energy E = 2.56 eV. If we transfer these results to examine our means of deposition, we find that at 650°C oxygen would relax in the lattice in a time $\tau_A \sim$ 0.02 sec, which is much less than a monolayer deposition time. The diffusion constant of oxygen can be obtained from the rate above[7] and, in the present case, can be used to show that oxygen can travel a significant distance during deposition (220 Å for a 3.5 μm thick film deposited at 20 Å/sec). This would tend to equilibrate the semiconductor layer rapidly. The barrier potential E is expected to be proportional to the bond strength which for hydrogen-to-silicon is less than half of that for oxygen. This can be used to show similarly that hydrogen, if it is trapped in the lattice, may simply have diffused out of the film at our deposition temperature. This would explain the lack of hydrogen in our infrared spectrum.

Another measurement to be compared to that of sputtered silicon is its x-ray diffraction. Experiments before annealing and crystallization of sputtered films reveal a halo structure which sharpens as the crystallizing phase change commences at elevated temperature (at $\sim$500 to 600°C over $\sim$3 hours for silicon).[4] Such a phase change has not been identified in our layers at comparable temperatures and times. The phase change can also occur at lowered temperature near an interface with a miscible metal ($\sim$540°C for Si-Ag).[8] This would have a predictable effect of modifying the rear interface reflectivity R_m and consequently the optical structure. In our present films, introduction of a chromium or silicon oxide interlayer $\sim$100 Å thick inhibits the effect.[9] The lack of Bragg structure which we obtain is indicative then of a silicon phase with no long range order. A simple Scherrer analysis[10] of some of our samples would suggest very small microcrystals of $\sim$10 Å linear dimension in a homogeneous substance; yet the probable heterogeneity of the film is likely to further disperse the x radiation diffracted from it.

One can ask whether impurity or void clustering occurs in the lattice. On a strictly statistical basis using a passive billiard ball picture, the distribution of a given defect having an average density N_0/V_0 in a volume V follows a Poisson distribution.[11] Thus the probability P_n for an n-defect cluster is

$$P_n = \frac{1}{n!}\left(\frac{N_0 V}{V_0}\right)^n e^{-N_0 V/V_0}. \tag{4}$$

The answer to the above query is then affirmative if the concentration is comparable to 1/V. If the impurities must be close enough to interact, then the volume of interest is $V \sim a_L^3$. This means that for an appreciable probability for two or more defects in the same volume V, the concentration must be quite high. This approximation breaks down as defects become overlapping in a given region (large n). However, taken literally, it would suggest that clustering of oxygen, even at our measured concentrations, would be relatively unimportant.

The infrared fringe structure of Figs. 1 and 2 was examined as a function of temperature for energies E < 0.3 eV, and the wavelength shifts of the first few orders of interference minima could be compared to temperature variations of refractive index of crystalline silicon films[12] using

$$\left(\frac{1}{n}\frac{dn}{dT} + \frac{1}{\ell}\frac{d\ell}{dT}\right)\Delta T = \frac{\Delta\lambda}{\lambda}. \tag{5}$$

The quantities in the brackets are here understood to be values for the temperature variation of the refractive index and the thermal expansion coefficient of a free surface and are $3.9 \times 10^{-5}(°C)^{-1}$ (Ref. 12) and $7.6 \times 10^{-6}(°C)^{-1}$ (Ref. 13), respectively. The first number is known to about 10%, and the second number is a correction of about 20% on the first. In a linear extrapolation between 300 K and 900 K (using data from Ref. 12), the left side of expression (5) yields 2.9×10^{-2}. Measured relative wavelength variations $\Delta\lambda/\lambda$ yield 2.6×10^{-2} to approximately the same accuracy as above. The refractive index changes can be accounted for by the temperature dependence of the electronic structure. The approximate agreement of the numbers determined may suggest that nearest neighbor interactions dominate temperature variations.

Several features of the temperature dependence of the fringe structure of the silicon film shown in Fig. 2 deserve comment. The resonance peak due to oxygen (Si-O) is only weakly temperature dependent above 300 K in comparison to fringe modes. The oxygen absorption peak height decreases by $\sim$10% in the 300 K to 900 K range. Since the linewidth increases by about this amount, the line strength remains approximately constant. The temperature dependence of the film fringe cavity minima is similar in each of the first few orders. The relative temperature change in fringe depth, a quantity proportional to absorption, increases as the fundamental edge of silicon is approached in energy. This suggests a mechanism for this residual absorption which we believe to be tied to existence of states below this threshold. The fact that the absorption increases with temperature suggests a process involving phonons and thus is proportional to phonon number. This assumes an unaltered interfacial metal reflectivity. Such an interpretation seems probable, since the same absorptive temperature dependence repeats as the films are temperature cycled.

Acknowledgment is given of partial support of this research by the National Science Foundation.

REFERENCES

1. B. O. Seraphin, "Semiconductors in Solar Energy Conversion, Proceedings of the XIII International Conference on the Physics of Semiconductors," Rome Italy, Aug. 30-Sept. 3, 1976, p. 97.
2. S. O. Sari, P. Hollingsworth Smith, and H. S. Gurev, Phys. Rev. B 15, 4817 (1977).
3. S. O. Sari, P. Hollingsworth Smith, and H. Oona, Bull. Am. Phys. Soc. 22, 436 (1977), and to be published.
4. M. H. Brodsky, R. S. Title, K. Weiser, and G. Pettit, Phys. Rev. B 1, 2632 (1970).
5. K. D. Masterson, in "Chemical-Vapor-Deposition Research for Fabrication of Solar Energy Converters," Annual Progress Report NSF/RANN/SE/GI-36731X/PR/74/4.
6. G. A. N. Connell and J. R. Pawlik, Phys. Rev. B 13, 787 (1976).
7. J. W. Corbett, R. S. McDonald, and G. D. Watkins, J. Phys. Chem. Solids 25, 873 (1964).
8. S. R. Herd, P. Cahudhari, and M. H. Brodsky, J. Non-Cryst. Solids 7, 309 (1972).
9. In Eq. (2) this has the effect of multiplying the term $(1-R_m)/2$ by a factor of n/n_I where n_I is the interlayer refractive index, if the interlayer is nonabsorbing and very thin compared to a wavelength.
10. J. Bouman, X-Ray Crystallography (North-Holland, Amsterdam, 1951), Chap. 4.
11. P. M. Morse, Thermal Physics (Benjamin, N. Y., 1965), pp. 157-158.
12. M. Cardona, W. Paul, and H. Brooks, J. Phys. Chem. Solids 8, 204 (1959).
13. Handbook of Chemistry and Physics (Chemical Rubber Company, Cleveland, 1961), 44th edition, p. 2329.

THE EFFECT OF SURFACE ROUGHNESS ON THE OPTICAL FUNCTIONS OF REAL METALS

J.P. Marton
McMaster University, Hamilton, Ontario, Canada, L8S 4M1

ABSTRACT

A rough surface introduces features in the optical functions of metals which are additional to those that appear in the case of a smooth surface. The features associated with rough surfaces are the conduction resonance in the real (ϵ_1) and imaginary (ϵ_2) parts of the dielectric function, and the shifted free electron plasmon in the loss function. In addition, if the metal has an interband transition near the conduction resonance, an extra interband feature appears in both ϵ_1 and ϵ_2 which is shifted in frequency from the interband transition due to splitting. Also, if the interband transition is close to the bulk plasma frequency, a split of the plasma peak will occur in the loss function. Examples for real metals are presented.

INTRODUCTION

It has been shown both theoretically[1,2] and experimentally[3,4] that for the purpose of optical characterization, the rough surface of a metal may be replaced by a thin composite film with certain well defined properties. If the scale of the surface roughness is on the order of $\lambda/10$ or less and if the surface is otherwise free of oxides or other contamination then a thickness d and a metallic volume fraction q may be assigned to the composite film. In this case, the complex dielectric function $\tilde{\xi}$ of the film is given by the well known Maxwell-Garnett (MG) Theory,[5]

$$\frac{\tilde{\xi}(\omega) - n^2}{\tilde{\xi}(\omega) + 2n^2} = q \frac{\tilde{\epsilon}(\omega) - n^2}{\tilde{\epsilon}(\omega) + 2n^2} \quad , \tag{1}$$

where n is the index of refraction of the matrix material in the composite (in the present case, it is air) and $\tilde{\epsilon}$ is the dielectric function of the parent bulk metal.

Optical measurements, such as spectral reflectivity or polarimetry may be now interpreted by the Fresnel equations in terms of a metallic substrate with a smooth plane surface having dielectric function $\tilde{\epsilon}$, covered by a film of thickness d of dielectric function $\tilde{\xi}$. The measurements thus yield values for d and q which in turn may be used to characterize the rough surface.

THEORY

In the simplest case, when the metal has no interband transitions in the spectral range of interest, aggregation leads to an optical feature called the optical conduction resonance[6] (OCR). It resembles an interband transition in the dielectric function, which has been mistakenly interpreted as such on a number of occasions.[7] Aggregation also shifts the bulk plasmon peak in the loss function. The frequencies at which the OCR occurs and at which the shifted bulk plasmon peak appears depend on the q value according to

$$\omega_R' = [\frac{\omega_p^2}{3}(1-q) - \frac{1}{\tau^2}]^{1/2} , \qquad (2)$$

$$\omega_R'' = [\frac{\omega_p^2}{3}(1+2q) - \frac{1}{\tau^2}]^{1/2} , \qquad (3)$$

where ω_R', ω_R'' and ω_p are respectively the OCR, shifted bulk plasma and bulk plasma frequencies and τ is the free electron relaxation time. The relevant optical functions for a sodium-like free electron metal at q = 0.7 is shown in Fig. 1, and the change of the OCR and plasma frequencies with q is shown in Fig. 2.

In the case where the metal has an interband transition in the frequency range of interest, the situation is more complex. For simplicity, we may substitute the interband transition with a Lorentzian oscillator and superimpose this on the free electron curves in the optical functions, as shown in Fig. 3. An analysis of the optical functions of an aggregated composite of that metal shows several features which are absent in the free electron case. First we find that the Lorentz feature is copied onto the aggregated functions practically unchanged; second, we see that if the oscillator is close to the OCR peak it shifts or splits it; and third, we observe the same shift or split to the loss function if the oscillator is close to the bulk plasma frequency. The features are illustrated in Fig. 4 for a sodium-like metal and its aggregated composite at q = 0.7.

REAL METALS

Real metals are neither free electron type nor have Lorentz oscillator-like interband transitions. However, their aggregated composite can still be analyzed by the method discussed above and their spectral features be understood by the simple Lorentzian model. Once the spectral features of the aggregated composite is understood, real metals with rough surfaces may be characterized by the values of d and q from the simple Fresnel and MG theories. In Fig. 5, the optical functions of bulk and aggregated Cs and Ag are shown. The bulk data were obtained from the work of Smith[8] and of Wang et al[9] for Cs and from the work of Johnson and Christy[10] for Ag. The main interband transitions are shown by arrows at the top

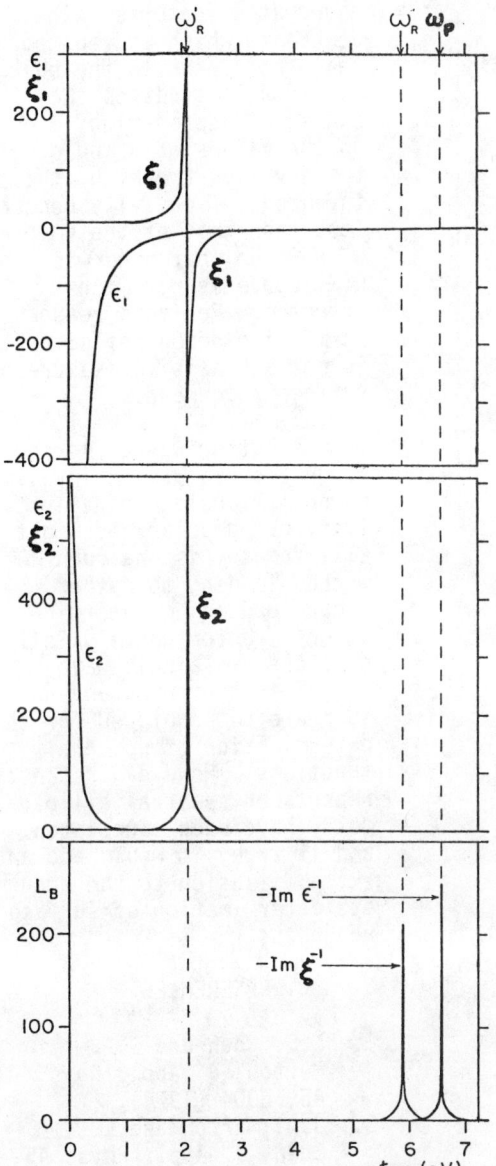

Fig. 1. The dielectric functions ϵ_1, ϵ_2 and loss function of a sodium-like free electron metal are plotted along with the corresponding functions ξ_1, ξ_2, and L for an aggregated composite at q = 0.7. The OCR at ω_R' and the shifted plasmon at ω_R' are evident. Their positions are q dependent.

of the figures and the shifts and splittings they cause in the OCR and plasma features of the aggregated functions are evident from the figures.

DISCUSSION AND SUMMARY

The optical functions of bulk metals with smooth surfaces are characterized by the evancescent region in the infrared, by the plasma edge in the ultraviolet, and by interband transitions. These lead to high reflectivity in the infrared, low reflectivity beyond the plasma edge, and reflectivity dips at the interband peaks.

The optical functions of aggregated metals were shown above to be characterized by the OCR and the shifted plasma peak as well as by the interband transitions. The latter was also shown to effect the two former features by shifting or splitting. These lead to low reflectivity in the infrared, high reflectivity between the OCR and the plasma edges, and low reflectivity beyond the plasma edge. The dips due to interband transitions in the high reflectivity region are copied from the bulk.

The optical functions of bulk metals with rough surfaces are characterized by a combination of the smooth bulk and the

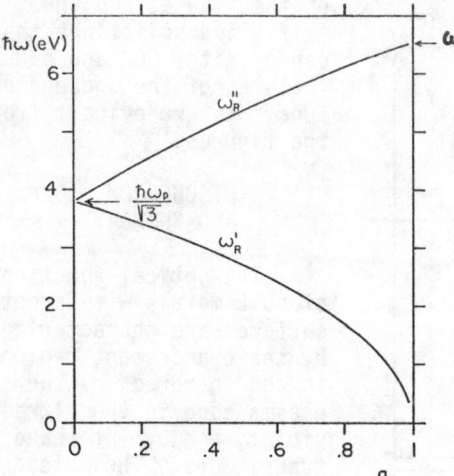

Fig. 2. The change of ω_R' and ω_R'' with q. At low q values, the two frequencies are the same.

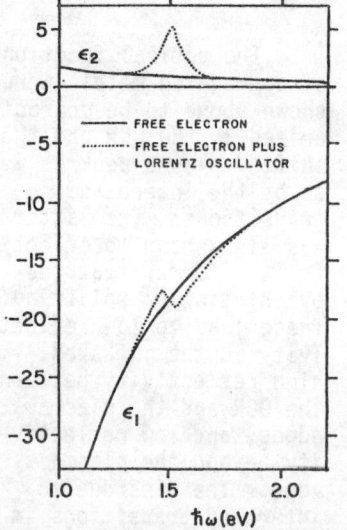

Fig. 3. Simulation of an interband transition by a Lorentzian oscillator superimposed on the free electron curve of a sodium-like bulk metal.

aggregated features. The degree to which aggregated features appear in the bulk functions depends on the degree of aggregation, i.e. on the values of d and q. If these values are high, the infrared reflectivity becomes less than that of the smooth bulk value, but otherwise the $R(\lambda)$ curve is not much effected. For this reason, normal incidence reflectivity is not a sensitive measure of surface roughness.

A better way of surface roughness evaluation involves nonnormal incidence reflectivity using polarized light. Ellipsometry is one such method leading to rather accurate d and q determination[2]. Unfortunately, ellipsometers are single wavelength instruments, and as such are not equipped for the determination of spectral functions. However, recently reports on spectral ellipsometry have been circulated, and if true, it would add an extra dimension to the accurate determination of surface roughness.

REFERENCES

1. E.C. Chan and J.P. Marton, J. Appl. Phys. 45, 5004 (1974).
2. J.P. Marton and E.C. Chan, J. Appl. Phys. 45, 5008 (1974).
3. J.P. Marton, Appl. Phys. Letters 18, 140 (1971).
4. J.P. Marton, J. Electrochem. Soc. 123, 370c (1976).
5. J.C. Maxwell Garnett, Philos. Trans. R. Soc.

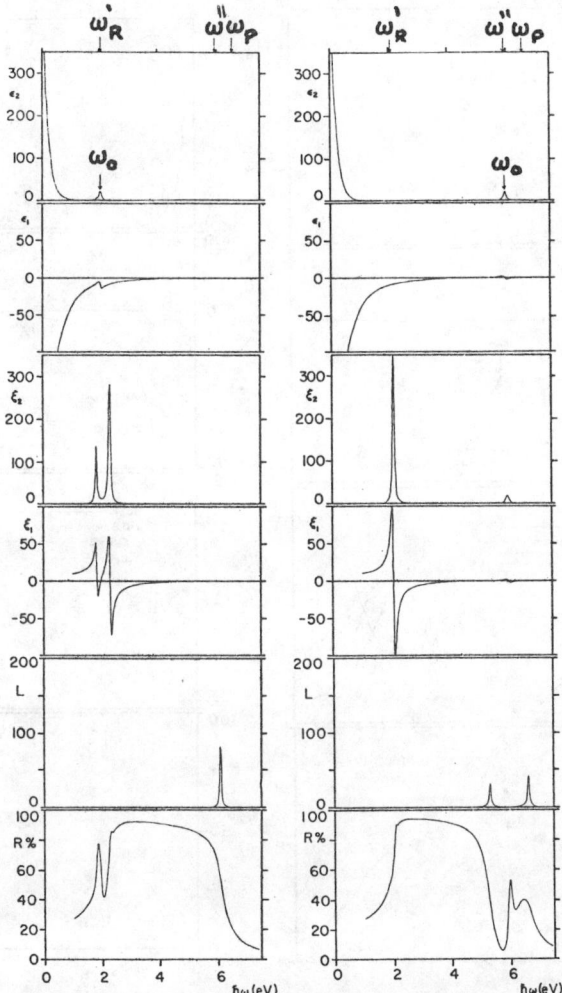

Fig. 4. The effect of an interband oscillator at ω_0 on the optical functions of a sodium-like aggregated composite at $q = 0.7$. At the left, ω_0 is close to the free electron OCR and it splits it, while on the right, ω_0 is close to the bulk plasma peak causing it to split. In the latter case, the OCR is not affected and the oscillator is copied onto ξ_1 and ξ_2. The reflectance is also affected. At left, the aggregation edge is split and at the right, the plasma edge is **split**.

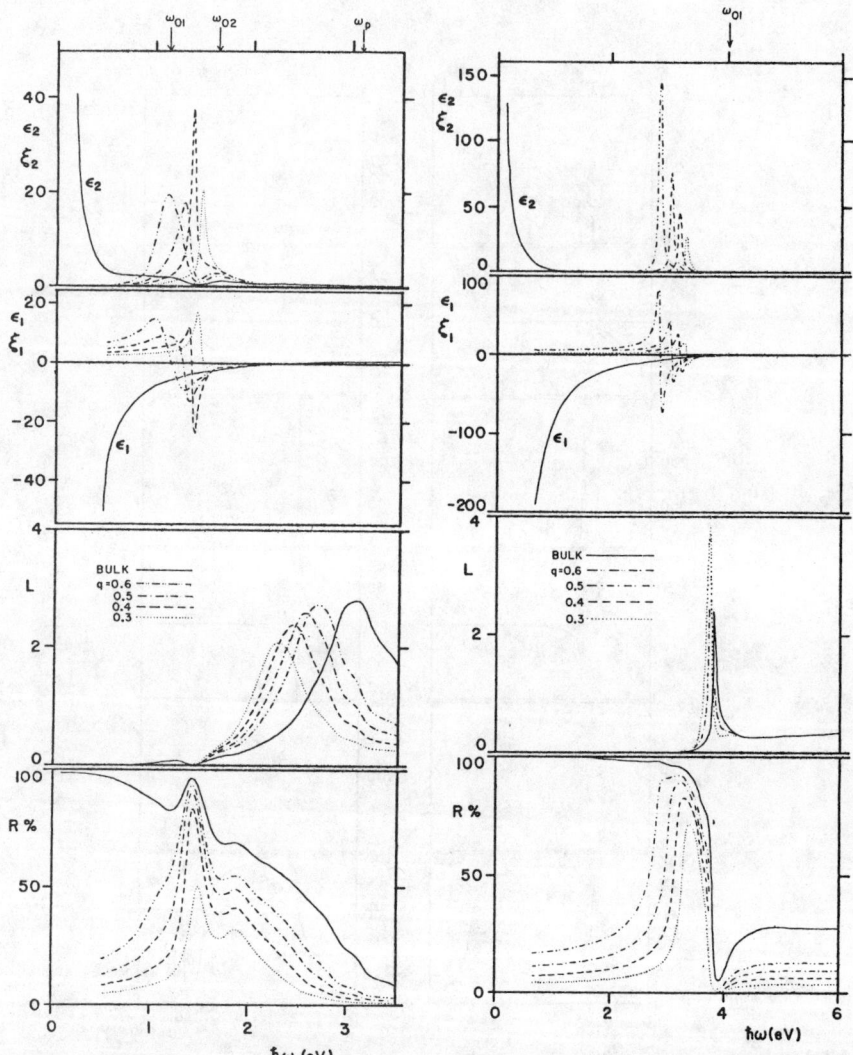

Fig. 5. Optical functions of bulk and aggregated Cs (left) and Ag (right). The main interband transitions in Cs affect the OCR, but not the plasma peak in L. Conversely, the main interband transition in Ag splits and shifts the plasma peak in L (in the figure only the lower energy peak is shown), but does not affect the OCR.

Lond. 203, 385 (1904); A205, 237 (1906).
6. J.P. Marton and J.R. Lemon, Phys. Rev. B4, 271 (1971).
7. H. Mayer and B. Hietel, in "Optical Properties and Electronic Structure of Metals and Alloys", F. Abeles ed. (Wiley, N.Y., 1966).

8. N.J. Smith, Phys. Rev. B2, 2840 (1970).
9. U.S. Wang, E.T. Arakawa, and T.A. Callcott, J. Opt. Soc. Am. 61, 740 (1971).
10. P.B. Johnson and R.W. Christy, Phys. Rev. B6, 4320 (1972).

FAR-INFRARED ABSORPTION BY ELECTRON-HOLE DROPS IN GERMANIUM*

T. Timusk and H.G. Zarate
Department of Physics, McMaster University
Hamilton, Ontario, Canada L8S 4M1

ABSTRACT

We report here on new measurements of the far-infrared resonance absorption of electron-hole drops in the region 1.5 to 50 meV, at 1.2°K. The absorption consists of a broad assymmetric line peaking at 9.3 meV, with a threshold at (2.9 ± 0.1) meV and a cut-off at (43 ± 2) meV. The observed lineshape is in excellent agreement with the results of recent theoretical calculations by Rose et. al.,[1] that include the effect of interband transitions between the heavy and light holes bands. For very small excitation levels an additional broadening due to scattering with the surface has been observed.

INTRODUCTION

The far-infrared absorption spectrum of electron-hole drops (EHD) has been measured by several investigators[2,3,4] over a restricted range of energies and excitation levels. Vavilov et al.[2] emphasized the region around the plasma peak in their early work but to test the recent theories[1,5] of the interband absorption careful measurements above 30 meV and below 4 meV are needed. We report here on results over the complete range 1.5 meV to 50 meV. Small discrepancies in the reported line shapes[2,3] have been attributed to a droplet size variation[1] at varying excitation levels. We present lines shapes measured over a wide range of excitation intensities. The corresponding absorption at the peak of the line varies from 0.03 to 1.6. The absorption is defined here as the quantity $\ln(I_o/I)$ where I_o and I are the incident and transmitted intensities.

THEORY

Two effects have to be taken into account in explaining the infrared absorption by EHD: the plasma oscillations of electrons and holes and interband transition between heavy and light hole bands. For small drops $r << \frac{\lambda}{\varepsilon_o}$ the lowest order term in the Mie expansion can be taken for the crossection for absorption:[6]

$$\sigma_{abs} = \frac{8\pi^2\sqrt{\varepsilon_o}}{\lambda} R^3 \text{Im}\left\{\frac{\varepsilon-1}{\varepsilon+2}\right\}. \qquad (1)$$

*Work supported in part by the National Research Council of Canada.

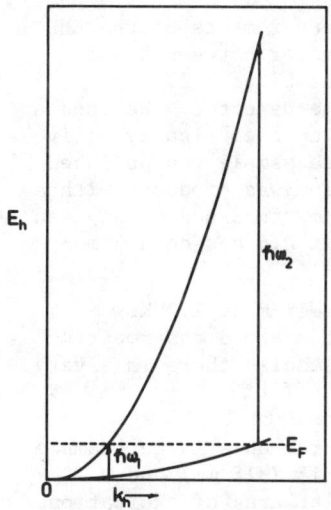

Fig. 1. The allowed transitions between the heavy and light holes branches, giving rise to the interband contribution of the EHD absorption.

The simplest form of ε assumes a free carrier model with finite damping γ due to electron-hole scattering. This mechanism gives rise to a broad resonance at $\omega_p/\sqrt{3}$.[2] However, the experimental line shape, particularly at higher frequencies, deviates from this model. A complete description of the absorption[1,5] includes the additional absorption due to interband transitions by holes (Fig. 1).[7] Vertical transitions are possible for holes between two limits $\hbar\omega_1$ and $\hbar\omega_2$ determined by the fermi energy of the holes and structure of the two bands. Using known band parameters for germanium and a fermi energy of the holes in the EHD of about 4 meV, an absorption rising sharply at 3 meV and tailing off at 44 meV is predicted. This contribution to ε has to be added to the free electron hole part and the final absorption calculated with e.q. 1. The result of such a recent calculation is shown as the solid curve in Fig. 2.[8]

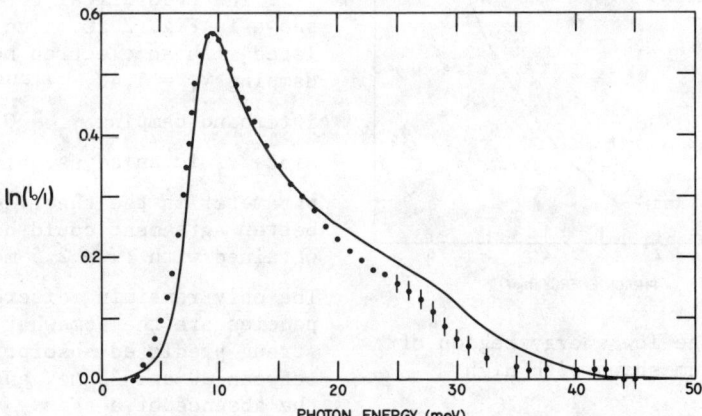

Fig. 2. Overall absorption of EHD in Ge at 1.2°K. The dots are experimental points and the solid line corresponds to theoretical results of Rose et. al. Note the observed threshold at 3 meV and the high frequency cut-off at 43 meV.

EXPERIMENTAL RESULTS AND DISCUSSION

The filled circles in Fig. 2 are our measurements of the EHD absorption. We used a Michelson interferometer between 5 and 50 meV and a lamellar grating interferometer between 1.5 and 8 meV. A Ge bolometer at 0.3°K was used as a detector. We used a germanium sample 1 X 6 X 15 mm in size with an electrically active impurity concentration of 2 X 10^{10} cm^{-3}. The sample was polished optically and etched with CP4. The excitation was produced with a CW YAG: Nd laser at powers between 3 and 50 mW focussed on an area 5 mm diameter to give a maximum absorption of 0.6 and on a 3 mm dia area for larger values of $\log(I_o/I)$.

The sample was immersed in superfluid helium at 1.2°K.

The EHD absorption at 1.2°K consists of a broad assymetric band peaking at 9.3 ± 0.1 meV. At low frequencies there is a very sharp onset at 3 meV and a less distinct cutoff at 43 ± 2 meV. Fig. 3 shows this region in more detail measured on a larger sample (2 X 15 X 15 mm) and a smaller area of excitation (1 mm dia). There is a distinct threshold at 2.75 meV which, when corrected for instrumental resolution and thermal broadening, yields an estimated absorption onset of 2.9 ± 0.1 meV.

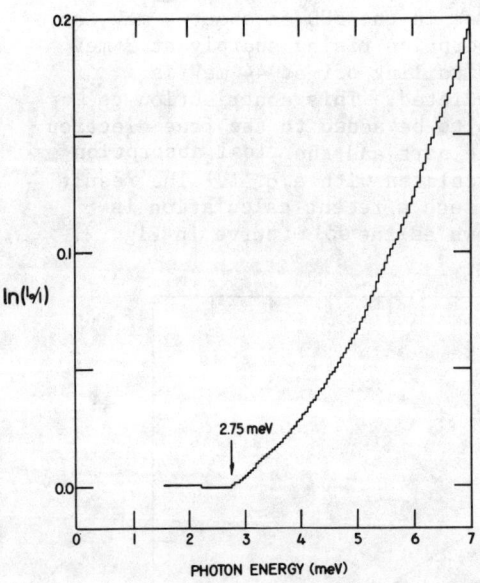

Fig. 3. The low energy region of the spectrum shown with high resolution.

The theoretical curve shown in Fig. 2 is calculated with an electron hole damping γ_1 = 1.48 meV and an interband damping $\gamma_2 \cong 0$. Since γ_1 is an adjustable parameter in the theory even better agreement could be obtained with γ_1 = 2.5 meV. The only remaining discrepancies are the somewhat strong predicted absorption between 20 and 35 meV and the absence of a sharp threshold at 3 meV in the theory. The theory assumes constant interband matrix element, but it is possible that far out in the [111] direction where the contribution at 30 meV is made, a more detailed calculation is needed.[8] At the low end the quadratic dependence of the absorption is the result of an assumed frequency independent γ_1. It is known, however, that γ_1 must go to zero near the fermi surface as the phase space for possible intraband scattering processes is reduced. Similarly, the interband threshold will be sharp since the final state hole lifetimes become very large near the fermi surface ($\hbar\omega_2$ in Fig. 1). A more complete calculation with γ_1 and γ_2 that vanish near the fermi surface should improve the agreement between experiment and theory at low frequencies.

SMALL PARTICLE EFFECTS

We observed no change in the EHD line shape for absorptions such that $0.5 < \ln(I_o/I)_{max} < 1.6$, but at lower excitation levels (Fig. 4), the line shape changes. The line broadens and both the peak and the threshold $\hbar\omega_1$ shift to lower energies. We suggest these effects are related to the increasing importance of the droplet surface as the excitation level and, as shown by a number of studies,[10] droplet size, is reduced.

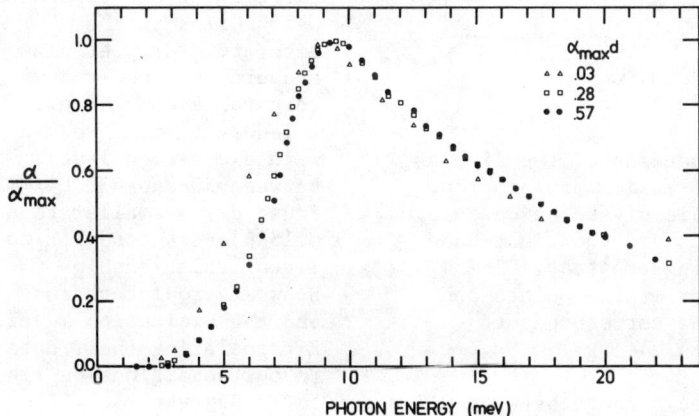

Fig. 4. Absorptions corresponding to different excitation levels normalized to maximum absorption — αd stands for $\ln(I_o/I)$.

The low energy half absorption point, we will call it E_1, shifts from 6.8 meV to 6.0 meV (Figure 5). We can interpret this effect as an additional damping term γ_s simple added to γ_1 to account for surface scattering. To estimate the order of magnitude of this effect we assume that

$$\gamma = \gamma_1 + \gamma_s = \gamma_1 + v_F/R_s \qquad (2)$$

where v_F is the fermi velocity and R the droplet radius. From the theoretical curves of Rose et. al. we can find how the low energy half maximum E_1 varies with γ. We find:

$$E_1 = [7.4 - 0.2\gamma] \text{meV} \tag{3}$$

With this equation we have labelled the γ axis in Fig. 5. Substituting from equation 2 we obtain finally:

$$E_1 = [6.88 - 0.074\ R^{-1}] \text{meV} \tag{4}$$

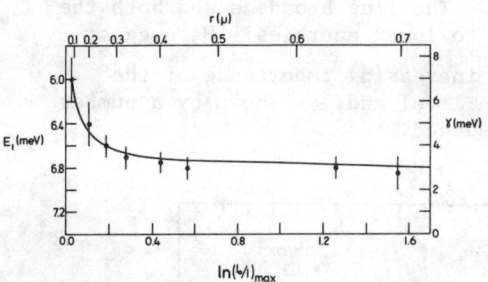

Fig. 5. Broadening of the line as a function of maximum absorption. The upper scale gives the calculated radii of drops and the right-hand one the damping constant. The dots are experimental points and the solid line corresponds to equation 4.

The solid line in Fig. 5 is the plot of this equation fitted to two points. It allows us to calculate a radius for the drops.

The radii obtained this way are very reasonable. While we cannot measure droplet size directly in our apparatus, we can find in the literature droplet size measurements under comparable conditions. Etienne et. al.[9] worked with excitation levels between 20 and 430 $\mu W/mm^2$ found drops smaller than 0.15 μ and a good fit to a linear relationship between droplet radius and the excitation level. Extrapolating their data to our conditions we find our sizes should range between 0.1 and 1.5 μ. Bagev et. al.[10] measured drops at 3.8 and 8 mW on a 900 μ spot from 2 to 4°K. Extrapolating their curves gives values smaller by a factor of two than what is shown in Fig. 5. Voos et. al.[11] found that above 10 mW excitation level on a very small spot at 2°K a droplet size of 2 μ was found independent of excitation level up to 100 mW. Extrapolating to our lower temperature we find that these conditions correspond to the upper level of our excitation.

In summary, it appears that at an an absorption of 0.3 or below, corresponding to a droplet radius of 0.3 μ, surface scattering becomes a significant mode of electron hole damping.

The plasma frequency of the smaller drops also moves to lower energies. For a 0.1 μ drop we estimate $\Delta\omega_{max}$ of 1.0 meV and a density reduction of 20%. This value is not unreasonable since the bohr radius of the exciton is about 0.01 μ and hence in a 0.1 μ

drop 30% of the volume is already in the low density surface region.

We would like to thank J.H. Rose, H.B. Shore and T.M. Rice for very valuable discussions.

REFERENCES

1. J.H. Rose, H.B. Shore and T.M. Rice; (preprint), see also J.H. Rose and H.B. Shore, Bull. Am. Phys. Soc. 20, 470 (1975), and T.M. Rice, Bull. Am. Phys. Soc. 20, 470 (1975).
2. V.S. Vavilov, V.A. Zayats and V.N. Murzin; Zh. Exper. Teor. Fiz. Pisma 10, 304 (1969); [Sov. Phys. JETP. Letters 10, 195 (1969)].
3. T. Timusk and A. Silin, Phys. Stat. Sol (b) 69, 87 (1975).
4. R.L. Aurbach, L. Eaves, R.S. Markiewicz and P.L. Richards; Solid State Comm. 19, 1023 (1976).
5. V.N. Murzin, V.A. Zayats and V.L. Kononenko; Fiz. Tverd. Tela 17, 2684 (1975) [Sov. Phys. Solid State, 17, 1785 (1975)].
6. D. Deirmendjian; Electromagnetic Scattering on Spherical Polydispersions; Elsevier, New York (1969).
7. M. Combescot and P. Nozieres; Solid State Comm. 10, 301 (1972).
8. J.H. Rose, H.B. Shore and T.M. Rice (Private Communication).
9. B. Etienne, C. Benôit à la Guillaume and M. Voos; Phys. Rev. Letters 35, 536 (1975).
10. V.S. Bagaev, N.V. Zamkovets, L.V. Keldysh, N.N. Sibel'din, and V.A. Tsvetkov; Zh. Eksp. Teor. Fiz. 70, 1501 (1976), [Sov. Phys. JETP, 43, 783 (1976)].
11. M. Voos, K.L. Shaklee and J.M. Worlock; Phys. Rev. Letters 33, 1161 (1974).

"INFERENCE OF INHOMOGENEOUS OPTICAL ABSORPTION ON THE NEAR-FIELD DYNAMICS OF LEDs"

A. Zehe
Universidad Autónoma de Puebla, Centro de Investigación
Apdo. Postal J-48, Puebla, Pue., México

ABSTRACT

The transient behavior of the near-field distribution generated by a time-dependent current applied to p-n junctions at forward bias is investigated. The configuration treated corresponds to that of a p^+pn GaAs diode in FABRY-PEROT structure. Within the framework of the assumed model, the spatial and temporal distribution of the near-field on the diode side faces are described, and these are in complete accord with the experimental findings. The main result is a physical picture of the extent of the so called 'Phase Inhomogeneity Effect' which turns out to be a valuable tool for investigating deviations of the minority carrier distribution from equilibrium.

INTRODUCTION

In recent years light generation in semiconductors, especially in p-n junctions, has been studied rather extensively[1,2]. A special modulation effect has been found with experiments in the radiation field of LEDs by use of high modulation frequencies of the minority carrier density[3]. This so called Phase Inhomogeneity Effect (PIE) reflects a non-stationary minority-carrier distribution in the recombination region, and is influenced considerably by the local optical properties of the light propagation area within the p-n structure.

It is the purpose of this paper to provide a physical picture of the formation of the PIE in the near-field of LEDs as a consequence of both the excess carrier and recombination dynamics, respectively, and the photon propagation through regions with rapidly changing optical properties. As a geometrical structure, we choose that of an edge emitting LED and investigate the near-field properties in a particular side face. The electrical and optical properties of this structure are those of a p^+pn GaAs LED, which is excited by a high-frequency sine-wave current. In order to describe the near-field dynamics of LEDs, the intensity distribution in a face perpendicular to the radiating areas is calculated supposing the emission strength of such layers to be proportional to the concentration of recombining minority carriers. Using a spatial and temporal distribution function of the carrier density, the physical nature of the PIE is discussed in terms of the inhomogeneous optical absorption properties of the light generation region.

THEORY

Consider a sandwich structure (Fig.1) of semiconducting material, where the upper part is p-type and the lower part n-type. In contrast to the case of inhomogeneous media with a varying dielectric constant and a magnetic permeability (e.g. dielectric film) we take here the special case of a homogeneous real part of the refractive index but an inhomogeneous optical absorption coefficient, $\alpha(r)$. Further we assume that the atomic oscillators feel no anisotropy (as e.g. in birefringent crystals) providing that $\alpha(r)$ is a scalar quantity.

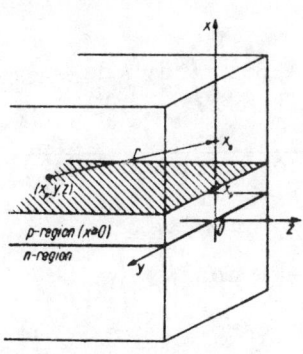

Fig.1. Diode model.
$x=x_p$ fixes a radiating layer within the p-region, $x=x_s$ is a surface point at (y=o, z=o)

In the lower part of the structure the optical and electrical properties are taken to have constant values, whereas above $x_p=o$ the optical absorption follows

$$\alpha(x_p) = A - B\cosh^{-2} ax_p \qquad (1)$$

where A, B, a, are constants. The interface between the p and n region is located at $x_p=o$.

LIGHT PROPAGATION

Usually light-field properties of a pn structure are described in the oscillation mode concept. This has been successfully done for the case of LASER operation[4,5,6]. In the spontaneous mode operation the light field of LEDs is formed by a large number of oscillation modes causing a considerable complexity of the physical model. Holding in mind that for p^+pn GaAs diodes the outside detectable light intensity, - although generated in the p region -, propagates, mainly through the low absorbing n side of the junction[7], the use of ray optics should be justified as a first approximation[8]. We will show that experimental results are matched quite well even by this approach, although a more sophisticated model should be desirable.

Now imagine that light is emitted from a y-z plane (Fig. 1) at x_p consisting of emitter points of equal strength providing a homogeneous radiating plane, embedded in the medium with an absorption function corresponding to (1). The intensity distribution $I(x_s)$ at the surface points $(x_s, y_s=0, z_s=0)$ in the x-y plane follows the eq.2, and along the line perpendicular to the junction plane results in the following expressions:

$$I(x_s) = I_o \int_{z_o}^{d} \int_{0}^{y} \exp\left\{-\alpha_\vartheta \cdot r_1 + (A-B)r_2\right\} dy\, dz \qquad (2)$$

$$\alpha_\vartheta = A - \frac{2B}{a\, r\, \cos\vartheta} \cdot \frac{\sinh(a\, r\, \cos\vartheta)}{\cosh(a\, r\, \cos\vartheta) + \cosh a(r\, \cos\vartheta + 2x_p)} \qquad (3)$$

$$y(z) = \frac{z}{\cos\beta_c}\left\{1 - \left(1 + \frac{(x_s-x_p)^2}{z^2}\right)\cos^2\beta_c\right\} \qquad (4)$$

$$z_o = |x_s - x_p|/\tan\beta_c$$

$$r_1 = r\frac{x_p}{x_s+x_p} = -r_2 + \left[(x_s-x_p)^2 + y^2 + z^2\right]^{1/2} \qquad (5)$$

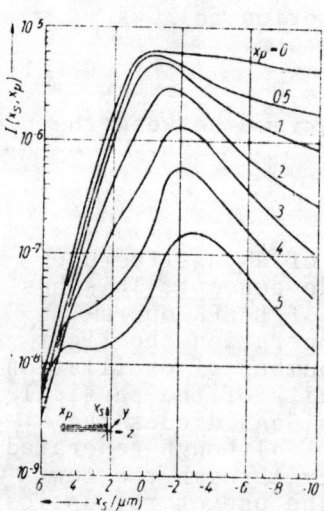

Fig.2. Light intensity $I(x_s,x_p)$ at a surface point (x_s, o, o) arising from a area source situated at x_p (x_p denotes the distance in the p-region)

where n is the semiconductor refractive index. Applying the following values: $d=150\,\mu m$, $n=3.6$, $A=0.51\,\mu m^{-1}$, $B=0.50\,\mu m^{-1}$, $a=0.2\,\mu m^{-1}$, the solution curves of (2) are illustrated in Fig.2. The remarkable feature is, that the intensity $I(x_s)$ arising from the emission plane at $x_p=o$ dominates that from either emission plane at x_p 0 over the whole range of values x_s. This is valid even for the case $x_s=x_p$, where the contribution of emission events perpendicular to the observation point x_s is smaller than that of the plane at $x_p=o$. Provided that y-z planes at different x_p-values are emitting at the same time, and assuming that the emission strength of those planes can be described by weighting factors $n(x_p,t)$, which have a spatial and temporal dependence, then the intensity $I(x_s,t)$ follows from (2) by a further integration to become:

$$I^*(x_s,t) \sim \int_{x'}^{x''} I(x_s) \cdot n(x_p,t) \, dx_p \qquad (6)$$

The upper and lower bound, x', x'', of this integration can easily be determined by the extent of the weighting factor $n(x_p,t)$ corresponding to x_p.

PHOTON DISTRIBUTION AND PHASE-INHOMOGENEITY EFFECT

The distribution $n(x,t)$ of the minority-carrier density beyond the depletion zone in the p and n region is usually determined from the time-dependent continuity equation. For the case of small signal excitation by a sinusoidal voltage is

$$n(x_p,t) \sim \exp(-x_p/L) + \frac{qu_o}{kT} \exp g \cdot \cos(\omega t - h) \qquad (7)$$

$$g = -\frac{x_p}{L}\left\{\frac{[1+\omega^2\tau^2]^{1/2}+1}{2}\right\}^{1/2} ; \quad h = \frac{x_p}{L}\left\{\frac{[1+\omega^2\tau^2]^{1/2}-1}{2}\right\}^{1/2}$$

keeping in mind the well known assumptions during its development[9] (ω- modulation frequency, u_o-amplitude of modulation voltage, L - diffusion length, τ - radiative life time.

The phase inhomogeneous state we define as follows (see Fig.3): The injection of minority carriers into the p-n junction of an edge emitter, which follows a sine function with an angular frequency ω, causes an oscillating luminescent light amplitude following the same rhythm in the near field. In relation to ω, wave surfaces S can be found in the light field of the diode which connect points of equal phase φ. A unit normal $\mathcal{n}_1(x_o,z_o)$ is introduced to provide a reference orientation for every other unit normal $\mathcal{n}(x,z)$ on a given wave surface S_i and we subject these vectors to the collinearity condition $[\mathcal{n}_1, \mathcal{n}_2] = 0$. If this condition is not fulfilled for every vector we speak of a phase - inhomogeneous state.

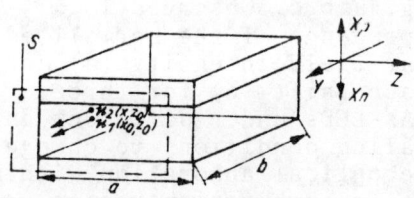

Fig.3. Definition of the phase inhomogeneous state

Solving (6) it turns out, that the phase relation between a modulating injection current and the modulated near-field intensity is independent of the coordinates $(x_s, y_s, 0)$ only up to a certain frequency cutoff. This limit depends essentially on the steepness of the optical absorption function, $\alpha(x_s)$, within the recombination region. Above this frequency limit the phase-relation, φ, proves particularly a dependence on the transverse coordinate, x_s.

RESULTS AND DISCUSSION

Both theoretical (c) and experimental results of the near-field intensity $I(x_s)$ and phase $\varphi(x_s)$ (a,b, respectively) are shown in Fig. 4 a,b,c. The only parameter changed during the calculation of (6) has been the steepness of the absorption function, $\alpha(x_p)$, within the recombination region.

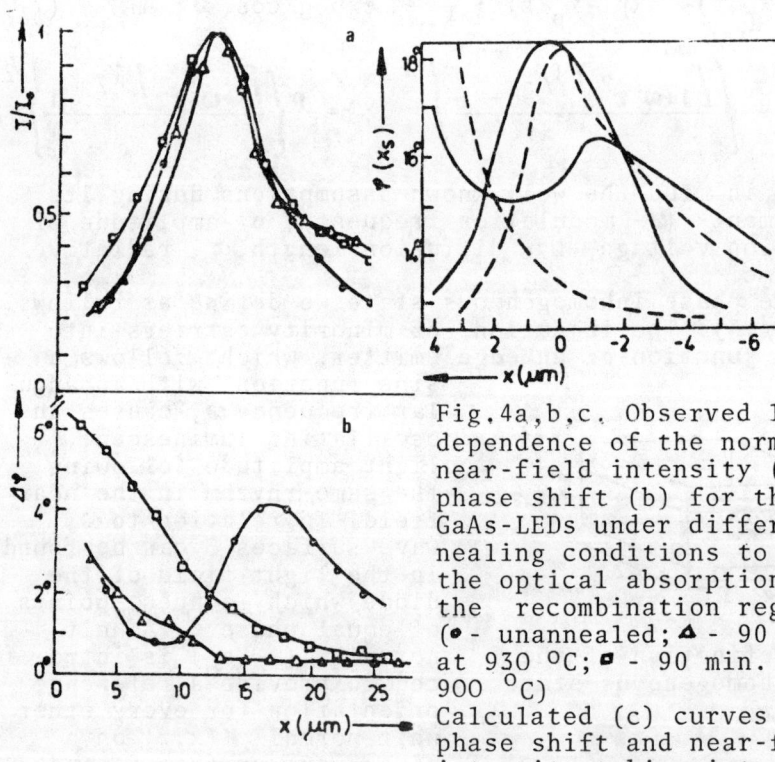

Fig. 4a,b,c. Observed local dependence of the normalized near-field intensity (a) and phase shift (b) for three GaAs-LEDs under different annealing conditions to change the optical absorption within the recombination region (● - unannealed; ▲ - 90 min. at 930 °C; ◘ - 90 min. at 900 °C)
Calculated (c) curves for phase shift and near-field intensity taking into account different absorption profiles around the recombination region.

The conclusion is appropriate that the PIE of electroluminescence diodes can be reduced by providing for a diminution of the contribution of the greatly delayed recombination processes in the near field. This is possible by giving the recombination area optical absorption properties which increase greatly in the direction of diffusion of the carriers. Such an optoelectronic structure is best achieved by steep doping gradients.

It turns out that the overlapping of the position functions of the recombination rate (light generation) and optical absorption (along light propagation) in the diode volume provides an explanation for the PIE and that phase inhomogeneity is inevitable when a set of conditions relating to $\omega\tau$ are fulfilled. The course of $\varphi(x_s)$ contains typical transient properties of the minority carriers which have not yet achieved equilibrium distribution with regard to time. The latter aspect particularly is becoming more important since the PIE is relatively open to accurate experimentation.

REFERENCES

1. A.A. Bergh and P.J. Dean, Proc. IEEE, 60, 156 (1972)
2. J.I. Pankove, Electroluminescence, Springer-Verlag Berlin-Heidelberg-New York, 1977.
3. A. Zehe, Ann. Phys. (Leipzig), 7. Folge, 29 (1973)1
4. F.K. Reinhart, I. Hayashi, and M.B. Panish, J.Appl. Phys. 42, 4466 (1971)
5. M.M. Antonoff, J. Appl. Phys. 35, 3623 (1064)
6. M.J. Adams, and M. Cross, Electron.Lett.7, 569, (1971)
7. A. Zehe, and K. Jacobs, Czech.J. Physics, B 23, 567 (1973)
8. A. Zehe, and G. Roepke, Ann. Phys. (G), 29, 351 (1973)
9. E.J. Cassignol; Halbleiter, Phil. Techn. Bibl.1966

CHARACTERIZATION OF PHOSPHORUS DOPED SILICON BY INFRARED REFLECTIVITY

R. Bennaceur
Faculté des Sciences de TUNIS, Departement de Physique,
Campus Universitaire le Belvédère TUNIS

ABSTRACT

A theoretical study of the propagation of E.M. waves in inhomogeneous media has been used to determine the profile of the free carrier concentration in semiconductors, with the aid of infrared reflectivity curves. The doping of ghe semiconductor was done by implantation. The profile of free carriers was approximated by a deformed Gaussian, whose parameters were found by comparison of the experimental and theoretical coefficients of reflectivity. The sensitivity of the reflectivity curve to the different parameters of the Gaussian was investigated. This method seems to be well adapted to the study of the heavily doped semiconductors (implanation doses greater than 10^{14} cm^{-2} for energy about 100 KeV).

INTRODUCTION

The study of inhomogeneously doped semiconductors in electronics, going from simple p-n junctions to integrated circuits seems to be of great importance. For the characterization of such structure different methods have been used : destructive, or non-destructive suchas the method of infrared reflectivity. Reflectivity has been used by Gusev, Streltsov, and Khaibullin[1] in which they approximate the profile by a step function. This method can give only an approximate value for the thickness of the layer. Subashiev and Kukharshi[2] have used the first order expansion of the solution of Maxwell's equations. This method can give only the value of the free carrier concentration and the slope of the free carrier distribution at the surface. A third method[3,4,5] consists of the approximations of the inhomogeneous semiconductor by a superposition of different homogeneous or inhomogeneous layers[9].

In the case of the semiconductor doped by ion implantation with the usual energies, the length of the inhomogeniety is of the order 10^3 Å. The wave length of the light in the region of the plasma reflectivity is longer than the inhomogeneous region and the previous methods are no longer valid except for that of reference 9. We will treat the general solution of Maxwell's equations and will apply this method for the determination of the free carrier distribution. The same method can be used for the study of other one dimensional inhomogenieties, for example, the inhomogeneous exciton gas at the surface of semiconductors.

EQUATION OF PROPAGATION OF THE ELECTROMAGNETIC WAVES

Consider the electric field of the polarized E.M. wave

$$E(u,t) = E_y(u) e^{i\omega t} \tag{1}$$

satisfying the wave equation

$$\frac{d^2 E_y(u)}{du^2} - \left(\frac{\omega}{c}\right)^2 \varepsilon(u) E_y(u) = 0 \tag{2}$$

with the Drude dielectric constant

$$\varepsilon_1(u) = \varepsilon_\infty \left[\left(1 - \frac{4\pi N(u) e^2}{m^* \varepsilon_\infty} \frac{1}{1+\omega^2 \tau^2(u)}\right) \right]$$

$$\varepsilon_2(u) = \frac{4\pi N(u) e^2}{m^* \varepsilon_\infty} \frac{\tau(u)}{1+\omega^2 \tau^2(u)} \tag{3}$$

with the relaxation time τ given by the empirical relation[6] :
for $N > 6 \times 10^{18}$ cm^{-3}

$$\tau(u) = \left| \left(0.75 - \frac{1}{6} \ln\left(\frac{N(u) - 6 \cdot 10^{18}}{10 e^1}\right)\right) \right| \times 10^{-15} \text{sec} \tag{4}$$

For $N \leq 6 \times 10^{18}$ cm^{-3}, $\tau(u) = 3 \times 10^{-14}$ sec.

TRANSFORMATION OF THE WAVE EQUATION (2)

We write the solution $E_y(u)$ as

$$E_y(u) = C_1 \exp\left(\int_0^u z_1(\xi) d\xi\right) + C_2 \exp\left(\int_0^u z_2(\xi) d\xi\right) \tag{5}$$

with

$$z_1 = w_1 + i \frac{\omega}{c} v_1 \quad \text{and} \quad z_2 = w_2 - i \frac{\omega}{c} v_2 \tag{6}$$

from (2) and (5) we obtain :

$$z'_i + z_i^2 + \left(\frac{\omega}{c}\right)^2 \varepsilon(u) = 0 \quad i = 1, 2 \tag{7}$$

Where the prime denotes a derivative.
When the semiconductor is homogeneous $z'_i = 0$ and

$$v_1 = v_2 = n \quad ; \quad w_1 = -w_2 = -k\frac{\omega}{c}$$

where n,k are respectively the refractive index and the extinction coefficient. The functions $v_i(u)$ and $w_i(u)c/\omega$ can be interpreted as the analogus of the refractive index and the excitation coefficient in the one-dimensional inhomgeneous media. We will see later the analogy between n, k and v_i, $w_i\frac{c}{\omega}$. From equations (6) and (7) we obtain the following equations:

$$\left. \begin{array}{l} w'_i + w_i^2 - (\frac{\omega}{c})^2 (v_i^2 - \varepsilon_1(u)) = 0 \\ v'_i + 2w_i v_i + (-1)^i (\frac{\omega}{c})^2 \varepsilon_2(u) = 0 \end{array} \right\} i = 1, 2 \qquad (8)$$

For the homogeneous case we obtain the usual relations for determining n and k.

APPROXIMATIONS OF WEAKLY (W.I.) AND STRONGLY (S.I.) INEHOMOGENEOUS MEDIA

The parameters that will be used in these approximations are the wave length of the light in the medium and the thickness of the inhomogeneous layer L. We define λ_e by:

$$\lambda_e = \frac{\lambda}{\sqrt{n^2 + k^2}}$$

The first approximation (W.I.) is obtained when:

$$|z'_i| \ll |z_i^2| \quad \text{or} \quad (\frac{\omega}{c})^2(\varepsilon(u)) \qquad (9)a$$

The inequatity (9)a is equivalent to:

$$\frac{\lambda_e}{2\pi} \ll L \qquad (9)b$$

This corresponds to the usual geometrical approximation. In the second approximation (S.I.),

$$|z'_i| \gg z_i^2 \qquad (10)a$$

From this inequality we obtain

$$\frac{\lambda_e}{2\pi} \gg L \qquad (10)b$$

We consider in more detail this two approximations : a) weakly inhomogeneous media.

In this case we can use z'_i as a perturbation with respect to z_i^2 (or $(\omega/c)^2 \varepsilon(u)$). In first order we obtain

$$w_i(u) = (-1)^{i+1} (\frac{\omega}{c}) k(u) - \frac{k'(u)k(u) + n'(u)h(u)}{2(n^2(u) + k^2(u))}$$

$$v_i(u) = n(u) + (-1)^{i+1} \frac{c}{\omega} \frac{k'(u)n(u) - n'(u)k(u)}{2\omega(n^2(u) + k^2(u))} \qquad (11)$$

We see that w_i and v_i depend locally on n and k.

b) Strongly inhomogeneous media, we obtain in this case when we neglect completely z_i^2 with respect to z'_i :

$$v_i(u) = (-1)^{i+1} (\frac{\omega}{c}) \int_L^u \varepsilon_2(\xi)d\xi + n(L)$$

$$w_i(u) = -(\frac{\omega}{c})^2 \int_L^u \varepsilon_1(\xi)d\xi + (-1)^{i+1} (\frac{\omega}{c})k(L)$$

We can do better approximations using perturbation theory in this case.

c) In general case we need numerical solution of the non-linear equation (7).

REFLECTIVITY COEFFICIENT OF THE INHOMOTENEOUS MEDIUM

By unsing the continuity relation of the electric field and his derivative we obtain :

$$R_1 = |\frac{1 - \frac{ic}{\omega} z_2(o)}{1 + \frac{ic}{\omega} z_2(o)}|^2 = \frac{(v_2(o)-1)^2 + (\frac{c}{\omega})^2 w_2^2(o)}{(v_2(o)+1)^2 + (\frac{c}{\omega})^2 w_2^2(o)} \quad (12)$$

The expression (10) is analogus in this form to the usual expression of the reflectivity coefficient in homogeneous media with

$$n \to v_2 \text{ and } k \to \frac{c}{\omega} w_2$$

In the case of the W.I. approximation R depends only on n(o), n'(o), k(o) and k'(o). We obtain the same reflectivity coefficient obtained by other methods by Subashiev and Kukharshi.

$$R° = \frac{(n + \Delta n - 1)^2 + (k + \Delta k)^2}{(n + \Delta n + 1)^2 + (k + \Delta k)^2} \quad (13)$$

where n + An and k + Ak represent the value of $v_2(o)$ and $\frac{c}{\omega} w_2(o)$ obtained from the relation (11). In this approximation we can obtain only the free carrier density N(o) and dN/du_o. In the case of the S.I. approximation we obtain.

$$v_2(o) = (\frac{\omega}{c}) \int_o^1 \varepsilon_2(u)du + n(1)$$

$$w_2(o) = (\frac{\omega}{c})^2 \int_o^1 \varepsilon_1(u)du - \frac{\omega}{c} k(1)$$

And we see that R depends on the whole inhomogeneous layer.

NUMERICAL SOLUTION OF THE MODIFIED MAXWELL EQUATION (7)

We have used the following boundary conditions : when the depth n is very large the profile is assumed to be very smooth, then we apply the W.I. approximation. For the following point we use the predictor-corrector of volume with 500 steps[7]. When

$u = 0$ we determine the reflectivity coefficient R.

APPLICATION OF THE METHOD FOR THE DETERMINATION OF THE PROFILE OF THE FREE CARRIER IN Si DOPED BY PHOSPHOROUS

a) Model of the free carrier profile.

The Gaussean profile predicted by the theory of ion-implanation of Lindhart-Scharff-Sshiott[8] cannot reproduce the real free carrier distribution. Experimental data concerning the distribution of the free carriers and the impurities show that the form of the profile is no longer Gaussian and is asymetric.

We will consider the free carrier distribution in the asymetric

$$N = N_{max} \exp{-\beta(u-d)^2} + N_o \text{ for } u \leq d$$
$$N = N_{max} \exp{-\alpha(u-d)^2} + N_o \text{ for } u \geq d \quad (15)$$

and we adjust the parameters N_{max}, α, β and d obtain the best fit between the theory and experimental reflectivity coefficients obtained by Geffroy[9].

b) Profil obtained by I.R. Reflectivity. Figure (1) shows the result obtained for two samples implanted with doses of 5×10^{15} ions /cm^2 and with energies of 40 and 90 keV. The reflectivity curves (1) and (2) correspond respectively to the profile (a) ($N_{max} = 3 \times 10^{20}$ cm^{-3} d = 0.11μ, $\alpha = 450\mu^{-2}$ and $\beta = 50\mu^{-2}$) and profile (b) ($N_{max} = 3 \times 10^{20}$, d = 0.04$\mu$, = 450$\mu^{-2}$ and $\beta = 50\mu^{-2}$). The determined profile reflects the variation of the implantation parameters. The concentration maximum changes form 0.04μ for 40kev to 0.11μ for 70kev. The L.S.S. theory predict depths of 0.049 and 0.085μ.

CONCLUSION

We have seen that with the use of the non-destructive method: the I.R. reflectivity, we can determine the profile of the free carrier.

This method gives good results for the heavily doped semiconductors (implantation doses greater than 10^{14} cm^{-2} for the energy of about 100kev). For the lower doses the reflectivity curve became flat and the ploting gives only approximate value of the parameters.

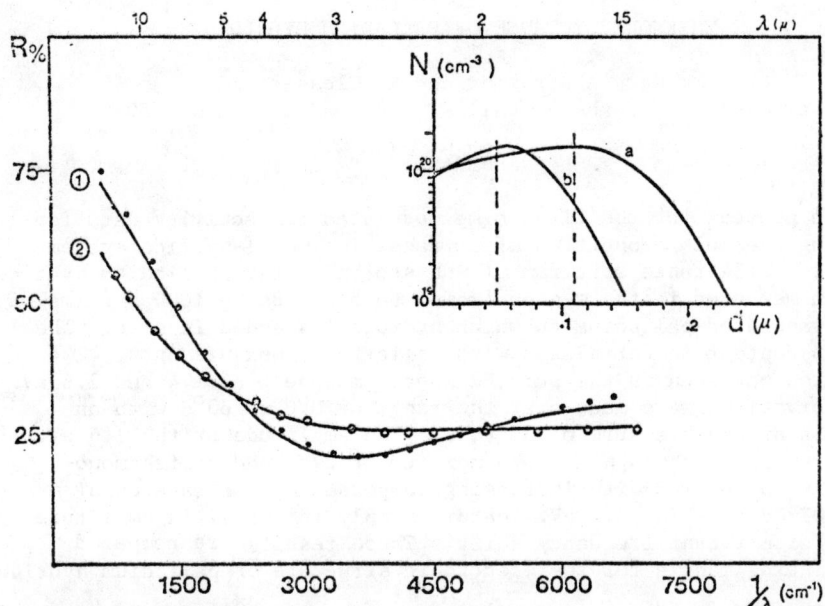

Figure 1 : Reflectivity curves of the Silicon Implanted by phosphore. Experimental curves : points correspond to energy 90 KeV and Dose : 5.10^{15} ions/cm^2 ; stars : energy 40 KeV and dose 5.10^{15} ions/cm^2. Curves (a) and (b) correspond to electron profils obtenued from curve (1) au (2).

REFERENCES

1. V.M. Gusev, L.N. Streltsov, I.B. Khaibullin, Sov. Phys. Semic. 5, 737 (1971).
2. V.D. Subashiev, A.A. Kukharski, Phys. Stat. Solidi 23, 447 (1967).
3. V.M. Evdokimov, A.A. Kukharski, L.M. Strel'rsov, v.B. Titov and I.B. Khaibullin, Sov. Phys. Semic. 4,5, (1970)
4. T. Abe and T. Kato Jap J. Appl. Phys. 4, 741 (1965).
5. L.M. Lampert J. Appl. Phys. 43 (11) 4612(1972).
6. R. Bennaceur, Thèse de doctorat es-Sciences, Université de TUNIS (1975).
7. W.J. Johnson, J.F. Gibbons, "Profected Range Statistics in Semiconductors," Stanford University Book Store (1969).
8. Y. Geffroy, Thèse 3° Cycle Université Pierre et Marie Curie, Paris (1975).
9. R.S. Stephens and G.D. Cody : will appear in Thin Solid Films 49 (1977).

THERMOREFLECTANCE OF PALLADIUM HYDRIDE

Gary A. Frazier[*] and R. Glosser
University of Texas at Dallas, Richardson, Texas, 75080

ABSTRACT

We present for the first time modulated reflectivity data representing the bulk properties of a transition metal-hydride system. The thermoreflectance spectrum of sub-stoichiometric palladium hydride has been measured in the region from 1 to 5 eV. using 1000 Å films of e-gun evaporated palladium to which hydrogen is added in situ. The hydrogen content is correlated with resistivity measurements. Two structures not seen in the pure Pd spectrum appear at 1.4 and 1.6 eV. The observations were made over the range of $0°C$ to $60°C$ with an ambient hydrogen pressure of 35 torr. The amplitude of the 1.6 eV. structure passes through a sharp maximum at $27°C$ and shifts monotonically to the red with increasing temperature at a rate $dE/dT = -3 \times 10^{-3} eV/°C$. The 1.4 eV. feature simply increases in amplitude with no significant frequency shift. These results are compared with existing models for the electronic structure of palladium hydride.

INTRODUCTION

A significant amount of theoretical and practical consideration has been given to the transition metal hydrides. Particular hydrides have been applied in nuclear and chemical physics as neutron moderators and hydrogenation catalysts. A basic problem with these materials is the embrittlement of the base metal due apparently to short range amorphism in the metal hydrogen bonds. That is, although the long range periodic structure of the host metal is maintained after hydride formation, the local substoichiometric distribution of hydrogens over available sites is random. Strain due to boundary discontinuities in hydrogen concentration induces microscopic embrittlement.

The band structure of stoichiometric and substoichiometric palladium hydride(PdH) has been studied based upon APW[1] and CPA calculations[2]. It has been shown that the simple rigid band model cannot be applied to PdH in that Fermi level shifts are also accompanied by the pulling down of previously unoccupied states below the Fermi level.

To our knowledge, no optical reflectivity data exists on any of the transition metal hydrides. Modulation spectroscopy has been useful in correlating observed optical transitions with band structure calculations. Fermi-level and critical point transitions can be assigned precise energies and modulation of symmetry parameters can determine the placement of these structures in the Brillouin zone. The stoichiometry of a metal hydride can be conveniently varied from nearly pure metal to nearly pure metal hydride. Thus the experimentalist can help determine the success of theory in predicting substoichiometric behavior based upon pure crystal models by monitoring the optical spectrum of the hydride as the hydrogen concentration is varied.

The purpose of this paper is to present preliminary data on palladium and palladium hydride using the technique of thermoreflectance. As a first step, two compositions i.e. Pd and $PdH_{.65}$ will be examined. Possible correlation with existing band structure models is discussed.

EXPERIMENT

The apparatus and instrumentation for the standard thermoreflectance experiment has been described elsewhere.[5] The notable feature of the equipment is the dewar used to cool the sample. A vertical glass column was sealed to a concentric Varian type vacuum flange to act as a reservoir for liquid N_2, dry ice etc. A 1"x 0.08" sapphire window was fused to the side wall of this tube to serve as an efficient thermal contact for the sample. In this geometry, the sample is in nearly direct contact with the coolant. This results in a very low thermal resistance for the substrate which is necessary for good modulation efficiency.[5]

The sample was prepared by vacuum evaporation of Pd wire (99.9995%) using a 4KW e-gun. A 1" x 3 mil quartz disc supported the sample during evaporation and was masked to produce 2 cm x 0.4 cm x 1000 Å films. The evaporation rate was 2.5 Å/sec at a base pressure of 8×10^{-8} torr. Contact to the film was made through indium soldered copper wire. The resistance of the Pd film, measured in situ, was determined to be 1.5 ohms/square which compares very well to the calculated bulk resistance[6] of 1.3 ohms/square for a 1000Å sheet. After evaporation the sample was attached to the dewar with silicone grease. The dewar was immediately placed in the reflectometer and the system pumped to better than 2×10^{-7} torr. The total time the sample was exposed to air was less than 5 minutes. The coolant for the samples used was uncirculated ice water. When necessary, hydrogen gas (99.9975%) was added via a pre-evacuated manifold. In all cases the modulation was fixed at 2 Watts/cm^2 averaged over a 50% duty square wave.

RESULTS

The thermoreflectance for Pd from 1.2 to 2 eV. is featureless within noise limits except for a slight D.C. offset. The vacuum chamber was filled with 35 torr of hydrogen gas and the scan repeated. Two large structures appeared at 1.4 and 1.6 eV with a $\Delta R/R$ of about 10^{-4}. The structures shifted in intensity initially, but the peak energies and relative intensity of the two features were very reproducible after 1 hour.

The Pd spectrum and PdH_x spectrum were reproduced with a second film under identical conditions.

The $\Delta R/R$ PdH$_x$ spectrum versus energy is shown in figure 1 for various temperatures. The amplitude and peak energy shifts for the two peaks are significant. $\Delta R/R$ vs T and Energy peak versus T were taken from figure 1, and are plotted in figures 2 and 3. It is seen that the low energy structure shows no significant energy shift and monotonically increases amplitude with temperature. The high energy peak has a definite red shift of about -3×10^{-3} eV/°C and also passes through a sharp $\Delta R/R$ maximum at 27°C. It should be noted that the E(T) inflection point in the high energy peak and its $\Delta R/R$ maximum occur at the same temperature.

ANALYSIS AND DISCUSSION

The electrical resistance of the PdH samples increased to 175% of their pure Pd value after hydrogen addition. This value, and the PdH P-C-T diagram for 35 torr H$_2$ at 30°C, predicts[7] a substoichiometric ratio of H to Pd of .65. This indicates that the hydride was in the so called β phase throughout the temperature range studied. The equilibrium value of x in PdH$_x$ should decrease from .66 to .62 as the temperature increases from 0°C to 50°C. The correlation of resistivity changes and new optical structure with hydrogen addition shows that our thermoreflectance results are due to bulk phenomenon and not just surface changes.

An examination of the band structure of Switendick[1] for Pd and stoichiometric PdH predicts three possible Fermi level transitions and no critical point transitions in the region of 1 to 2.5 eV. Two of the transitions could originate along Δ from the two highest valence bands, below the Fermi level, to the Fermi level. Similarly, a third possibility is from the highest valence band along Q, below the Fermi level, to the Fermi level. The initial state for all three transitions have a d state character and only small energy modification of these states is expected as hydrogen is added. Since only stoichiometric PdH band structure has been published, it is not possible to accurately follow these predicted features to the structure of PdH$_{.65}$. It is possible that the large band shifts which occur in the formation of PdH allow transitions which cannot be predicted with any accuracy. However, a study of the direction of band motion and the Fermi level shift predicts that the three PdH structures mentioned earlier should all show a red shift as hydrogen concentration is decreased. For a static H$_2$ pressure, as in our experiment, this would mean that peaks should have a red shift with temperature increase.

Given the idea that the only possible transitions near 1.5 eV are d state to Fermi level transitions, there is good agreement between predicted energies and the observed structure at 1.4 and 1.6 eV. The 1.6 eV peak has a red shift of the correct magnitude while the 1.4 eV peak remains fixed. The lack of motion of this feature is anomalous to the band structure generalizations made. We could account for a zero temperature dependence if an energy difference between two bands were passing through a shallow maximum or minimum centered around the PdH transition energy. Transitions between parallel bands, or bands

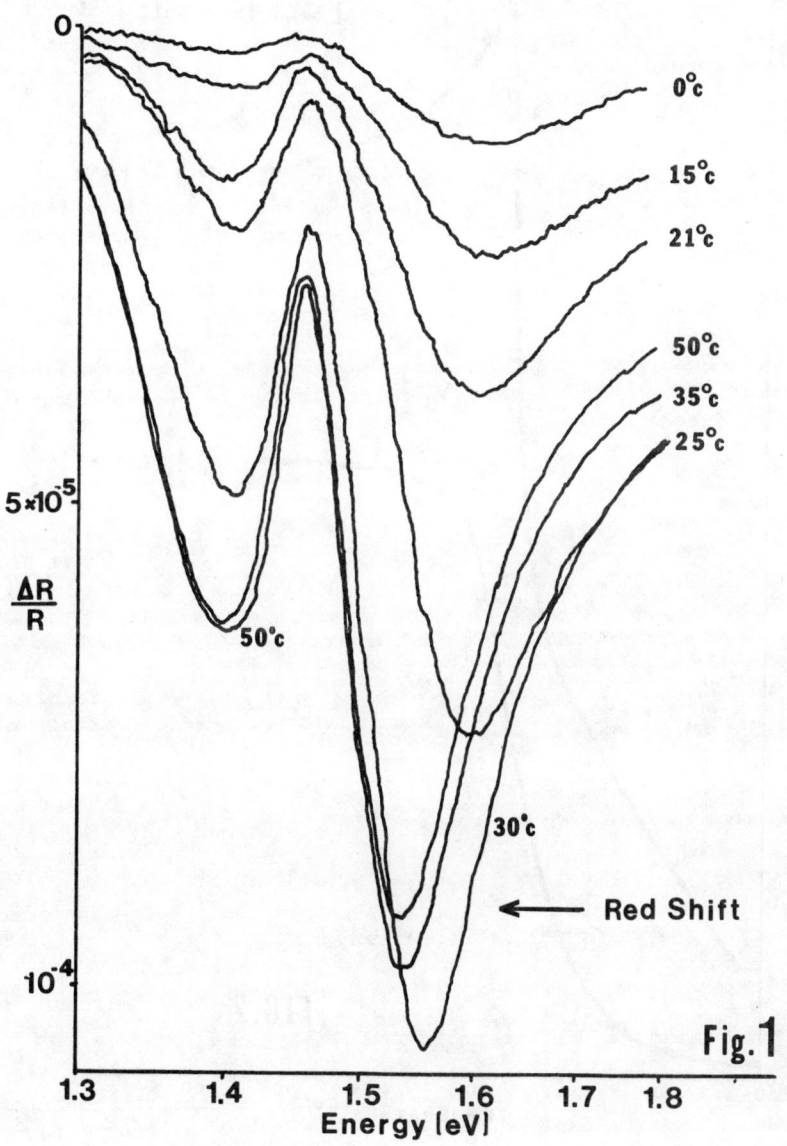

Fig. 1

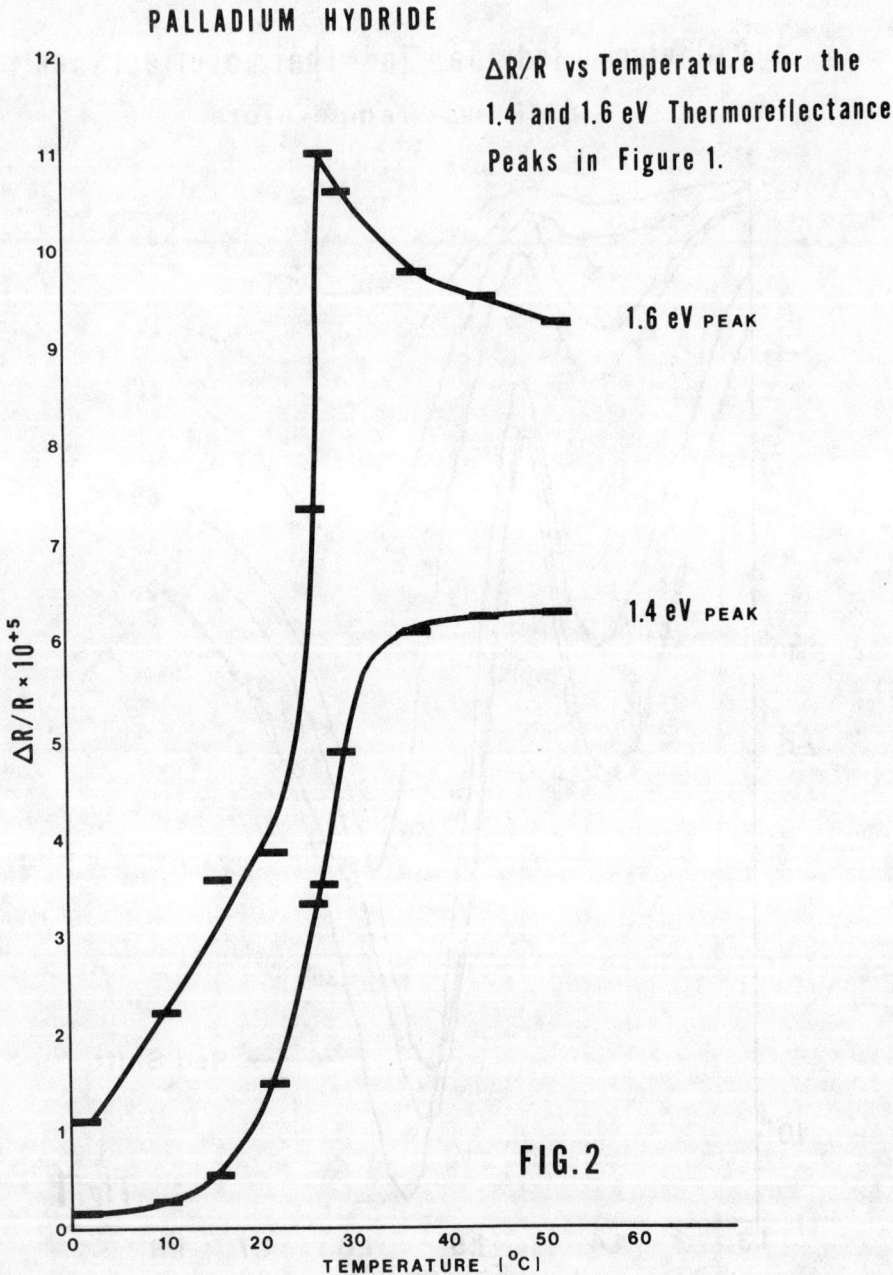

FIG. 2

which shift their energy in oppostion to the Fermi level shift can cancel out any concentration dependence of the transition energy. Further work in which the hydrogen concentration is slowly varied should help resolve this uncertainty.

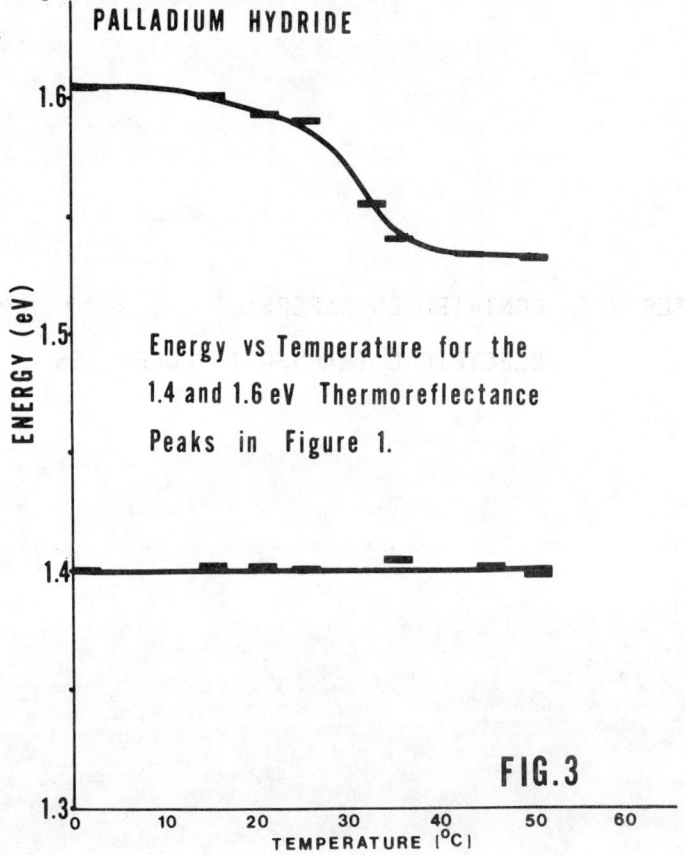

FIG.3 Energy vs Temperature for the 1.4 and 1.6 eV Thermoreflectance Peaks in Figure 1.

* Supported by University of Texas at Dallas Organized Research Funds

REFERENCES

1. A. C. Switendick, Ber Bunsenges 76, 535 (1972).
2. J. S. Faulkner, Phys. Rev. B 13, 2391 (1976).
3. A. Y. Yu and W. E. Spicer, Phys. Rev. 169, 497 (1968).
4. D. E. Eastman, J. K. Cashion and A. C. Switendick, Phys. Rev. Lett. 27, 35 (1971).
5. M. Cardona, Modulation Spectroscopy, Solid State Physics, Supplement 11, Acedemic Press (1969).
6. Weast, Handbook of Chemistry and Physics, 52nd Ed.,The Chemical Rubber Co. (1977).
7. F. A. Lewis, The Palladium Hydrogen System, Acedemic Press (1967)

CHAPTER V: CONTRIBUTED PAPERS:
ELECTRICAL TRANSPORT PROPERTIES

MEAN POTENTIAL AND FIELD IN A RANDOM DIELECTRIC

Kenneth S. Mendelson and Debra Schwacher
Physics Department, Marquette University
Milwaukee, Wisconsin 53233

ABSTRACT

A Green's function method is used to study the electrostatic potential and field in an infinite random dielectric containing a random space charge. The permittivity is assumed to be locally and statistically isotropic and statistically homogeneous. The space charge density is assumed to be statistically independent of the permittivity. Then the mean potential and electric field are determined by the mean space charge density and the non-local effective permittivity of the random medium. If the mean quantities do not vary appreciably over distances characteristic of the variation in the permittivity, the non-local effective permittivity can be replaced by a constant effective permittivity.

INTRODUCTION

The problem to be discussed in this paper is the determination of the mean electrostatic potential and field in an infinite random dielectric. The potential $\phi(\vec{r})$ satisfies the equation

$$\nabla \cdot \epsilon(\vec{r}) \nabla \phi(\vec{r}) = -\rho(\vec{r}) \tag{1}$$

where $\epsilon(\vec{r})$ is a random scalar function that is statistically homogeneous and isotropic, and $\rho(\vec{r})$ is a random function that is statistically independent of ϵ. In any given sample taken from a statistical ensemble ϵ is a piecewise continuous function and $\rho \to 0$ as $\vec{r} \to \infty$. In a given sample the potential is a continuous function for which the normal component of $\epsilon(\vec{r}) \nabla \phi(\vec{r})$ is continuous across any surface of discontinuity of ϵ and which satisfies the boundary condition $\phi \to 0$ as $\vec{r} \to \infty$.

FORMAL CALCULATION OF THE MEAN POTENTIAL

A formal solution for the potential can be obtained by means of a Green's function $g(\vec{r},\vec{r}')$ which satisfies the equation

$$\nabla \cdot \epsilon(\vec{r}) \nabla g(\vec{r},\vec{r}') = -\delta(\vec{r}-\vec{r}'), \tag{2}$$

ISSN: 0094-243X/78/349/$1.50 Copyright 1978 American Institute of Physics

where $\delta(\vec{r}-\vec{r}')$ is the Dirac delta function, and is subject to the boundary condition

$$g(\vec{r},\vec{r}') \to 0 \text{ as } |\vec{r}-\vec{r}'| \to \infty \quad . \quad (3)$$

The Green's function is continuous in each of its arguments and symmetric to interchange of its arguments.[1]

To obtain the potential we multiply Eq. (1) by $g(\vec{r},\vec{r}')$ and integrate over $\vec{r}$, which yields

$$\int g(\vec{r},\vec{r}')\nabla\cdot\varepsilon(\vec{r})\nabla\phi(\vec{r})d^3r = -\int\rho(\vec{r})g(\vec{r},\vec{r}')d^3r \quad . \quad (4)$$

We integrate the left hand side twice by parts noting that, because of the boundary conditions, the integrals over the surface at infinity vanish. Then using Eq. (3), the properties of the Dirac function and the symmetry of g to interchange of its arguments, we have

$$\phi(\vec{r}') = \int\rho(\vec{r})g(\vec{r}',\vec{r})d^3r \quad . \quad (5)$$

Since the charge density is independent of the permittivity, it is also independent of g. Thus the mean potential is given by (renaming the variables)

$$<\phi(\vec{r})> = \int <\rho(\vec{r}')><g(\vec{r},\vec{r}')>d^3r' \quad . \quad (6)$$

THE EFFECTIVE PERMITTIVITY

The mean Green's function will be calculated in terms of the effective permittivity $\hat{\varepsilon}(\vec{r}-\vec{r}')$ of the dielectric. This quantity is defined by[2]

$$<\vec{D}(\vec{r})> = <\varepsilon(\vec{r})\vec{E}(\vec{r})> = \int\hat{\varepsilon}(\vec{r}-\vec{r}')<\vec{E}(\vec{r}')>d^3r' \quad . \quad (7)$$

In reference 1 it was shown that

$$\hat{\varepsilon} = <\varepsilon(1+G\varepsilon)><1+G\varepsilon>^{-1} \quad (8)$$

where

$$G_{ij}(\vec{r},\vec{r}') = -\frac{\partial^2}{\partial x_i \partial x_j'} g(\vec{r},\vec{r}') \quad . \quad (9)$$

The product of two operators A and B means

$$[AB]_{ij}(\vec{r},\vec{r}') = \sum_k \int A_{ik}(\vec{r},\vec{r}'')B_{kj}(\vec{r}'',\vec{r}')d^3r'' \quad . \quad (10)$$

The permittivity operator is explicitly

$$\varepsilon_{ij}(\vec{r},\vec{r}') = \varepsilon(\vec{r})\delta_{ij}\delta(\vec{r}-\vec{r}') \qquad (11)$$

and the unit operator is

$$1_{ij}(\vec{r},\vec{r}') = \delta_{ij}\delta(\vec{r}-\vec{r}') . \qquad (12)$$

We introduce a zero-order Green's function $g^o(r-r')$ which satisfies the equation

$$\varepsilon_o \nabla^2 g^o(\vec{r}-\vec{r}') = -\delta(\vec{r}-\vec{r}') \qquad (13)$$

and the boundary condition (3). Here ε_o is a definite, but arbitrary, constant. In fact

$$g^o(\vec{r}-\vec{r}') = \frac{1}{4\pi\varepsilon_o |\vec{r}-\vec{r}'|} . \qquad (14)$$

It is shown in reference 1 that

$$G = (1-G^o\delta\varepsilon)^{-1}G^o \qquad (15)$$

where

$$\delta\varepsilon(\vec{r}) = \varepsilon(\vec{r}) - \varepsilon_o \qquad (16)$$

and G^o is given by an equation similiar to (9).

By a direct calculation

$$1 + G\varepsilon = (1-G^o\delta\varepsilon)^{-1}(1+G^o\varepsilon_o) = G\, G^{o-1}(1+G^o\varepsilon_o).$$

On substituting this into Eq. (8), and noting that $G^{o-1}(1+G^o\varepsilon_o)$ is not a random operator, we obtain

$$\hat{\varepsilon} = <\varepsilon G><G>^{-1} \qquad (17)$$

from which

$$<G> = \hat{\varepsilon}^{-1} <\varepsilon G> . \qquad (18)$$

CALCULATION OF MEAN QUANTITIES

On writing in the arguments, Eq. (18) becomes

$$\left\langle \frac{\partial^2}{\partial x_i \partial x_j'} g(\vec{r},\vec{r}') \right\rangle = \int \hat{\varepsilon}^{-1}(\vec{r}-\vec{r}'') \left\langle \varepsilon(\vec{r}'') \frac{\partial^2}{\partial x_i'' \partial x_j'} g(\vec{r}'',\vec{r}') \right\rangle d^3r''. \qquad (19)$$

Because $g(\vec{r},\vec{r}')$ is continuous in its variables, the order of differentiation and averaging can be interchanged,[3] to give

$$\frac{\partial^2}{\partial x_i \partial x_j'} <g(\vec{r},\vec{r}')> = \int \hat{\epsilon}^{-1}(\vec{r}-\vec{r}'') \frac{\partial}{\partial x_j'} <\epsilon(\vec{r}'') \frac{\partial}{\partial x_i''} g(\vec{r}'',\vec{r}') > d^3 r''. \tag{20}$$

Since the material is statistically homogeneous and isotropic, it follows from symmetry that $<\epsilon(\vec{r}) \nabla g(\vec{r},\vec{r}')>$ has the form [4]

$$<\epsilon(\vec{r}) \nabla g(\vec{r},\vec{r}')> = f(|\vec{r}-\vec{r}'|)(\vec{r}-\vec{r}'). \tag{21}$$

Integrate Eq. (2) over a sphere centered at the point $\vec{r}'$. By means of Gauss's divergence theorem this yields

$$-1 = \oint \epsilon(\vec{r}) \nabla g(\vec{r},\vec{r}') \cdot \frac{(\vec{r}-\vec{r}')}{|\vec{r}-\vec{r}'|} dS. \tag{22}$$

Taking the mean of this equation and using Eq. (21) gives

$$-1 = 4\pi |\vec{r}-\vec{r}'|^3 f(|\vec{r}-\vec{r}'|) \tag{23}$$

from which

$$<\epsilon(\vec{r}) \nabla g(\vec{r},\vec{r}')> = \frac{1}{4\pi} \nabla \frac{1}{|\vec{r}-\vec{r}'|} = \epsilon_o \nabla g^o(\vec{r}-\vec{r}'). \tag{24}$$

We substitute Eq. (24) into Eq. (20) and note that, from symmetry, $<g(\vec{r},\vec{r}')>$ is a function only of $\vec{r}-\vec{r}'$. Thus

$$\frac{\partial^2}{\partial x_i \partial x_j'} <g(\vec{r},\vec{r}')> = -\frac{\partial^2}{\partial x_i' \partial x_j'} <g(\vec{r},\vec{r}')>$$

and, with a similar interchange on the right, Eq. (20) becomes

$$\frac{\partial^2}{\partial x_i' \partial x_j'} <g(\vec{r},\vec{r}')> = \epsilon_o \int \hat{\epsilon}^{-1}(\vec{r}-\vec{r}'') \frac{\partial^2}{\partial x_i' \partial x_j'} g^o(\vec{r}''-\vec{r}') d^3 r''. \tag{25}$$

The derivatives on the right are now with respect to a parameter, not the variable of integration, and can be taken out of the integral. Then integrating twice with respect to $\vec{r}'$ and noting that, since $<g(\vec{r},\vec{r}')>$ is a function only of $|\vec{r}-\vec{r}'|$, the arbitrary functions of $\vec{r}$ appearing in the integrations vanish, we obtain

$$<g(\vec{r},\vec{r}')> = \epsilon_o \int \hat{\epsilon}^{-1}(\vec{r}-\vec{r}'') g^o(\vec{r}''-\vec{r}') d^3 r''. \tag{26}$$

Substitution of Eq. (26) into Eq. (6) gives for the mean potential

$$<\phi(\vec{r})> = \int d^3r'' <\rho(\vec{r}'')> \varepsilon_o \int d^3r' \hat{\varepsilon}^{-1}(\vec{r}-\vec{r}')g^o(\vec{r}'-\vec{r}''). \tag{27}$$

On interchanging the order of integrations, this becomes

$$<\phi(\vec{r})> = \varepsilon^* \int \hat{\varepsilon}^{-1}(\vec{r}-\vec{r}')\phi_o(\vec{r}')d^3r' \tag{28}$$

where

$$\phi_o(\vec{r}') = \frac{\varepsilon o}{\varepsilon^*} \int <\rho(\vec{r}'')> g^o(\vec{r}'-\vec{r}'')d^3r'' \tag{29}$$

and ε^* is an arbitrary constant which, for convenience, we will take to be

$$\varepsilon^* = \int \hat{\varepsilon}(\vec{r}-\vec{r}')d^3r' . \tag{30}$$

This is the effective permittivity associated with constant-mean fields. By differentiation one finds that ϕ_o satisfies the equation

$$\varepsilon^* \nabla^2 \phi_o(\vec{r}) = -<\rho(\vec{r})>, \tag{31}$$

that is, ϕ_o is the potential in a homogeneous and isotropic dielectric having permittivity ε^* and charge density $<\rho(\vec{r})>$.

Since $\phi(\vec{r})$ is a continuous function, we can interchange differentiation and averaging to obtain

$$<\vec{E}(\vec{r})> = -<\nabla\phi(\vec{r})> = -\varepsilon^* \int [\nabla\hat{\varepsilon}^{-1}(\vec{r}-\vec{r}')]\phi_o(\vec{r}')d^3r'. \tag{32}$$

We replace ∇ by $-\nabla'$ and integrate by parts noting that as $\vec{r}' \to \infty$, $\hat{\varepsilon}^{-1}(\vec{r}-\vec{r}') \to 0$ (see next section) and $\phi_o(\vec{r}') \to 0$ so the integral over the surface at infinity vanishes. Thus we have

$$<\vec{E}(\vec{r})> = \varepsilon^* \int \hat{\varepsilon}^{-1}(\vec{r}-\vec{r}')\vec{E}_o(\vec{r}')d^3r' \tag{33}$$

where $\vec{E}_o = -\nabla\phi_o$.

USE OF REPRESENTATIVE VOLUME AVERAGES

The mean values used in the preceding discussion are ensemble averages. In applications one would like to replace the ensemble average at the point $\vec{r}$ by an average over a representative volume around $\vec{r}$. Beran and McCoy[5] have shown that this can be done if there are two distinct length scales associated with the problem. One scale, denoted by ℓ, is associated with the variations of the permittivity and would be given by some correlation length. The second, denoted by L, is associated with the variation of the mean charge density, potential or electric field and would be given by the shortest significant wavelength in a Fourier expansion of any of these quantitites. If $L \gg \ell$, one can find a representative volume with dimensions that are much larger than ℓ, but much smaller than L. Ensemble averages can be replaced by averages over such a representative volume.

Under the same conditions one can simplify Eqs. (28) and (33). The function $\hat{\varepsilon}^{-1}(\vec{r}-\vec{r}')$ has a maximum at $\vec{r}=\vec{r}'$ and is essentially zero for $|\vec{r}-\vec{r}'| > \ell$. Thus the significant contribution to the integrals in Eqs. (28) and (33) comes from a region of dimension ℓ over which, if $L \gg \ell$, ϕ_o and $\vec{E}_o$ are essentially constant and can be taken out of the respective integrals. Then since, by definition,

$$\int \hat{\varepsilon}^{-1}(\vec{r}-\vec{r}')\hat{\varepsilon}(\vec{r}'-\vec{r}'')d^3r' = \delta(\vec{r}-\vec{r}'')$$

which implies, on integrating over $\vec{r}''$, that

$$\varepsilon^* \int \hat{\varepsilon}^{-1}(\vec{r}-\vec{r}')d^3r' = 1,$$

we have

$$<\phi(\vec{r})> = \phi_o \qquad (34)$$

and

$$<\vec{E}(\vec{r})> = \vec{E}_o . \qquad (35)$$

REFERENCES

1. K.S. Mendelson, J. Appl. Phys., to be published 12/77.
2. The fields in this definition satisfy a homogeneous equation rather than Eq. (1). This paper proves that the same effective permittivity can be used with more general fields.
3. P.G. Saffman, Studies in Appl. Math. <u>50</u>, 93-101 (1971).
4. G.K. Batchelor, <u>The Theory of Homogeneous Turbulence</u>, (Cambridge U. P., Cambridge, England, (1953), sec. 3.3.
5. M.J. Beran and J.J. McCoy, Q. Appl. Math. <u>28</u>, 245-258 (1970).

MICROSCOPIC FIELDS AND CURRENTS IN D.C. ELECTRICAL CONDUCTIVITY*

R. S. Sorbello
University of Wisconsin-Milwaukee, Milwaukee, Wi. 53201

ABSTRACT

Kubo response functions are considered for the microscopic electric field and current in a metal which is subjected to a uniform external field. The functions are evaluated for the model system of weakly scattering impurities in a jellium background. Both the microscopic field and current around an impurity are very localized and show dipolar-like angular variation.

INTRODUCTION

Microscopic electric fields and currents in the neighborhood of localized scatterers are quantities of fundamental interest in studies of electron transport. However, with the notable exception of the semi-classical treatment by Landauer twenty years ago,[1] surprisingly little work has been published on the subject. Recently, however, there has been renewed activity in this area.[2] This activity has been motivated in large part by studies of electromigration, or atomic migration in the presence of electric fields and currents.[3] The microscopic electric field is, in fact, the driving force for electromigration.[4]

An expression for the microscopic field can be obtained immediately from the formula given by Kumar and Sorbello[5] for the force $\vec{F}$ on an ion in the presence of an external d.c. field $\vec{E}$. Specializing their formula to the case of a test charge Q at position $\vec{X}$, we can then find the local microscopic electric field $\vec{E}_L(\vec{X})$ whose definition is $\vec{E}_L(\vec{X}) = \lim_{Q \to 0} \vec{F}/Q$. The result is[4,5]

$$E_L(\vec{X}) = \vec{E} - \partial \Phi(\vec{X})/\partial \vec{X} \qquad (1)$$

Here Φ is the microscopic potential defined by

$$\Phi(\vec{X}) = \langle \phi(\vec{X}) \rangle \qquad (2)$$

where $\phi(\vec{X}) = -\sum_j e/|\vec{X}-\vec{x}^j|$ is the potential operator at position $\vec{X}$ due to the electrons. The sum is over all electrons j, their positions being denoted by x^j. The brackets in Eq. (2) represent the expectation value in the quantum mechanical state of the system.

*Research sponsored by the Air Force Office of Scientific Research under Grant No. AFOSR-76-3082.

The corresponding expression for the microscopic current is simply the expectation value of the current operator. That is,

$$\vec{J}(\vec{X}) = \langle \vec{j}(\vec{X}) \rangle \qquad (3)$$

where $\vec{j}(\vec{X})$ is the particle-current operator at position $\vec{X}$. Within second-quantization, we have $\vec{j}(\vec{X}) = \sum_{\vec{k},\vec{q}} a^{\dagger}_{\vec{k}} a_{\vec{k}-\vec{q}} (\vec{k}-\vec{q}/2) \exp(i\vec{q}\cdot\vec{X})$ where the creation and annihilation operators are $a^{\dagger}$ and a, respectively.

LINEAR RESPONSE THEORY

We shall evaluate expressions (2) and (3) using Kubo's linear-response theory.[6] An exact expression for Φ to first-order in $\vec{E}$ can be shown to be[7]

$$\Phi(\vec{X}) = \lim_{\omega\to 0} \frac{i}{\omega} \langle\langle \phi(\vec{X}); \vec{v} \rangle\rangle \cdot e\vec{E} \qquad (4)$$

The double-brackets in Eq. (4) define the usual correlation function $\langle\langle A; B \rangle\rangle = -i\int_0^\infty e^{i\omega t} \langle [A(t),B(0)] \rangle_0 dt$ where $\langle ... \rangle_0$ represents the equilibrium statistical average and $A(t)$ is in the Heisenberg representation. $\vec{v}$ is the total velocity operator and e is the charge of the proton. In analogy with Eq. (4), one can also obtain an expression for $\vec{J}(\vec{X})$ by replacing ϕ by $\vec{j}$ and Φ by $\vec{J}$ in Eq. (4).

Following Sham's technique we evaluate Eq. (4) diagrammatically for a system of an impurity in a jellium background.[8] We allow for electron-electron interaction by screening all bare electron-ion interactions through the dielectric function $\varepsilon(q)$. Typical diagrams are shown in Fig. 1. The dashed lines ending in an × represent the screened electron-impurity interaction $V(q)$. The dashed line ending with an arrow represents the screened electric potential

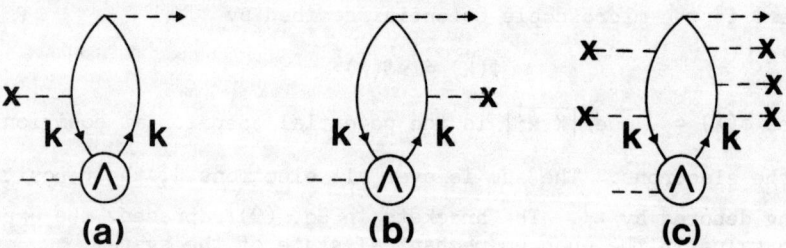

Fig. 1. Contributions to the response function in first-order (a,b) and fifth-order (c).

$\phi_s(q)\exp(-i\vec{q}\cdot\vec{X})$, where $\phi_s(q) = -4\pi e/q^2\varepsilon(q)$. The current vertex is denoted by Λ. At this vertex the electron-hole pair labelled $\vec{k}$ is created by the presence of $\vec{v}\cdot e\vec{E}$ in Eq. (4). The diagrams (a) and (b), which are first-order in V can be evaluated using the rules given by Sham for long mean-free-path. We obtain

$$\Phi(\vec{X}) = \frac{e^2\tau\vec{E}\cdot\vec{X}}{\pi^2 X} \int_0^{2k_F} \frac{V(q)}{\varepsilon(q)} j_1(qX) dq \qquad (5)$$

where τ is the relaxation time, j_1 is the Bessel function, k_F is the Fermi wavevector, and $X = |\vec{X}|$. The impurity is taken to be at $\vec{X} = 0$.

The evaluation of $\vec{J}(\vec{X})$ from the appropriate generalization of Eq. (4) is also easily obtained to first-order in V. Again, the diagrams of Fig. 1 apply except that the dashed line ending in an arrow represents the matrix element of $\vec{J}(\vec{X})$. We find that the first-order contribution to the current can be written as

$$\vec{J}(\vec{X}) = \vec{\nabla} \times \vec{\nabla} \times g(X)\vec{E} \qquad (6)$$

where $\vec{\nabla} = \partial/\partial\vec{X}$ and

$$g(X) = (\frac{e\tau}{8\pi^3}) \frac{1}{X} \int_0^\infty V(q) q^3 [\varepsilon(q)-1] \sin qX \, dq$$

If $V(q)$ is taken to be a pseudopotential, then $g(X)$ turns out to be, apart from a factor of $e^2\tau$, the pseudo-charge-density at X.[9]

NUMERICAL CALCULATION

In our calculations we used Shaw's local model-potential[10] for $V(q)$. We took values for the parameters appropriate to an $A\ell$ interstitial in an $A\ell$ matrix. Results for the first-order (linear in V) radial components of $\vec{J}$ and of $\vec{E}_L$ are shown in Fig. 2. The radial component of $\vec{J}$ is labelled J_r and the radial component of $\vec{E}_L$ is labelled E_r. The position, which is denoted by r, is measured from the impurity along a vector parallel to $\vec{E}$. That is, we have set $\vec{X} = r\vec{E}/|\vec{E}|$ in our equations. The radial components for an arbitrary position $\vec{X}$ can be obtained from the curves in Fig. 2 by multiplying them by $\cos\theta$ where θ is the angle between $\vec{X}$ and $\vec{E}$. This is characteristic of dipolar fields.

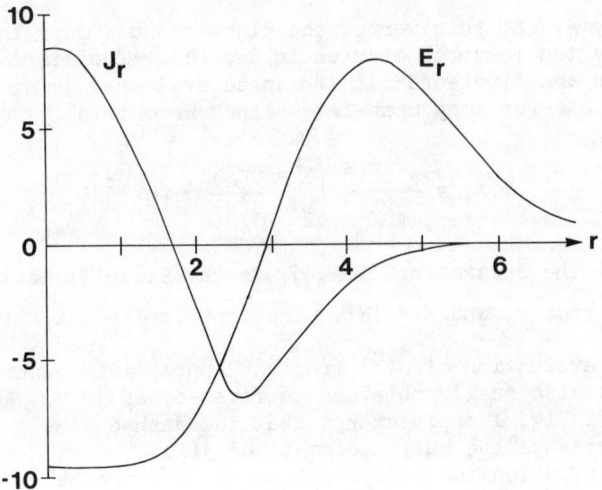

Fig. 2. Radial components of electric field and current in arbitrary units (r is in a.u.).

Note that J_r falls-off very rapidly. Asymptotically ($k_F r \gg 1$), Eqs. (5) and (6) lead to $E_r \sim r^{-2} \cos 2k_F r$ and $J_r \sim r^{-4} \sin 2k_F r$. If one extends the above calculations to second-order in V one finds corrections to J_r and E_r which may be significant for large r. In second-order we find $E_r \sim r^{-3}$ and $J_r \sim r^{-2}$ without the Friedel oscillations.

CONCLUSION

We have found that localized fields and currents exhibiting dipolar-like angular characteristics are present around localized scatterers. We note from Fig. 1 that $\vec{J}$ and $\vec{E}_L$ do not obey a local Ohm's law, although the macroscopic averages of $\vec{J}$ and $\vec{E}_L$ certainly satisfy such a law for a distribution of impurities. When typical numbers are used in Eqs. (5) and (6), we find that for dilute impurities J_r at $r = 0$ can be comparable to the macroscopic current while E_r at $r = 0$ can be several times larger than $\vec{E}$. The latter remains true even near the melting point, where τ is smallest.

We point out that the microscopic fields and currents which we have obtained do not persist indefinitely for very large r. Rather, for $r \geq \ell$ where ℓ is the mean-free-path, one obtains a decaying behavior like $\exp(-r/2\ell)$ due to the finite quasi-particle lifetime in the propagators of Fig. 1.[11] It is thus not clear how long-range

($r \gg \ell$) macroscopic fields can be attributed to localized dipoles centered on individual impurities as in Landauer's model.[1] However, in the $r \gg \ell$ region one needs to evaluate the Λ-vertex more carefully. In the simplest picture, this would lead to replacement of $\vec{E}$ by the macroscopic electric field in the local region of interest. This picture de-couples the long-range macroscopic field from the microscopic variations in the region of interest. The net effect is to replace the Λ-vertex by the semi-classical distribution function $f(\vec{k},\vec{X})$ appropriate to the region of interest. Such an approximation (but without the $\vec{X}$-dependence explicitly displayed) has been suggested by Sham[8] and Schaich.[12] We intend to apply this approximate picture to impurities near extended inhomogeneities such as surfaces and interfaces.

REFERENCES

1. R. Landauer, IBM J. Res. Dev. **1**, 223 (1957).
2. R. Landauer, Phys. Rev. B **14**, 1474 (1976); A. K. Das and R. Peierls, J. Phys. C **6**, 2811 (1973); H. E. Rorschach, Annals of Physics **98**, 70 (1976).
3. For the most recent bibliography on electromigration, see R. S. Sorbello, in <u>Electro- and Thermo-transport in Metals and Alloys</u>, edited by R. E. Hummel and H. B. Huntington (The Metallurgical Society of AIME, New York, 1977).
4. R. S. Sorbello and B. Dasgupta, submitted to Physical Review.
5. P. Kumar and R. S. Sorbello, Thin Solid Films **25**, 25 (1975).
6. R. Kubo, J. Phys. Soc. Japan **12**, 570 (1957).
7. Eq. (4) can be obtained directly from Eqs. (9) and (12) of Ref. 5 if one uses ϕ in place of $\partial V/\partial \vec{X}$ in Ref. 5.
8. L. Sham, Phys. Rev. B **12**, 3142 (1975).
9. W. A. Harrison, <u>Pseudopotentials in the Theory of Metals</u> (Benjamin, New York, 1976).
10. R. W. Shaw, Jr., Phys. Rev. B **5**, 4742 (1972). Values for the parameters were chosen as follows: $Z = 3$; $R_{OMP} = 2.24$ a.u.; $k_F = 0.9273$ a.u.
11. A. A. Abrikosov, L. P. Gorkov, and I. E. Dzyoloshinski, <u>Methods of Quantum Field Theory in Statistical Physics</u> (Prentice-Hall, Englewood Cliffs, N.J., 1963), p. 332.
12. W. L. Schaich, Phys. Rev. B **13**, 3360 (1976).

MINIMUM METALLIC CONDUCTIVITY IN GRANULAR METAL FILMS

B. Abeles[†] and Ping Sheng
RCA Laboratories, Princeton, NJ 08540

ABSTRACT

Granular metal films, consisting of fine mixtures of immiscible metals and insulators, are found to exhibit a minimum metallic conductivity, σ_m, which varies inversely as the average metal grain size d. It is shown that, as a consequence of electron localization on metal grains due to the electrostatic charging energy associated with the grain, $\sigma_m \simeq e^2/2z\hbar d$ (z is the number of nearest neighbor grains), in order of magnitude agreement with observations.

Metal-nonmetal transitions resulting from a change in conduction mechanism from metallic to hopping between localized states have been studied in a variety of physical systems. Of particular theoretical interest is the magnitude of the minimum metallic conductivity, σ_m, at which the transition occurs. Minimum metallic conductivity in heavily compensated semiconductors[1] and in Si inversion layers[2] has been discussed within the framework of the Anderson model[3] in which the localization of electrons results from random fluctuations of potentials. It has recently been proposed by Licciardello[4] that Anderson localization is responsible for the minimum metallic conductivity observed in two-dimensional granular metals such as ultrathin metal films. In this paper we propose, however, that σ_m in granular metals is a direct consequence of electron localization on individual grains due to the electrostatic charging energy E_c associated with a small metal grain.

Granular metals consist of a fine mixture of mutually immiscible metals and insulators.[5] As a function of volume fraction of metal, x, the electrical resistivity ρ exhibits the characteristic behavior shown in Fig. 1 for the Ni-SiO$_2$ system.[5] At 4.2 K, the sharp increase in resistivity near x = 0.5 corresponds to the breaking up of the metal continuum into isolated metal grains about 40 Å in size. The fact that the resistive transition is smeared out even at 4.2 K is due to percolation effects associated with a distribution of grain sizes and tunnel barrier thicknesses in the films.[6] The inset shows that the temperature coefficient of resistivity (TCR) exhibits an abrupt change in sign from positive to negative indicating a transition in the character of electronic transport from metallic to thermally activated tunneling between isolated metal grains. We define the minimum metallic conductivity as one at which the TCR changes sign (σ_m = 60 Ω^{-1} cm^{-1} in Fig. 1). In the following we show how σ_m follows from simple considerations of the effect of the charging energy on electron localization.

Consider two neutral neighboring grains separated by a tunnel barrier. Disregarding for the moment the charging energy, it is

[†]Exxon Research and Engineering Company, Linden, NJ 07036

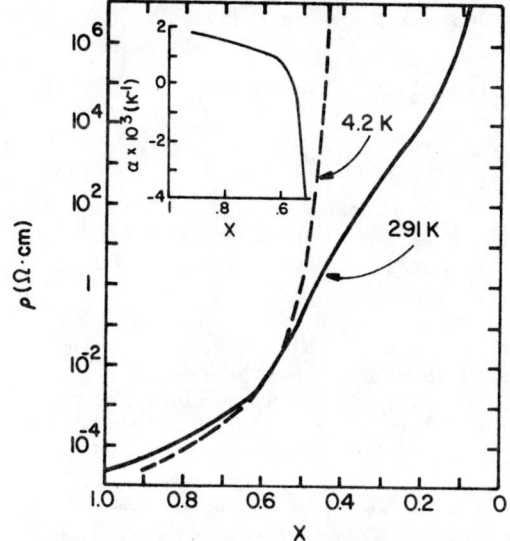

Fig. 1. Resistivity ρ as a function of volume fraction of metal x for granular Ni-SiO$_2$ films measured at 4.2 K and 291 K;[5] film thickness is about 5000 Å. Inset shows the temperature coefficient of resistivity $\alpha = [1-(\rho(4.2)/\rho(291))]/(291-4.2)$ as a function of x.

well known that due to coherent tunneling of the electron, each energy level is split by an amount $\delta E = \hbar/\tau$, where τ is the tunneling time.[7] The tunneling time τ can be expressed as[7] $\tau = RC$ where R is the zero bias tunneling resistance between two neighboring grains and C is the capacity of the two grains given by $C \sim \varepsilon d$ (ε being the effective dielectric constant of the granular metal and d being the grain size). Next we consider the effect of charging energy. The transfer of an electron from one neutral grain to another neighboring neutral grain, resulting in a pair of positively and negatively charged grains, requires a charging energy $E_c = e^2/2C$. For grain size of 50 Å and $\varepsilon = 10$, we have $E_c \simeq 30$ meV. The criterion for the localization of electrons is given by comparing the interaction energy δE with the charging energy E_c. We would expect that when $\delta E \leq E_c$, electrons are localized. A completely analogous situation arises in the Hubbard model[8] where localization for a half-filled parabolic band sets in when $\Delta/I = 1.15$, where Δ, the bandwidth, corresponds to δE and I, the intra-atomic repulsion between two electrons, corresponds to E_c. Using the criterion $\delta E = E_c$ for the onset of electron localization, we obtain

$$\frac{e^2}{2C} = \frac{\hbar}{\tau} = \frac{\hbar}{RC}. \qquad (1)$$

In the case where each grain is coupled to z nearest neighbors, R in Eq. (1) is replaced by R/z, and in that case it follows that

$$\frac{1}{R} = \frac{e^2}{2z\hbar}. \qquad (2)$$

To calculate the conductivity of an array of grains requires detailed knowledge of the spatial distribution of the grains. In the case of square lattice in two dimensions and a simple cubic lattice in three dimensions with lattice constant D, we have from Eq. (2):

$$\sigma_m = \begin{cases} \dfrac{e^2}{2z\hbar} & \text{two dimensional } (z = 4), \quad (3a) \\ \dfrac{e^2}{2zD\hbar} & \text{three dimensional } (z = 6). \quad (3b) \end{cases}$$

Eq. (3) was obtained by noting that in a unit cube there are $1/D^2$ (1/D for a unit square) resistors in parallel, each having a resistance of R/D.

In Table I are listed the observed minimum metallic conductivities for several three-dimensional granular metal systems.[5] In the cases where the average grain size d is available, the values of σ_m are calculated from Eq. (3b) corresponding to a simple cubic lattice and using the approximation $D \simeq d$. While the calculated values are all lower than the experimental values, they agree within an order

TABLE I. Volume fractions of metal, x_o, average grain size, d, and minimum metallic conductivities, σ_m, of granular metals

	x_o	d (Å)	σ_m (Ω^{-1} cm^{-1}) exp.	calc.
Al-SiO$_2$	.65	< 20	1000	> 100
Ni-SiO$_2$	.56	40	60	51
Au-Ta$_2$O$_5$	.55		100	
Au-SiO$_2$	.50		500	
Au-Al$_2$O$_3$	.45	60	300	35
Nb-Al$_2$O$_3$	.50	< 20	770	> 100
W-Al$_2$O$_3$	.53	30	142	67
W-Al$_2$O$_3$*	.48	130	20	15
Ag-SiO$_2$	.40		300	

*Annealed at 900°C for 6 hours.

of magnitude or better. The agreement is considered satisfactory in view of the fact that we have used a much simplified model which does not take into account a distribution in grain sizes and percolation effects. Although there is considerable scatter in the data, Table I shows a definite trend for σ_m to decrease with increasing d, in agreement with prediction. This behavior is seen particularly convincingly in the case of $W-Al_2O_3$ system where an increase in d by annealing is accompanied by a proportional decrease in σ_m.

In the case of two-dimensional granular metals with $z = 4$, Eq. (3a) yields the value $\sigma_m = 0.125\ e^2/\hbar = 3.0 \times 10^{-5}\ \Omega^{-1}$. This value is in good agreement with the minimum metallic conductance observed in ultrathin films.[9-11] It should be noted that the calculation of Licciardello et al.[12] based on the Anderson model leads to $\sigma_m = 0.12\ e^2/\hbar$, which is essentially the same as our result for two-dimensional systems. However, the authors did not take into account the charging energy but only considered the level splitting W in each grain caused by the quantization of motion. For an isolated grain,[13] $W \simeq E_F/n\Omega$ (E_F is the Fermi energy, n is the electron density, and Ω is the grain volume.) In the case of a 50 Å grain, $W \simeq 1$ meV, which is much smaller than E_c ($\simeq 30$ meV). Since the same coupling energy between two grains is assumed in both models, this would suggest that when $E_c \gg W$, it is the effect of the charging energy which is responsible for electron localization. Furthermore, we note that the temperature dependence of the granular metal conductivity in the nonmetallic regime is given by [5,14] $\ln \sigma \propto T^{-\frac{1}{2}}$, consistent with hopping conductivity in a granular system characterized by a distribution of charging energies, with the level splitting W not playing a significant[5] role. In conclusion we note that other physical phenomena in granular metals that depend on the character of the electron wave function (localized or extended) undergo drastic changes in the vicinity of the minimum metallic conductivity; examples are superconductivity and ferromagnetism in granular metals.[5]

REFERENCES

1. F.R. Allen, R.H. Wallis and C.J. Adkins, Proc. 5th Intern. Conf. on Amorphous and Liq. Semiconductors (Taylor & Francis, London, 1974), p. 865.
2. N. Mott, M. Pepper, S. Pollitt, R.H. Wallis and C.J. Adkins, Proc. R. Soc. Lond. A345, 169 (1975); D.C. Tsui and S.J. Allen, Jr., Phys. Rev. Lett. 34, 1293 (1975); A. Hartstein and A.B. Fowler, J. Phys. C8, L249 (1975).
3. P.W. Anderson, Phys. Rev. 109, 1492 (1958).
4. D.C. Licciardello, presented at the Mid-winter Conf. on Electronic and Magnetic States in Disordered Materials, Laguna Beach, Calif., Jan. 11-15, 1977; to be published.
5. B. Abeles, Applied Solid State Science, Vol. 6, ed. R. Wolfe (Academic Press, N.Y., 1976), p. 1; B. Abeles, P. Sheng, M.D. Coutts and Y. Arié, Adv. Phys. 24, 407 (1975).
6. B. Abeles, H.L. Pinch and J.I. Gittleman, Phys. Rev. Lett. 35, 247 (1975).
7. K.K. Thornber, T.C. McGill and C.A. Mead, J. Appl. Phys. 38, 2384 (1967); E.O. Kane, Tunneling Phenomena in Solids, eds. E. Burstein and S. Lundqvist (Plenum Press, N.Y., 1969), p. 1. Experimental verification of $\tau = RC$ can be found in, for example, G. Lewicki and C.A. Mead, Phys. Rev. Lett. 16, D939 (1966); and

C.A. Mead in Tunneling Phenomena in Solids, eds. E. Burstein and S. Lundqvist (Plenum Press, N.Y., 1969), p. 127.
8. J. Hubbard, Proc. Roy. Soc. Lond. A281, 401 (1964).
9. W.F.G. Swann, Phil. Mag. 28, 467 (1914).
10. N.T. Liang, Y. Shan and S.Y. Wang, Phys. Rev. Lett. 37, 526 (1976).
11. T. Andersson, J. Phys. D9, 973 (1976).
12. D.C. Licciardello and D.J. Thouless, Phys. Rev. Lett. 35, 1475 (1975).
13. R. Kubo, J. Phys. Soc. Japan 17, 975 (1962).
14. P. Sheng, B. Abeles and Y. Arié, Phys. Rev. Lett. 31, 44 (1973).

THE DIELECTRIC CONSTANTS AND LOSSES OF EXPLOSIVE MIXTURES BELOW 100 kHz

W. D. Gregory, R. Dunn, L. Capots, and L. Morelli
Physics Dept. Georgetown Univ., Washington D. C.

ABSTRACT

Explosives form a class of materials characterized by a highly polar substance (the explosive per-se) in an inert (non-explosive) background (clay, wood shavings, plastic, etc.). We have measured the relative dielectric constant and the losses as a function of frequency for seven types of explosives over a three decade frequency range. We find that the data are suggestive of the results obtained in the "effective medium" theory.[1]

[1] B. E. Springett, Phys. Rev. Lett. 31, 1463 (1973).

CONDUCTIVITY OF THICK FILM (CERMET) RESISTORS AS A FUNCTION OF METALLIC PARTICLE VOLUME FRACTION*

G. E. Pike
Sandia Laboratories, Albuquerque, NM 87115

ABSTRACT

Thick film resistors (TkFR) are composites of metallic particles embedded in a glass matrix. Conduction is through interconnected particle chains. For each of five systems studied the conductivity varies over many decades with particle volume fraction v_m. It follows a power law dependence, $G = K(v_m - v_{m,c})^t$, where K, $v_{m,c}$ and t are constants for a given system. There are two interesting features of this relation for TkFRs: $v_{m,c}$ is very low (0.02 or less) and t is rather high ($3 < t < 7$). These values of $v_{m,c}$ and t differ considerably from those of percolative systems in which $v_{m,c}$ is much higher and the critical exponent $t \sim 1.6$. Nevertheless, a model is proposed for which the conductance is indeed governed by percolation. This model takes explicit account of the resistor formation process and the observed microstructure. It is essentially a modified bond percolation model in which the metallic particles are <u>partial</u> <u>bonds</u> in the lattice connecting the interstices of the relatively larger, close-packed glass particles.

INTRODUCTION

The term thick film resistor (TkFR) refers to a class of resistive materials, used widely in the electronics industry, which are prepared according to a well established technology.[1] Preparation starts with an "ink" composed of glass and metallic powders suspended in a viscous organic fluid. The glass powders for the TkFR's discussed in this paper are 1 to 10 μm particles of various lead borosilicates. The metallic particles are either RuO_2 or $Bi_2Ru_2O_7$ and they are submicron. The organic fluid is present mainly to permit the mixture to be screen printed onto an alumina substrate, after which it is air dried at an elevated temperature. Finally the printed substrate is slowly fired (heated) to a temperature well above the glass transition temperature (typically 30°C per min. until a final temperature near 850°C is reached). The finished resistor is 10 to 20 μm thick and has a dark, glassy appearance. However, selectively removing a few microns of glass with an acid etch reveals a network of connected metallic particles as shown in Fig. 1 for a $Bi_2Ru_2O_7$ TkFR.[2]

---

*Prepared for the U.S. Energy Research and Development Administration under Contract AT(29-1)-789.

FORMATION AND STRUCTURE

Vest and co-workers have made an extensive study of the microstructure formation processes in TkFRs.[3-5] They find that the microstructure evolves in a series of steps. As the temperature T increases, the first change from a powder mixture is the initiation of sintering between adjacent glass particles. Neck growth proceeds by Newtonian viscous flow. This occurs at T ~ 500°C where the metallic particles can be neither wet significantly by the glass nor sintered together. Constrained to be on the surface of the glass, these smaller particles are pushed into pores and channels between the glass particles as the glass sintering goes to completion. At some higher temperature the glass begins to wet the metallic particles with the result that surface tension draws adjacent metallic particles closer together.

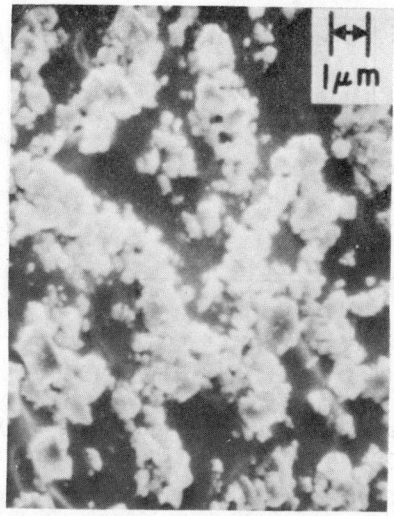

Fig. 1. Scanning Electron Micrograph (SEM) of a slightly etched, $10^6 \Omega/\square$ D-1200 TkFR.

Finally at the highest temperatures the still solid metallic particles begin to sinter by the liquid-phase process, which involves a thin interparticle layer of molten glass. Once this stage has been achieved, the essential features of the TkFR are relatively unchanged by either higher firing temperatures or longer firing times.[3,6]

ELECTRICAL

Although most commercial TkFR systems have compositions which are highly proprietary, there are some studies in which compositional information is available.[2,3,6,7] From these we have obtained results for the dependence of sheet conductivity, G, on the volume fraction of metallic particles v_m (relative to glass plus metallic particles; i.e., the organic volume was not considered). Figure 2 shows the results for two RuO_2 TkFR systems plotted as log G vs log $(v_m - v_{m,c})$. For the data of Ref. 7 we have taken $v_{m,c}$ to be zero. However, for the results of Ref. 3, $v_{m,c}$ was chosen as 0.021. This is close to the v_m value that Vest claims is a lower limit for making reproducible TkFRs with his system,[3] and it also yields a very linear graph. In each of these systems the many-decade variation of (sheet) conductivity is achieved by varying only one parameter: the concentraction v_m of metallic particles in the "ink". Similar results are seen in Fig. 3 for some $Bi_2Ru_2O_7$ TkFRs. The D-1200 and D-1400 are commercial

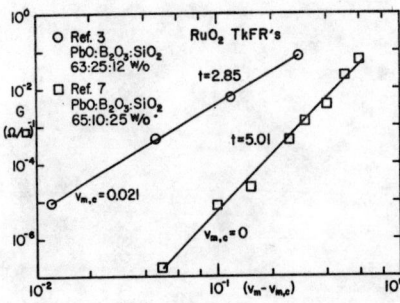

Fig. 2. Sheet conductivity G vs RuO_2 volume fraction v_m.

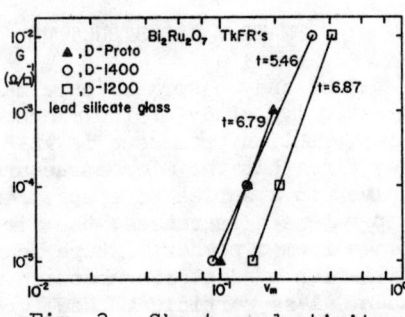

Fig. 3. Sheet conductivity G vs $Bi_2Ru_2O_7$ volume fraction v_m.

systems for which analysis of the principal constituents has been performed.[2,6,8] The D-Proto was custom formulated from $Bi_2Ru_2O_7$ and $PbO(SiO_2)_2$ plus a standard organic.[2,6,8] For all these systems $v_{m,c}$ was taken to be zero. Figures 2 and 3 show that the conductivity is accurately described by the form

$$G = K(v_m - v_{m,c})^t, \tag{1}$$

where the appropriate value of t for each system is entered on the figures. This is the first time that this functional form has been demonstrated for TkFRs.

DISCUSSION

An important aspect of the TkFR systems discussed above is that they all have the same conduction mechanism. It has recently been shown that conduction through these TkFRs is along the sintered chains of metallic particles.[2,6] The major impedance there is due to electron tunneling through the thin interparticle glassy layer remaining from the liquid-phase sintering. However, the finite conductivity of the metallic particles is also important. This is most easily seen from the temperature dependence of TkFR conductivity as shown in Fig. 4 for a typical system.[2,6] As the temperature is raised above several hundred Kelvin, the conductivity begins to decrease and its dependence approaches the (small) temperature dependence of the metallic particle conductivity. From Fig. 4 it is also apparent that the variation of TkFR conductivity with v_m is not due to different thicknesses of the interparticle tunnel barriers. For example, if the barriers in the $10^5 \Omega/\square$ materials were 100 times more resistive than those in the $10^3 \Omega/\square$ material, then the particle resistance in the former would totally negligable. Hence the contribution of the metallic particle to the TkFR temperature dependence of conductivity would also be negligable. This is contrary to the experimental results which show a metallic contribution for all compositions. Thus the conductivity variations cannot be ascribed to substantially different tunnel barriers.

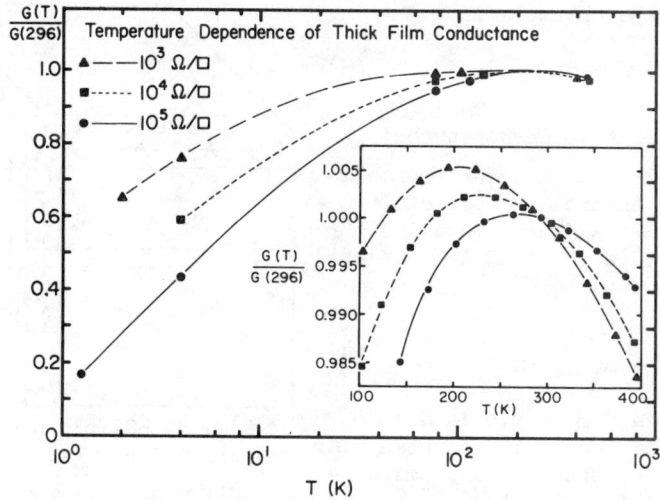

Fig. 4. Normalized conductivity vs temperature for the D-Proto system.

The functional dependence of Eq. (1) is reminiscent of that for conduction in a resistor lattice just above the percolation threshold,[9,10]

$$G = K'(p - p_c)^t, \qquad (2)$$

where p is the fraction of bond positions occupied by a resistor. However if one identifies v_m with p, then the values of $v_{m,c} < 0.02$ are considerably smaller than the published value of $p_c = 0.25$ for the simple cubic (sc) lattice[9] or $p_c \sim 0.10$ to 0.15 for various degrees of bond correlation on this same lattice.[9,10] Also regardless of correlation the value of $t \approx 1.6$ seems to characterize the 3-D resistor lattices while the TkFRs exhibit values of $3 < t < 7$.[9,10] Nevertheless, since the TkFR conductivity is increasing much faster that v_m, and since the conductivity at all values of v_m is controlled by similar chains of metallic particles, it seems that <u>connectivity</u> changes among chain segments must be largely responsible for the conductivity variations. Since this is exactly the situation in resistor networks for $p \gtrsim p_c$, one should ask whether the values of $v_{m,c}$ and t for TkFRs can be reconciled with p_c and t for the networks.

Recall that during formation the smaller metallic particles could not penetrate the larger glass particles, and that the glass sintering squeezed the metallic particles into pores and channels. Assume, for a rough estimate of $v_{m,c}$, that the glass particles sinter into cubes of side ℓ_1 with the spherical metallic particles of diameter ℓ_2 constrained to occupy rectangular channels of dimensions $\ell_1 \times \ell_2 \times \ell_2$ along the cube edges (see Fig. 5). Then the fractional volume of the channels is

$$v_2 = 3\ell^2 - 2\ell^3 \qquad (3)$$

where $\ell = \ell_2/\ell_1$. These channels form a sc lattice and the probability that a channel is completely occupied by $1/\ell$ adjacent metallic particles (i.e., a bond is complete) is

$$p = [v_m/(0.5 v_2)]^{1/\ell} \qquad (4)$$

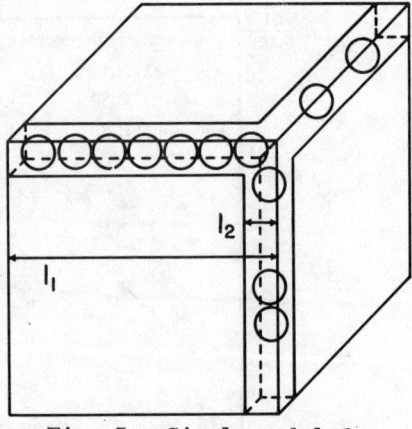

Fig. 5. Simple model for estimating $v_{m,c}$.

where the factor of 0.5 is due to spherical particles in a rectangular channel. For the TkFR systems discussed here ℓ is 0.1 to 0.2, so taking $\ell = 0.15$ and $p_c = 0.25$ and using Eqs. (3) and (4) one finds $v_{m,c} \sim 0.025$. Thus the condition that the metallic particles are <u>a priori</u> excluded from a large volume of the TkFR material permits <u>a small</u> amount of them to span the material as connected chain segments, each $\sim 1/\ell$ particles long.

The model above may be used only to demonstrate that $v_{m,c}$ can be rather small. It cannot be used to describe the conductivity much above $v_{m,c}$. This is because the channels would become completely filled (p = 1) for a small additional volume fraction of metallic particles. If p is nearly unity and the channels simply swell in cross-section as v_m increases, then connectivity would be unimportant and the conductivity would simply be proportional to v_m. This is not observed, and thus a channel model in which connectivity is strongly determined by the fraction of metallic particles seems to be incapable of explaining the conductivity data for $v_m > v_{m,c}$.

There is an alternative possibility that the channel connectivity is governed primarily by the glass sintering process. However, a satisfactory explanation of how this can happen has not yet been developed, and this aspect of the problem remains unsolved.

Although the detailed reasons for the large conductivity changes of TkFRs over a wide range of v_m are not known, this paper has shown that a large power law behavior is common to several systems, that connectivity changes dominate the conductivity variations, and that small volume fractions of metallic particles can reasonably be expected to form chain segments spanning the TkFRs.

ACKNOWLEDGEMENTS

The author is indebted to Dr. P. F. Carcia for first noting the functional dependence of G on v_m for the commercial systems, to Prof. R. W. Vest for permission to use his unpublished data, and to Dr. C. H. Seager for several useful discussions.

REFERENCES

1. M. L. Topfer, Thick Film Microelectronics, Fabrication, Design, and Applications (VanNostrand Reinhold, NY, 1971).
2. G. E. Pike and C. H. Seager, J. Appl. Phys., $\underline{48}$ (Nov. 1977).
3. R. W. Vest, Conduction Mechanisms in Thick Film Microcircuits, Final Technical Report, ARPA Order #1642 (Purdue Research Foundation, Ind., 1975).
4. G. L. Fuller, Ph.D. Thesis, Purdue Univ. (1972), unpublished.
5. A. N. Prabhu, Ph.D. Thesis, Purdue Univ. (1975), unpublished.
6. G. E. Pike and C. H. Seager, Electrical Properties of DuPont Birox and Cermalloy Thick Film Resistors II: Conduction Mechanisms, SAND76-0558 (Sandia Laboratories, Albuquerque, NM 1977).
7. H. C. Angus and P. E. Gainsbury, Electronic Components, $\underline{9}$, 84 (1968).
8. The D-1200 and D-1400 are commercial systems manufactured by the E. I. DuPont de Nemours & Co., Inc., Photo Products Div., Wilmington, Delaware, 19898. The D-Proto system was kindly supplied by Dr. P. F. Carcia of that company.
9. S. Kirkpatrick, Rev. Mod. Phys., $\underline{45}$, 574 (1973).
10. I. Webman, J. Jortner, and M. H. Cohen, Phys. Rev. B, $\underline{11}$, 2885 (1975).

PERCOLATION PROCESSES IN MIXTURES
OF CONDUCTING AND INSULATING PARTICLES[+]

H. Ottavi,[*] J. Clerc, G. Giraud, J. Roussenq,
E. Guyon,[**] and C. D. Mitescu[***]
Groupe de Physique, Université de Provence, Centre Saint-Jérôme
13397 Marseille, Cedex 4

ABSTRACT

We have measured the electrical conductivity of mixtures of hard, insulating and conducting particles, as a function of the fraction p of conductors. We observe the existence of a threshold p_c, below which the mixture is not conducting - a phenomenon of *site percolation* in an irregular lattice. The value of this threshold ($p_c \approx 0.3$), relatively insensitive to particle shape (spherical, oval, or baroque) or to applied pressure, is interpreted as being due to a relatively constant average number of nearest neighbor contacts ($z \approx 6-7$), reminiscent of "loose-packing". While the above results have been obtained with macroscopic mixtures (particle size $\approx$5mm), we report as well on experiments on microscopic powders, currently in progress.

We find, on the other hand, that pressure and the shape of the container do affect the conductance function $G(p)$, above the threshold. Specifically in the region $p \approx 1$, the slope of the normalized conductance $G(p)/G(1)$ varies inversely as the applied pressure. We interpret this as a superposition of a *bond percolation* phenomenon on the original *site* problem, where the applied pressure governs the probability p_B of existence of good bonds.

INTRODUCTION

Computer calculations of the conductance of percolation networks have been limited to rather small samples (<10^4 elements)[1] and size effects are of importance in the region of the percolation threshold.[2] We have undertaken a program of analog simulations using assemblies of insulating and conducting (metallically coated but of the same material as the insulating) particles. The electrical conductance measured between the conducting parallel ends of vertical cylinders filled with such a random

---

[+]Work supported in part by D.G.R.S.T. contract #736 PE.
[*]Also: Laboratoire d'électronique, Université de Provence, Centre Saint-Jérôme, 13397 Marseille.
[**]Also: Physique des Solides, Université Paris-Sud, 91405 Orsay.
[***]Permanent Address: Department of Physics, Pomona College, Claremont, California 91711.

ISSN: 0094-243X/78/372/$1.50 Copyright 1978 American Institute of Physics

mixture, normalized as $G(p)/G(p=1)$, is determined as a function of the fraction, p, of conducting objects. Pressure has been applied between these end plates in order to ensure electrical contact between particles in mechanical proximity.

Figure 1 gives the results obtained in three different containers holding about 10^5 spheres (radius 5.0±0.1mm); each data point represents the average of 10 independently mixed samples.

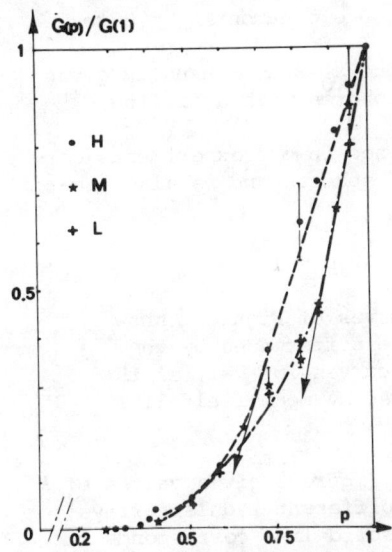

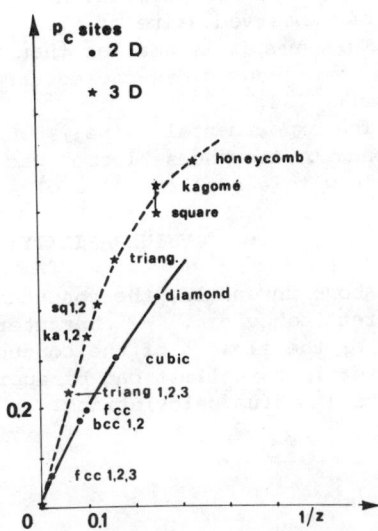

Fig. 1. The normalized conductance as a function of the fraction p of conductors, for three containers of varying aspect ratio: H(h=47 cm, ϕ=16 cm); M(h=21 cm, ϕ=24 cm); L(h=13.5 cm, ϕ=30 cm). (The arrows represent the tangents used to calculate the vulnerability K).

Fig. 2. Critical site percolation thresholds as a function of $1/z_0$ for various regular lattices. The continuous lines correspond to the single empirical function:

$$p_c = \sin(8.2/D\sqrt{2Dz_0 + z_0^2})$$

THRESHOLD MEASUREMENTS

The conductance vanishes below a critical threshold p_c=0.29±.02. This value is the same for all three containers and is equal to the result obtained in preliminary experiments[3] using irregularly shaped (ovoidal) objects of roughly the same size. We associate the constancy of p_c to that of the average number of electrical contacts between particles.

By assuming a filling process such that, as each new particle is added, it sets in a fixed position supported by three other objects, we can deduce an average coordination number z_0 of 6 for any convex-shaped body.

Figure 2 gives the value of p_c for *site percolation*, on different regular networks measured as a function of $1/z_0$. The data points lie on two curves corresponding respectively to dimension D = 2 and 3. If we assume that such an empirical relation applies to our non-periodic array, we deduce a number z_0 = 6 to 7 from our observed value of p_c.

This number is smaller than the value $z_0 \approx 8$ to 9 usually given for an amorphous close-packed array of spheres with a filling factor $f \approx 0.64$.[4]

The experimental value $f \approx 0.60$ obtained in our experiments corresponds to amorphous "loose" packing of spheres and is also compatible with a smaller value of z_0.

VULNERABILITY: SLOPE AT $p = 1$

Above threshold, the conductance curves of Figure 1 show different behaviors. We characterize the differences by the value of the slope K of the conductance curves for $p=1$, as the behavior in this limit can be approximated by mean-field-like effective medium calculations.

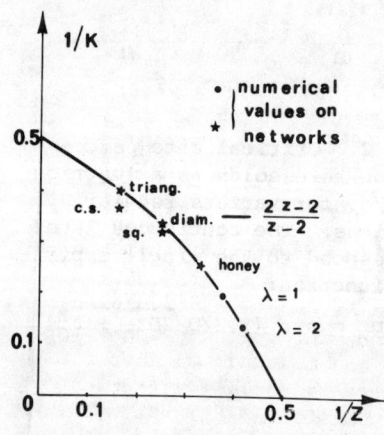

Fig. 3. Inverse of the vulnerability K vs. $1/z_0$. The stars represent a number of regular networks of different dimensionality. The dots correspond to special networks with intermediate sites.

Figure 3 gives values of K for different regular arrays. The solid line corresponds to the relation

$$K = (2z_0 - 2)/(z_0 - 2) \quad (2)$$

which obtains exactly for a certain class of arrays, independent of D.

Coming back to the results of Figure 1, we deduce a value $z_0 \approx 6$ for the container H ($K \approx 2.5$). However, for the two other containers a smaller value $z \lesssim 3$ would be deduced from the vulnerability, $K \approx 5$. We attribute the difference between the determination of z_0 to the smaller pressure used with the two containers M and L, of larger area. We have indeed checked that reducing the pressure applied on the cylinder H, increases the value of K (decreases z_1 from the value

$z_1 = 6 = z_0$), but does not change the threshold (keeps the value z_0 fixed).

We interpret the difference between the two determinations z_0 and z_1 as follows: near threshold all electrical contacts contribute to percolation and the connectivity $z_0 \simeq 6$ is the total average number of electrical contacts of a conducting sphere (*site problem*). On the other hand, near $p=1$ the "good" electrical contacts are dominant and the fraction of these contacts decreases with the pressure. The $p=1$ limit corresponds to a bond percolation problem. Further work is in progress on the effect of uniaxial pressure including an observed anisotropy of conductivity.

BEHAVIOR NEAR THRESHOLD

The critical behavior of the conductivity just above p_c should be

$$G(p) \propto (p - p_c)^\mu \qquad (2)$$

A value $\mu=1.6\pm.1$ has been deduced from numerical simulations of the bond problem on arrays[1] of size 25x25x25, while Adler et al[5] report $\mu=2.0$ for *site percolation* on a 16x16x16 lattice.

Due to the large scatter of our data for the three containers, and considering the uncertainty in p_c, we can only infer from our results an inequality $1.2<\mu<1.8$.

POWDERS

The above mentioned simulations have been restricted to a number ($\simeq 10^5$) hardly larger than the values used in computer experiments. We have undertaken studies on small size (10 to 100μ) coated and uncoated glass spheres, such that samples of 10^7 spheres can be obtained. Preliminary experiments indicate a threshold value $p_c=0.251\pm0.002$. The value is insensitive to the particle-size distribution (10 to 100μ). It is markedly smaller than the result for the larger objects. From this, applying the relationship of Figure 2, we would infer a coordination value $z_0 \simeq 8$ which is compatible with that of z_1 inferred from a study of the vulnerability. We tentatively attribute the difference from the results of the larger spheres to the closer packing obtained on this microsphere system (effect of a larger pressure and/or the effect of the distribution in the size of the particles.

We show in Figure 4a some preliminary data on the normalized conductivity $G(p)/G(1)$ for a sample of these particles (h≈φ≈1 cm). We observe a region near p_c ($\Delta p=p-p_c \simeq 0.05$) where the measured conductivities fluctuate considerably (on a scale too small to be evident in the figure). When we plot, in Figure 4b, the same conductivity data on a logarithmic scale against Δp, we find, outside that small region, a very good power-law behavior, extending over a remarkably wide range of Δp. We thence deduce a conductivity exponent $\mu=1.70\pm0.02$, which is not only consistent with previous estimates[1], but which agrees within experimental error with what we believe may be a more precise value μ=1.72±0.03, resulting from recent numerical calculations by a novel Monte-Carlo method[6].

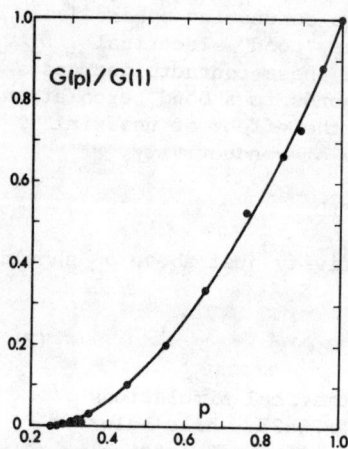

Fig. 4a. Normalized conductance $G(p)/G(1)$ vs. p for microsphere sample. {$G(1) = 1110 \ \Omega^{-1}$; $p_c = 0.251\pm0.002$}

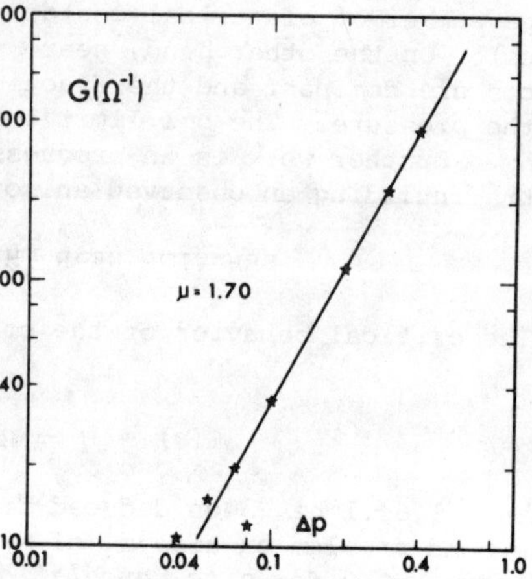

Fig. 4b. Logarithmic plot of conductance $G(p)$ vs. Δp ($=p-p_c$); note power-law behavior with exponent μ=1.70±0.02.

REFERENCES

1. S. Kirkpatrick, Rev. Mod. Phys. 45: 574 (1973)
2. J. Roussenq, J. Clerc, G. Giraud, E. Guyon, H. Ottavi, Journ. de Phys. Lett., 37: L99 (1976)
3. J. Clerc, G. Giraud, J. Roussenq, Compt. Rend. Acad. Sci. Paris 281B: 227 (1975)
4. J. L. Finney, Nature, 266: 309 (1977)
5. D. Adler, L.P. Flora, S.D. Senturia, Sol. St. Comm. 12, 9 (1973)
6. C.D. Mitescu, H. Ottavi, J. Roussenq, AIP Conf. Proc. (see paper immediately following, same conference)

DIFFUSION ON PERCOLATION LATTICES: THE LABYRINTHINE ANT[+]

C. D. Mitescu,[*] H. Ottavi,[**] J. Roussenq
Groupe de Physique, Université de Provence, Centre Saint-Jérôme
13397 Marseille, Cédex 4

ABSTRACT

In a recent conjecture, de Gennes has suggested that the parameters of a random walk on a percolation lattice (an ant in a labyrinth) should disclose information on the transport properties, especially the conductivity, of a percolation system in the critical region. By scaling arguments, we infer a number of linear relations between several exponents of the random walk and the critical exponents - μ, β, ν - of the percolation problem.

We report on Monte-Carlo calculations of such random walks, on *site* lattices above the threshold p_c, in two and three dimensions. We infer $\mu = 1.72 \pm 0.03$ for a simple cubic lattice and $\mu = 0.99 \pm 0.02$ for two-dimensional lattices; while the latter agrees with some recent predictions, we think here the critical region may be masked by size effects.

INTRODUCTION

In a recent discussion[1] of percolation-related phenomena, de Gennes suggested that the meanderings of an ant in a labyrinth - a random walk on a percolation lattice - should yield specific information on critical behavior, most significantly the critical exponent μ which characterizes the conductivity $\sigma(p) \propto |p - p_c|^\mu$.

We have previously reported[2] on a preliminary Monte-Carlo study on a square site lattice, below the threshold p_c, which seemed to lend substance to the original conjectures.

We discuss here more extensive studies, this time above p_c, still on site lattices - square and triangular in two dimensions and simple cubic in three.

The problem is treated in the following manner: The ant "parachutes" onto a randomly selected active site which we will call the *local origin*. It then attempts to move along one of the lattice directions, chosen at random, toward another site, making the step only if that site is active, or trying again another direction if it is blocked. For each successful step, labeled sequentially N, we compute R_N^2, the square of the displacement from the local origin. We then average

---

[+]Work supported in part by D.G.R.S.T. contract #736 PE.
[*]Permanent address: Department of Physics, Pomona College, Claremont, California 91711.
[**]Also: Laboratoire d'électronique, Université de Provence, Centre Saint-Jérôme, 13397 Marseille.

ISSN: 0094-243X/78/377/$1.50 Copyright 1978 American Institute of Physics

R_N^2, for a large number (800-1000) of choices of local origin; and plot the resulting $\langle R_N^2 \rangle$ as a function of N (proportional also to the time of travel of the ant).

RANDOM-WALK PARAMETERS

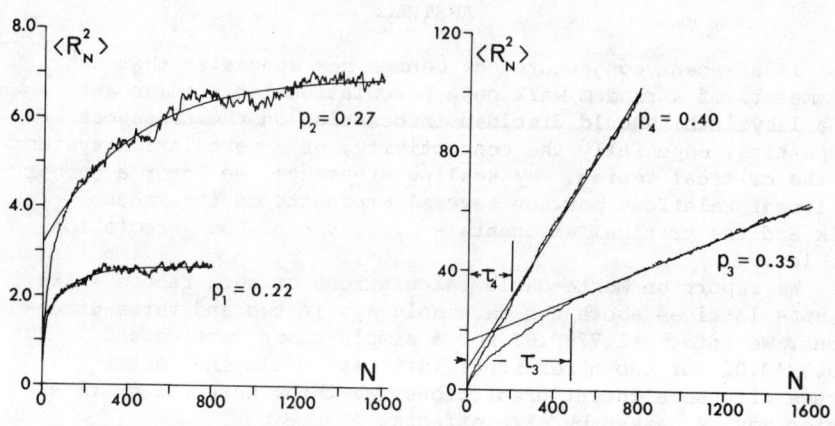

Fig. 1a and 1b. Variation of mean-square displacement $\langle R_N^2 \rangle$ with number of steps N, for several different site probabilities on a simple cubic lattice (p_c=0.312). In Fig. 1a, the smooth curves are least-square fits: $A - B \exp(-N/\theta)$, while in Fig. 1b the diffusion constant D is the slope of the linear region. (Note the different ordinate scales on the two figures!)

Figures 1a and 1b illustrate the observed behavior. For $p < p_c$ (Fig. 1a), we see that after a very narrow transition region $\langle R_N^2 \rangle$ approaches an asymptotic value $\langle R_\infty^2 \rangle$ with a single exponential time-constant, θ; both $\langle R_\infty^2 \rangle$ and θ are observed to increase rapidly as $p \to p_c$. For $p > p_c$ (Fig. 1b), we note a transition region of extent τ before $\langle R_N^2 \rangle$ reaches a typical, asymptotically linear behavior with slope D. This slope (Einstein diffusion constant) is much inferior to its value on a corresponding perfect lattice, and increases rapidly as $|p-p_c|$ increases, while the transition region τ rapidly diminishes as we depart from p_c.

On the basis of universality and dimensional arguments, we have suggested[2] that $\langle R_\infty^2 \rangle$ should scale as ξ^2, the square of the coherence distance $\propto |p-p_c|^{-2\nu}$, θ as $|p_c-p|^{\beta-\mu-2\nu}$, while de Gennes[3] has shown rigorously that D is proportional to the conductivity $\sigma(p)$; he has also suggested that τ might scale as $\xi^2/D \propto |p-p_c|^{-2\nu-\mu}$. A small amount of controversy exists about the first quantity $\langle R_\infty^2 \rangle$ which Stauffer[4] on the basis of scaling arguments has suggested should rather follow $|p-p_c|^{\beta-2\nu}$.

Our calculations have been carried out on a 48x48x48 3D lattice (simple cubic) and on 128x128 2D lattices (square and triangular), all with periodic boundary conditions. We have also a limited number of results, in two dimensions, for 256x256 lattices and are currently extending these to 400x400; in three dimensions, however, we are at the limit of the current capability of our programs. While our data base here is considerably larger than in our original observations, it is still at this stage limited by the fact that each curve of $<R_N^2>$ requires 20-50 minutes of calculation on a PDP 11/45 computer; accordingly, our results must be regarded as somewhat preliminary. A number of features, however, do stand out:

SIZE EFFECTS

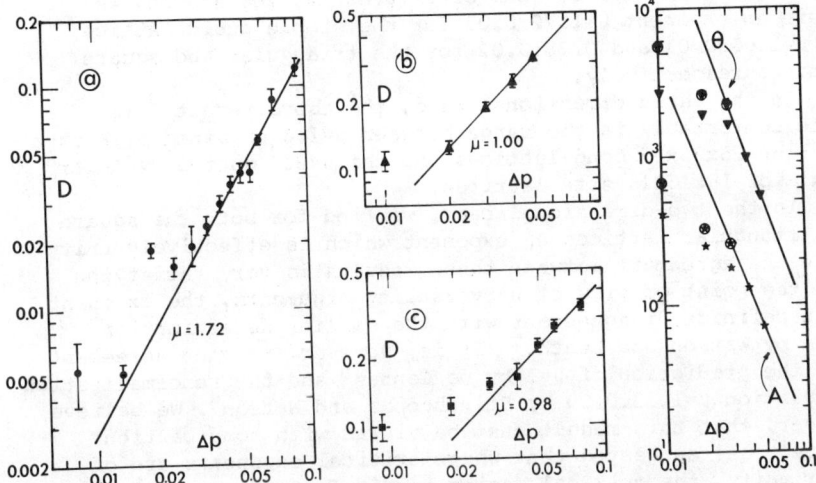

Fig. 2a, 2b, and 2c. Logarithmic plot of the diffusion constant D vs. $\Delta p = |p-p_c|$ for cubic, triangular, and square lattices respectively (p_c=0.312, 0.500, 0.591). The exponents µ are indicated on each figure.

Fig. 3. Variation of θ and A=$<R_\infty^2>$ with Δp on a triangular lattice. Circled data are for 256-wide lattice, remainder for 128.

The variation of the diffusion slope D with $\Delta p=|p-p_c|$ for each of the lattices investigated can be seen on Figures 2a, 2b, and 2c; the error bars correspond to the statistical scatter in the slopes obtained for the same probability, p. In each case, we observe systematic deviations from linear (power-law) behavior as Δp becomes small ($|\Delta p|$<0.035 for our 48x48x48 lattice and <0.025 for the 128x128 lattices). We attribute this deviation to crossover or size effects due to the finite dimensions of our lattices. While we now have too little data to prove this conclusively in the region above p_c, the effect is quite striking for $p<p_c$. We see in Figure 3 a few data points corresponding to the parameters $A(=<R_\infty^2>)$ and θ for triangular lattices 128x128 and 256x256. In each case, for small values of Δp the points for the larger lattice lie nearer a straight-line extrapolation of the dependence at larger values of Δp, than those of the smaller one.

We therefore suggest that a great deal of circumspection should be exercised in the derivation of critical exponents from finite size samples.

CRITICAL EXPONENT OF THE CONDUCTIVITY: μ

From the linear regions of Figures 2a, 2b, and 2c, we obtain an exponent $\mu=1.72\pm0.03$ for the simple cubic lattice, and $\mu=1.00\pm0.01$ and 0.98 ± 0.02 for the triangular and square lattices respectively.

In the three dimensional case, the above result falls quite comfortably in the range between $\mu=1.6$ obtained by Kirkpatrick[5] on 25x25x25 bond lattices and the $\mu=2.0$ quoted by Adler[6] et al for 16x16x16 site lattices.

In the two-dimensional case, we find for both the square and triangular lattices an exponent which is effectively unity. While the agreement between these results is very satisfying from the point of view of universality arguments, the exponent is in definite disagreement with the earlier determination $\mu=1.4$ by Watson and Leath.[7] It is, however, in good agreement with the prediction of $\mu=1$ by de Gennes[8] and the renormalization calculation $\mu=1.13\pm0.09$ by Stinchcombe and Watson[9]. We believe, however, that this result must be viewed with some caution. Pfeuty[10] has suggested that where critical exponents are of order unity, the critical region may in fact be extremely narrow and close to the threshold, before mean-field effects become dominant. We also emphasize that it is precisely the region very near p_c where the size effects discussed earlier become quite dominant. It should finally be noted that the values of D for the two-dimensional lattices become, unlike the 3D case, quite large even reasonably near p_c and "saturation" effects may also come in. The studies on larger lattices appear therefore to be quite crucial.

Because of the still limited data and the large statistical scatter observed in the other parameters: τ, $<R_\infty^2>$ and θ, we prefer to leave the discussion of the critical behavior of these parameters to a forthcoming more comprehensive paper.

ACKNOWLEDGMENTS

We are grateful for stimulating discussions with P.G. de Gennes, P. Pfeuty, D. Stauffer, and E. Guyon, and we thank the latter for much enthusiastic support and advice as well. One of us

(CDM) wishes to thank the Université de Provence for a six-week visiting appointment in the summer of 1977.

REFERENCES

1. P. G. de Gennes, La Recherche 7: 919 (1976)
2. C. D. Mitescu and J. Roussenq, Compt. Rend. Acad. Sci. Paris 283A: 999 (1976)
3. P. G. de Gennes, private communication.
4. D. Stauffer, J. Stat. Phys., (to be published)
5. S. Kirkpatrick, Rev. Mod. Phys. 45: 576 (1973)
6. D. Adler, L. P. Flora, S. D. Senturia, Sol. St. Comm. 12: 9 (1973)
7. B. P. Watson and P. L. Leath, Phys. Rev. B9: 4893 (1974)
8. P. G. de Gennes, Jour. de Phys. Lett., 37: L1 (1976)
9. R. B. Stinchcombe and B. P. Watson, J. Phys. (GB) C9: 3221 (1976)
10. P. Pfeuty, private communication.

AN APPROXIMATE CALCULATION OF THE DIMENSIONALITY DEPENDENCE OF THE RESISTOR LATTICE CONDUCTIVITY EXPONENTS

Paul M. Kogut and Joseph P. Straley
University of Kentucky, Lexington, Ky. 40506

ABSTRACT

Position space renormalization is discussed as a means of studying the resistor lattice conductivity exponents, t and s. We propose a decimation procedure which yields the approximate dimensionality dependence of the conductivity exponents and present the exponents in the context of other dimensionality-dependent theories. A bicritical exponent, v, is defined for the case of a lattice containing conductors, superconductors and insulators and its dimensionality dependence is determined to a low order of approximation. We argue that the exponent v describes unique critical behavior which is probably independent of t and s.

INTRODUCTION

Let us define some quantities of interest. Consider a resistor lattice containing links of conductivity chosen at random via the distribution function

$$P(\sigma) = p_a \delta(\sigma) + p_b \delta(\sigma - b) + p_c \delta(\sigma - \infty) \tag{1a}$$

where p_a, p_b and p_c are respectively the probabilities that a given link has conductivity 0, b or ∞ and $p_a + p_b + p_c = 1$. Associated with this three-phase lattice is a critical value, p^*, such that, if $p = p_b + p_c$, $\epsilon_1 = p - p^*$ and $\epsilon_2 = p^* - p_c$, the bulk conductivity, Σ, obeys:

$$\Sigma = 0 \quad \text{for} \quad \epsilon_1 \leq 0 \tag{1b}$$

$$\Sigma \sim \epsilon_1^t \quad \text{for} \quad \epsilon_1 \text{ small and positive} \tag{1c}$$

$$\Sigma \sim \epsilon_2^{-s} \quad \text{for} \quad \epsilon_2 \text{ small and positive} \tag{1d}$$

$$\Sigma = \infty \quad \text{for} \quad \epsilon_2 \leq 0 \tag{1e}$$

provided the point (p, p_c) is not too close to (p^*, p^*). As seen from fig. 1., if (p, p_c) is close to (p^*, p^*), we can define another exponent, v, through:

$$\Sigma \sim r^v \tag{1f}$$

for r small and positive, where r is the distance in concentration space from the bicritical point defined by $(p, p_c) = (p^*, p^*)$. Finally, for a lattice with links present (or absent) with probability p_1 (or $1 - p_1$), it is possible to define a correlation length expon-

ISSN: 0094-243X/78/382/$1.50 Copyright 1978 American Institute of Physics

ent, ν, which gives the divergence of the mean cluster size, ξ, through

$$\xi \sim (p^* - p_1)^{-\nu} \qquad (1g)$$

for p_1 just below p^*.

The ν, t and s exponents have been determined by many methods, including exact solutions on a limited number of lattices[1,3,4], effective medium theories[2,5], Monte Carlo solutions of Kirchhoff's equations[6,14,15], and position space renormalization techniques.[7-9] The last of these lends itself to a systematic incorporation of the variable of dimensionality, D, which avoids the need for vast amounts of computer space and time, as we will show. The underlying theory of position space renormalization, as applied to the resistor lattice is amply discussed elsewhere.[7-9]

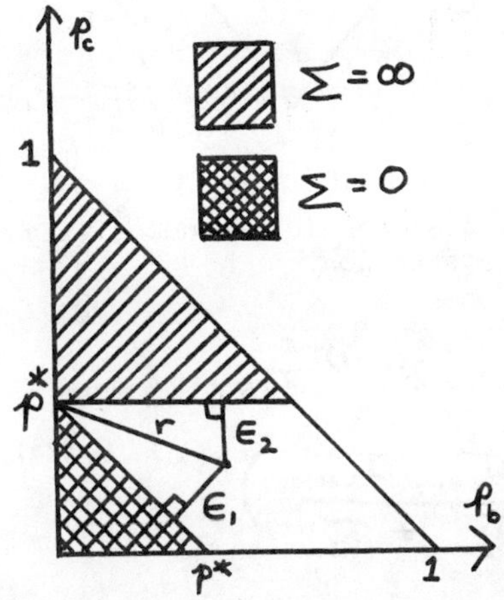

DETAILS OF THE METHOD

We introduce a D-dimensional exponent theory by considering a 3D simple cubic (SC) lattice. Decimation by removal of every other site of a 3D SC lattice generates a face-centered cubic (FCC) lattice. We therefore perform a second decimation, removal of the facial sites of the FCC lattice, and thus generate a new SC lattice whose lattice constant is a factor B=2 larger than the original SC lattice. Fig. 2. shows the unit cell of this doubly-decimated lattice. Paths like EG and GF arise from the first decimation. The second yields 4 paths like EGF, each of which consists of two conductors in series. A graphical representation of the conductivity of any lattice link after one (or two) decimations(s) is given in fig. 3., with N=2 (or N=4). A similar double decimation has recently been applied to 3D spin glasses.[18] This procedure neglects all but two-step conducting paths at each step of the decimation process and neglects correlations among links in each decimated cell and those of neighboring cells.

Fig. 1. Phase diagram showing ϵ_1, ϵ_2 and r in the concentration plane.

The extension to D dimensions is accomplished by considering the D-dimensional analogue of fig. 2. Instead of 4 conducting paths equivalent to EGF, there are N=2(D-1) such paths. And fig. 3 gives a graphical representation of the conductivity of any lattice link after one (or two) decimation(s) when N=2 (or N=2(D-1)).

Suppose $P^{(0)}(\sigma)$ is the conductivity distribution on an undecimated D-dimensional SC lattice. Then fig. 3 and the procedure outlined above imply a transformed distribution $P^{(2)}(\sigma)$, where

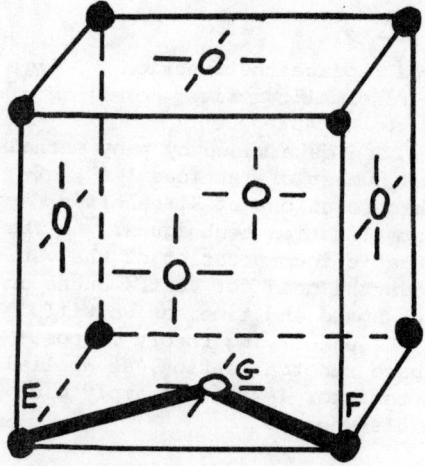

Fig. 2. Unit cell after two decimations. Open circles are clipped sites.

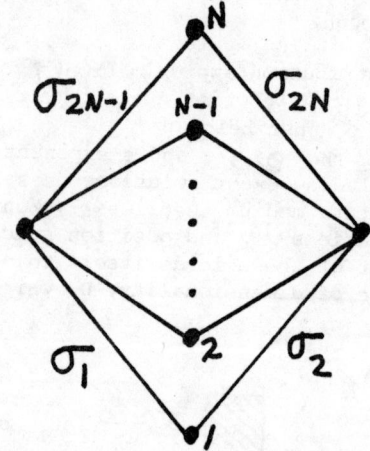

Fig. 3. Graphical conductivity representation.

$$\mathcal{P}^{(i)}(\sigma) = \int d\sigma_1 \cdots d\sigma_{2N}\, \mathcal{P}^{(i-1)}(\sigma_1) \cdots \mathcal{P}^{(i-1)}(\sigma_{2N}) \\ \times \delta\left(\sigma - \frac{\sigma_1 \sigma_2}{\sigma_1 + \sigma_2} - \cdots - \frac{\sigma_{2N-1}\sigma_{2N}}{\sigma_{2N-1}+\sigma_{2N}}\right) \quad (2a)$$

where i=1 implies N=2 and i=2 implies N=2(D-1). If $\mathcal{P}^{(i-1)}(\sigma)$ has the form of equ. (1a),

$$\mathcal{P}^{(i-1)}(\sigma) = p_a^{(i-1)} \delta(\sigma) + p_b^{(i-1)} \delta(\sigma - b^{(i-1)}) + p_c^{(i-1)} \delta(\sigma - \infty) \quad (2b)$$

then direct substitution into equ. (2a) yields

$$\mathcal{P}^{(i)}(\sigma) = p_a^{(i)} \delta(\sigma) + p_b^{(i)} f^{(i)}(\sigma, b^{(i-1)}) + p_c^{(i)} \delta(\sigma - \infty) \quad (3a)$$

where

$$p_b^{(i)} f^{(i)}(\sigma, b^{(i-1)}) = \sum_{I=1}^{N} \binom{N}{I} \left(1 - (p^{(i-1)})^2\right)^{N-I} \\ \times \sum_{K=0}^{I} \binom{I}{K} (2 p_c^{(i-1)})^K (p_b^{(i-1)})^{2I-K} \delta\left(\sigma - \frac{(K+I) b^{(i-1)}}{2}\right) \quad (3b)$$

$$p_c^{(i)} = 1 - \left(1 - (p_c^{(i-1)})^2\right)^N \equiv T_N(p_c^{(i-1)}) \quad (3c)$$

$$p^{(i)} = T_N(p^{(i-1)}) \; ; \; p^{(i-1)} = p_b^{(i-1)} + p_c^{(i-1)} \quad (3d)$$

$$p_a^{(i)} = 1 - p^{(i)} \tag{3e}$$

Simplification results if we make what we shall call the "distorted distribution approximation" (DDA). We replace the function $f^{(i)}$ by a single δ-function centered about the renormalized first moment of the function $f^{(i)}$. That is, make the following replacement:

$$f^{(i)}(\sigma, b^{(i-1)}) \longrightarrow \delta(\sigma - b^{(i)}) \tag{4a}$$

where
$$b^{(i)} = \frac{\int \sigma f^{(i)}(\sigma, b^{(i-1)}) d\sigma}{(N/2)} \tag{4b}$$

which may be calculated from equ. (3b) if we define

$$p_b^{(i)} = p^{(i)} - p_c^{(i)} \tag{4c}$$

Equ. (4c) renormalizes the sum of the generated concentrations to unity. The factor N/2 in equ. (4b) serves to force the doubly decimated link conductivities to the value $b^{(0)}$ in the case of an initial conductivity distribution with $p_a = p_c = 0$ and $p_b = 1$. Thus equ. (3a) has the same structure as equ. (2b) and repetition of the transformation proceeds without difficulty, particularly because equs. (3) and (4) lead to the exact relation

$$\frac{b^{(i)}}{b^{(i-1)}} = \frac{(4 p_c^{(i-1)} p_b^{(i-1)} + (p_b^{(i-1)})^2)(1 - (p_c^{(i-1)})^2)^{N-1}}{p_b^{(i)}} \tag{5}$$

Critical points, p_{SC}^* and p_{FCC}^*, respectively for the SC and FCC lattices, are given by the fixed point equations

$$p_{SC}^* = T_N(T_2(p_{SC}^*)) \tag{6a}$$

$$p_{FCC}^* = T_2(T_N(p_{FCC}^*)) \tag{6b}$$

which possess the interesting properties $T_2(p_{SC}^*) = p_{FCC}^*$ and $T_N(p_{FCC}^*) = p_{SC}^*$. From equ. (5), we can compute the ratio $b^{(2)}/b^{(0)}$ as a function, say $\gamma(p_a(0), p_b(0), p_c(0))$, of the initial concentrations. Define

$$\lambda_\varrho = d/dp \, (T_N(T_2(p)))\big|_{p=p_{SC}^*} \tag{7a}$$

$$\lambda_t = \gamma(1-p_{SC}^*, p_{SC}^*, 0) = p_{SC}^* p_{FCC}^* \tag{7b}$$

$$\lambda_s = \gamma(0, 1-p_{SC}^*, p_{SC}^*) = \frac{(1+3p_{SC}^*)(1+3p_{FCC}^*)}{(1+p_{SC}^*)(1+p_{FCC}^*)} \tag{7c}$$

$$\lambda_v = \gamma(1-p_{SC}^*, 0, p_{SC}^*) = \frac{16 p_{SC}^* p_{FCC}^* (1-p_{SC}^{*2})(1-p_{SC}^*)}{\lambda_\varrho (1-p_{FCC}^{*2})} \tag{7d}$$

where we have used equ. (5). Then if B is the ratio of the lattice constant after decimation to that before, the homogeneous function representation of the conductivity distribution function[8,9] implies: (i) $\lambda_\nu = B^{1/\nu}$, (ii) $\lambda_t = B^{-t/\nu}$ and (iii) $\lambda_s = B^{s/\nu}$. If, in addition, ν is independent of t and s and the homogeneity assumption valid near the bicritical point defined by $(p^{(0)}, p_c^{(0)}) = (p_{SC}^*, p_{SC}^*)$, we have: (iv) $\lambda_v = B^{-v/\nu}$.

RESULTS AND DISCUSSION

The values determined by the method of the previous section, and others from the literature for comparison, are given in table I. Several positive statements can be made.[16] (i) By reversing the order of the two steps of our decimation procedure, we find that the theory predicts the same exponents for a FCC lattice as for a SC lattice, in

Table I: Critical points and exponents

D	p_{SC}^*	p_{FCC}^*	ν	ν_{lit}	t	t_{lit}	s	s_{lit}	v
2	.618[a]	.618[b]	.818[a]	1.34[c]	1.14	1.1[f]	1.34	1.1[f]	0.00
3	.459	.378	.683	.82[c]	1.73	1.7[g]	.912	.7[f]	.682
4	.390	.281	.639	.66[d]	2.04		.747		1.01
5	.349	.228	.617	.52[d]	2.25		.653		1.23
6	.320	.195	.602	.5[d,e]	2.41	3[h,e]	.589	0[h,e]	1.40
7	.299	.171	.592	.5[d,e]	2.54	3[h,e]	.543	0[h,e]	1.53

a: first calculated in ref. 7.
b: The FCC critical point is artificial in 2 dimensions since a single decimation generates a square lattice.
c: ref. 10.
d: ref. 11.
e: Literature values are assumed to take on the mean field (Bethe lattice) values for D $\geq$ 6.
f: ref. 6.
g: ref. 12; also see refs. 6, 15 and 17.
h: ref. 4.

agreement with the universality principle. (ii) We can reproduce the table I exponents by numerically studying the values to which the ratio $b^{(2)}/b^{(0)}$ converges, after many applications of equ. (5), for various choices of $(p_a^{(0)}, p_b^{(0)}, p_c^{(0)})$ in the appropriate critical regime. This verifies the homogeneity assumptions and indicates therefore that ν is independent of t and s. (iii) To check the error introduced by the DDA, we have determined t and s by a

Monte Carlo simulation of the renormalized cell and have found the DDA t and s values to lie within the Monte Carlo (1000 samples) error bars. (iv) We have also determined t and s by a Monte Carlo method which includes N extra links which connect each of the central nodes marked "1" to "N" in fig. 3 to the adjacent pair of central nodes. This led to a slightly depressed value of t and a slightly elevated value of s. The former is consistent with the findings of Stinchcombe and Watson[8], who also studied the effects of inclusion of higher order conducting paths.

On the other hand, our critical point estimates our all too high (although we note that they do obey the rigorous inequality[13] $p_{SC} \geq 1/(2D-1)$). For the 2D "simple cubic" lattice, p_{SC}^* is exactly .5 and, in 3 dimensions, it is well established[13,17] that $p_{SC}^* = .25$ and $p_{FCC}^* = .12$. In addition, Kirkpatrick has noted[11] that one may expect ν, t and s to take on the mean field values for all $D \geq 6$, a condition our exponents fail to obey. In the infinite-dimensional limit, we find $\nu = .5$, $t = \infty$, $s = 0$ and $v = \infty$. Since the mean field Bethe lattice model yields $\nu = .5$, $t = 3$ and $s = 0$, we find that our exponents give the correct infinite dimensional limits for ν and s but not for t and v.

ACKNOWLEDGMENT

We wish to thank the US National Science Foundation for supporting this research.

REFERENCES

1. R. B. Stinchcombe, J. Phys. C 7, 179 (1974).
2. S. Kirkpatrick, Phys. Rev. Lett. 27, 1722 (1971).
3. J. P. Straley, J. Phys. C 9, 783 (1976).
4. J. P. Straley, J. Phys. C 10, 3009 (1977).
5. R. Landauer, J. Appl. Phys. 23, 779 (1952).
6. J. P. Straley, Phys. Rev. B 15, 5733 (1977).
7. A. P. Young and R. B. Stinchcombe, J. Phys. C 8, L535 (1975).
8. R. B. Stinchcombe and B. P. Watson, J. Phys. C 9, 3221 (1976).
9. J. P. Straley, J. Phys. C 10, 1903 (1977).
10. A. G. Dunn, J. W. Essam and D. S. Ritchie, J. Phys. C 8, 4219 (1975).
11. S. Kirkpatrick, Phys. Rev. Lett. 36, 69 (1976).
12. K. Onizuka, J. Phys. Soc. Jap. 39, 527 (1975).
13. V. K. Shante and S. Kirkpatrick, Adv. Phys. 20, 325 (1971).
14. G. E. Pike and C. H. Seager, Phys. Rev. B 10, 1421 (1974).
15. S. Kirkpatrick, Rev. Mod. Phys. 45, 574 (1973).
16. P. M. Kogut, unpublished calculations.
17. I. Webman, J. Jortner and M. H. Cohen, Phys. Rev. B 11, 2885 (1975).
18. B. W. Southern and A. P. Young, J. Phys. C 10, 2179 (1977).

RENORMALIZATION GROUP APPROACH FOR THE CONDUCTIVITY OF RANDOM CONDUCTANCE LATTICES

J. Bernasconi
Physics Dept., University of California, Los Angeles, Calif. 90024
and Brown Boveri Research Center, Baden, Switzerland

ABSTRACT

A new real-space renormalization group approach for the conductivity of random conductance lattices is presented. It can be regarded as an extension of the bond-probability renormalization recently proposed by Reynolds, Klein and Stanley.[1] We apply our transformation to the 2-dim square and the 3-dim simple cubic lattices with a binary distribution of conductances, $\rho(\sigma)=p\delta(\sigma-\sigma_1)+(1-p)\delta(\sigma-\sigma_2)$. It is shown that our approach is not only applicable near critical points, but represents a very good approximation to the lattice conductivity $\sigma(p)$ for all values of p and σ_2/σ_1. In particular, it produces the exact slopes of $\sigma(p)$ at p=0 and p=1, and in two dimensions satisfies the selfdual symmetry of the square lattice.

For the percolation conductivity (case $\sigma_1=1$, $\sigma_2=0$) we propose a new approximate solution to the renormalization relations which is very accurate for all values of p, where $p_c \leq p < 1$. The conductivity exponents t and s at the percolation threshold are estimated to take the values t=s≈1.33 in 2 dimensions and t≈2.14, s≈0.76 in 3 dimensions. Apart from the 3-dimensional s-exponent, our values are appreciably larger than the best numerical estimates (t=s≈1.1+0.05 in 2 dimensions and t≈1.7 + 0.05, s≈0.7+0.05 in 3 dimensions). Based on our renormalization group analysis we shall, however, present some arguments that the numerical approaches might underestimate the true exponents in certain cases.

[1] P.J. Reynolds, W. Klein and H. E. Stanley, J. Phys. C10, L167 (1977).

ISSN: 0094-243X/78/388/$1.50 Copyright 1978 American Institute of Physics

represent a magnetic field, the temperature, the pressure etc.

The system will be specified by a potential $\Phi(\underline{r}, z(\underline{r}), t)$. The order parameter $z(\underline{r})$ is determined by the equilibrium condition

$$\delta \Phi = 0 \tag{1}$$

where δ is the variation with respect to z.

In general Eq.(1) has an infinite number of solutions. A stable solution for fixed t must satisfy the condition

$$\delta^2 \Phi > 0 . \tag{2}$$

In general there are many stable solutions and the state of the system depends on its previous history.

If the system becomes unstable by changing t, it is postulated that it moves in the direction with

$$\delta^3 \Phi = \text{minimum} \tag{3}$$

THE MODEL

To be more specific we use the form

$$\Phi = \int \phi_o (\underline{r}, z(\underline{r}), t) d^3\underline{r} + \Phi_{int} . \tag{4}$$

The inhomogeneity of the system is represented by the dependence of $\phi_o(\underline{r}, z(\underline{r}), t)$ on $\underline{r}$.

In the following the interaction term is represented by the molecular field approximation

$$\Phi_{int}^{MF} = \frac{\gamma}{2} \int d^3\underline{r} \ (z(\underline{r}) - \bar{z})^2 \tag{5}$$

where

$$\bar{z} = \frac{1}{V} \int z(\underline{r}) d^3 r \tag{6}$$

is the space averaged order parameter.

The ansatz (5) introduces in a simple way the coupling of the order parameter z of different regions of the system. For magnetic systems γ is related to the Blochwall energy, for a metal-insulator-transition γ may be related to the elastic energy caused by locally different volume expansions.

Details of the relationship between the ansatz (5) and a particular system will not be discussed in this paper. In the following we will derive an equation for the hysteresis based on the model represented by Eqs.(4), (5) and (6).

THE HYSTERESIS OF AN INHOMOGENEOUS SYSTEM

J. BERNASCONI*
Dept. of Physics, University of California, Los Angeles, Ca

S. STRAESSLER and H.J. WIESMANN
Brown Boveri Research Center, CH-5405 Baden, Switzerlan

ABSTRACT

A simple model is introduced to study the hysteresis of a phase transition in an inhomogeneous system e.g. in a composite material. The inhomogeneity is described by a thermodynamic poten tial which depends on some space dependent order parameter and is also an explicite function of the space vector $\underline{r}$. The coupling between different regions is approximated by a molecular field ansatz. In particular we study the influence of the coupling strength and the degree of inhomogeneity on the width and spread of the hysteresis.

INTRODUCTION

All first order phase transitions show hysteresis effects. The reason is that there exist potential barriers against nucleation and growth of one phase in another phase.

Well known examples are: crystallisation, condensation, marten sitic transformations, metal insulator transitions, the magnetic hysteresis etc. In all cases the hysteresis is determined by the inhomogeneity of the system. This is particularly true in the case of crystal anisotropy where micromagnetics predicts the existence of a coercive force for a homogeneous system, which is in complete disagreement with experiments[1]. This discrepancy is usually referred to as "Brown's paradox". To resolve this paradox the stability criteria must be solved for an inhomogeneous magnet, a problem which is extremely difficult for realistic models[2]. In this paper we want to present a general solution to the problem of the hysteresis in an inhomogeneous system within a molecular field approximation. The justification of this approximation for specific applications will not be discussed in this paper.

We want to formulate the problem in terms of a space dependent scalar order parameter $z(\underline{r})$. This order parameter may be the angle of the magnetisation, the distortion in the case of a martensitic transformation, the density change in the case of a metal insulator transition, crystallisation or condensation. The hysteresis is then characterised by the behavior of the order parameter as a function of some external parameter t. Depending on the situation, t may

CONSTRUCTION OF THE HYSTERESIS EQUATION

The equilibrium condition (1) for the system represented by Eqs.(4), (5) and (6) leads to

$$\frac{\partial \phi_o}{\partial z} - \gamma(z(\underline{r})-\bar{z}) = 0 \ . \tag{7}$$

This equation must be solved simultaneously with Eq.(6). We assume that ϕ_o is such that Eq.(7) has for fixed $\bar{z}$ at most three real solutions

$$z_{-1} < z_o < z_{+1} \ .$$

Because of the particular form of ϕ_{int} it will turn out that only $z_\nu(\underline{r})$ with $\nu = \pm 1$ are stable. The system can be specified by a function $\sigma(\underline{r},t)$ where

$$\sigma(\underline{r},t) = \pm 1 \quad \text{if} \quad z(\underline{r}) = z_{\pm 1} \ . \tag{8}$$

The stability condition (2) for the particular choice of $\sigma(\underline{r},t)$ can be written in the form

$$1 - \gamma \int \frac{d^3 r}{F(\underline{r},z(\underline{r}),t)} > 0 \tag{9a}$$

if

$$F = \frac{\partial^2 \phi_o}{\partial z^2} - \gamma > 0 \ . \tag{10}$$

If the condition (10) is not satisfied for all values of $\underline{r}$ (9a) must be replaced by the condition

$$F > 0 \ . \tag{9b}$$

Let us suppose that for sufficiently low values of t the system is in the phase $\sigma(\underline{r}) \equiv -1$. If the value of t is increased we will reach a point where $\delta^2 \Phi > 0$ is no longer satisfied. The equation for this critical value of t is then given by equating the left hand side of Eq.(9) equal to zero. If the value of t is further increased, those values of $\sigma(\underline{r})$ must be changed which according to Eq.(3) give

$$\frac{\partial^3 \phi_o}{\partial z^3} = \text{minimum} \ . \tag{11}$$

To represent the basic structure of the hysteresis it is convenient to introduce the global order parameter

$$\rho(t) = \frac{1}{V} \int \sigma(\underline{r},t) d^3\underline{r} . \qquad (12)$$

AN APPLICATION

The potential ϕ_o is expanded in powers of the order parameter:

$$\phi_o = \sum_{n=1}^{N} a_n(\underline{r},t) z^n(\underline{r}) . \qquad (13)$$

In order to be able to describe a first order transition we must choose $N \geq 4$. For convenience we take

$$a_4 = \frac{1}{4}, \; a_3 = 0, \; a_2 = \frac{3}{2}, \; a_1 = t-p(\underline{r}) . \qquad (14)$$

The space dependent parameter $p(\underline{r})$ represents the inhomogeneity of the system. Its probability distribution $w(p)$ is assumed to be given by

$$w(p) = \frac{1}{P} \left| \theta(p + \frac{P}{2}) - \theta(p - \frac{P}{2}) \right| . \qquad (15)$$

Without interaction term ($\gamma = 0$) Eqs.(7) and (9b) lead to the following solution for $\rho(t)$:
For increasing values of t

$$\rho_+(t) = \begin{cases} -1 & t < 2 - \frac{P}{2} \\ \frac{2}{P}(t-2) & 2 - \frac{P}{2} < t < 2 + \frac{P}{2} \\ 1 & t > 2 + \frac{P}{2} \end{cases} \qquad (16)$$

For decreasing values of t

$$\rho_-(t) = \rho_+(t+4) . \qquad (17)$$

For $\gamma \neq 0$ it is possible to represent the hysteresis in the region $-1 < \rho < +1$ in the form

$$t = t(\rho) . \qquad (18)$$

For values

$$\gamma < \Gamma(P) \qquad (19)$$

where Γ is given by Eq.(21),

Eqs. (7) and (9b) lead to

$$t = \bar{t}(\rho) - \gamma \bar{z}(\rho) \qquad (20)$$

$$\bar{t}(\rho) = \frac{\rho P}{2} + 2(1 - \frac{\gamma}{3})^{3/2}$$

$$\bar{z} = \frac{3}{P}(1 - \frac{\gamma}{3})^2 \left| \frac{1}{4}(f^4(\zeta_1) - g^4(\zeta_2)) - \frac{1}{2}(f^2(\zeta_1) - g^2(\zeta_2)) - \frac{9}{4} \right|$$

$$f(\zeta) = (|\zeta| + (\zeta^2-1)^{1/2})^{1/3} + (|\zeta| - (\zeta^2-1)^{1/2})^{1/3}$$

$$g(\zeta) = \begin{cases} -f(\zeta) & \zeta < -1 \\ 2\cos\alpha & -1 < \zeta < 0 \\ -\cos\frac{1}{3}\alpha - \sqrt{3}\sin\frac{1}{3}\alpha & \zeta > 0 \end{cases}$$

$$\zeta_{\frac{1}{2}} = \frac{1}{2} \frac{\pm P/2 + \bar{t}(\rho)}{(1-\frac{\gamma}{3})^{3/2}}$$

$$\alpha = \arctan\sqrt{\frac{1}{\zeta_2^2} - 1}$$

$\Gamma(P)$ is given by

$$\Gamma(P) = \frac{6(\eta-1)}{4 + 2\eta + 3g(\eta)} \qquad (21)$$

where

$$\eta = 1 - \frac{P}{2} \frac{1}{(1-\frac{\gamma}{3})^{3/2}}.$$

The hysteresis for 3 pairs of values (P,γ) is given in Fig.1. We find from Eq.(20) that the hysteresis of the inhomogeneous system is reduced to a step function hysteresis for $\gamma > 0.3P$. This value of the interaction strength γ also corresponds to maximum reduction of the width h of the hysteresis with respect to γ. In Fig.2 we plot h as a function of P for this case.

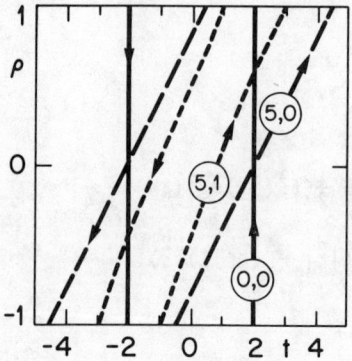 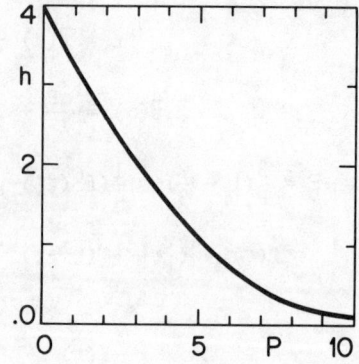

Fig. 1. Hysteresis for different values of (P,γ).

Fig. 2. Width h of the hysteresis as a function of P for γ = .3P.

SUMMARY

We have constructed a procedure represented by Eqs.(7), (9) and (11) to find the hysteresis of a general class of models represented by Eqs.(4), (5) and (6). In particular we have found an analytical solution (Eq.20)) for the hysteresis equation of the model defined by Eqs.(13), (14) and (15).

REFERENCES

1. W.F. Brown, Jr., Revs. Modern Phys. 17, 15 (1945).
2. For a review see e.g. J.D.Livingston, Proc.1972 Conf. on Magnetism and Magnetic Materials, A.I.P. Conf.Proc. 10, 643 (1973), and A. Aharoni, Rev.Mod.Phys. 34, 227 (1962).

---

*Permanent address: Brown Boveri Research Center, CH-5405 Baden, Switzerland.

ELECTRODE EFFECTS ON GEOMETRICAL MAGNETORESISTANCE

J. B. Sampsell and J. C. Garland
The Ohio State University, Columbus, Ohio 43210

ABSTRACT

Analytic solutions for the potential distribution of conducting flat plates are given for several electrode configurations. The solutions are believed to be asymptotically correct in the limit of an infinitely strong magnetic field. Transverse magnetoresistance data are shown for indium samples in order to illustrate the practical consequences of different electrode geometries.

Spurious measurement of the Hall Effect and magnetoresistance of metals and semi-conductors can be easily induced by a non-uniform current distribution in the specimen of interest.[1] In practice, current distortion effects can arise from many sources, including voids, strain fields, lattice defects, and surface imperfections. The problem of current distortions induced by such volume inhomogenieties has been considered elsewhere by the authors;[2] in this note we are specifically concerned with electrode and geometry effects pertaining to flat plate conductors in strong magnetic fields. This problem has been discussed recently by Jensen and Smith,[3] who obtained approximate solutions using the method of images. Their work showed the extent to which geometrical magnetoresistance in flat plate conductors can be considered as an end effect. The purpose of this note is to point out that the boundary conditions appropriate to this problem make it possible to obtain simple analytic expressions for the electrostatic potential in the limit $H \to \infty$.

We consider a thin conducting plate, presumed to lie in the x-y plane, which has a free electron conductivity and which carries a current $\vec{J}$ given by $\vec{J} = \sigma (-\vec{\nabla}\Phi)$. The non-zero elements of σ are $\sigma_{xx} = \sigma_{yy} = \sigma_0 \gamma$, $\sigma_{xy} = -\sigma_{yx} = \sigma_0 \gamma \beta$, and $\sigma_{zz} = \sigma_0$, where σ_0 is the zero field conductivity, $\beta = \omega_c \tau$ is the effective magnetic field oriented in the z-direction, and $\gamma = (1 + \beta^2)^{-1}$. In a strong magnetic field $\beta \gg 1$, the Hall angle in the medium approaches $\pi/2$, and the current density $\vec{J}(\vec{r})$ becomes directed along equipotential lines. We consider the perimeter of the sample to be divided into four regions, two consisting of those sections bounded by perfectly conducting current electrodes, and the remaining two consisting of the unbounded regions. In the high field limit the electrostatic potential along each of these regions is uniform. It is easily shown, in fact, that each electrode must be at the same potential as one of the two adjacent unbounded regions. However, the particular region to be paired with a given electrode depends on the direction of the magnetic field; reversing the field direction reverses the pairing. In this situation the electrostatic potential in the limit $\beta \to \infty$ may be expressed as an elementary solution to Laplace's equation. For a rectangular sample of length L and width w this solution is given by:

$$\Phi = \sum_{k=1}^{\infty} A_k^1 \sinh \frac{k\pi x}{w} + A_k^2 \sinh \frac{k\pi(x-\ell)}{w} \sin \frac{k\pi y}{w}$$

$$+ \sum_{m=1}^{\infty} A_m^1 \sinh \frac{m\pi y}{\ell} + A_m^2 \sinh \frac{m\pi(y-w)}{\ell} \sin \frac{m\pi x}{\ell} \quad (1)$$

where the constants are chosen to match the piecewise uniform potential along each of the four sides of the sample. For the geometry shown at the top of Figure 1, the constants are:

$$A_k^1 = A_k^2 = \begin{cases} \dfrac{-2V}{k\pi \sinh \frac{k\pi \ell}{w}} & k \text{ odd} \\ 0 & k \text{ even} \end{cases}$$

$$A_m^1 = A_m^2 = \begin{cases} \dfrac{2V}{m\pi \sinh \frac{m\pi w}{\ell}} & m \text{ odd} \\ 0 & m \text{ even} \end{cases} \quad (2)$$

for one direction of $\vec{H}$ and:

$$A_k^1 = A_k^2 = \begin{cases} \dfrac{-4V}{k\pi \sinh \frac{m\pi \ell}{w}} & k = 2, 6, 10 \\ 0 & \text{otherwise} \end{cases}$$

$$A_m^1 = A_m^2 = \begin{cases} \dfrac{-2V}{m\pi \sinh \frac{m\pi w}{\ell}} & m \text{ odd} \\ 0 & m \text{ even} \end{cases} \quad (3)$$

for the other direction of $\vec{H}$. In equations 2 and 3, V is taken to be the difference in potential between the current electrodes. The calculated equipotential lines are shown in Figure 1, and it should be noted that the potential distribution is strongly dependent upon the direction of the magnetic field.

Consider now the practical consequences of this current electrode arrangement on a conventional transverse magnetoresistance measurement, where the current through the specimen is fixed and potential differences are measured between two voltage electrodes (shown in Figure 1 as small arrows). For one magnetic field direction, there is a large component of electric field projected along

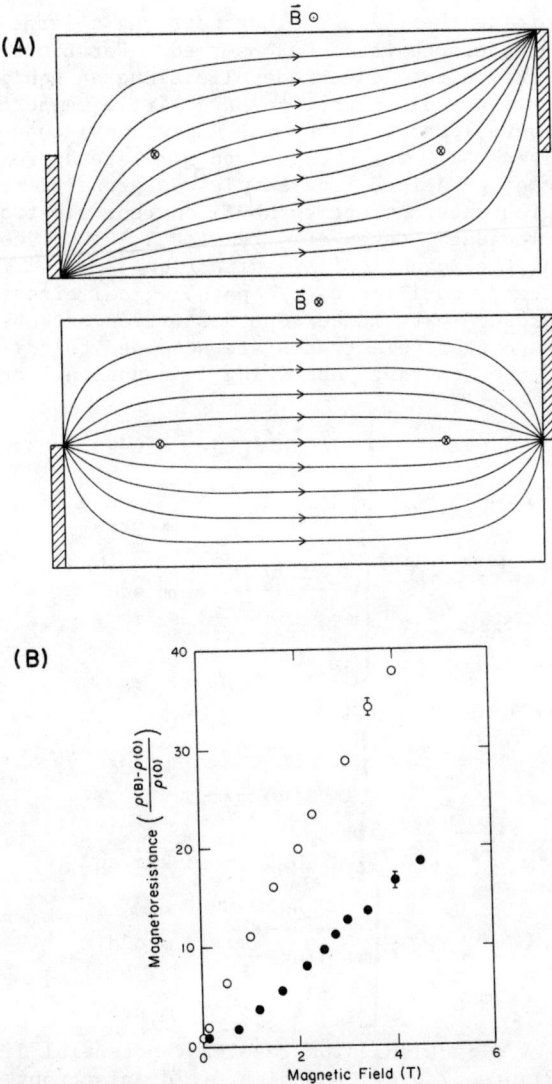

Fig. 1 (a) Diagram of equipotential lines in a thin plate sample for two directions of $\vec{H}$ in the limit $\beta \to \infty$. High conductivity current electrodes are shown as hashed areas. The scaling between equipotentials is V/10. (b) Magnetoresistance data for two directions of $\vec{H}$ in a thin indium sample for $\beta \approx 100$. Superconducting current electrodes were attached to approximate the geometry of a.

the line joining the voltage electrodes and a large (spurious) linear magnetoresistance would be observed. For the reverse field direction, the voltage electrodes lie along an equipotential line and the specimen would appear to have little magnetoresistance. The graph of Figure 1 shows transverse magnetoresistance data obtained for a specimen which was designed to simulate approximately this electrode configuration. The sample was prepared from an indium plate in which superconducting Nb-Ti current electrodes had been embedded. The data, taken at 4.2K with $\beta_{max} \approx 100$, show clearly the asymmetry of the magnetic field direction.

In Figure 2, we illustrate a pathological electrode geometry chosen to produce extreme current distortion effects. For this case, in which the current electrodes are attached to adjacent sides of the sample, the constants appearing in equation 1 are given by:

$$A_k^1 = A_k^2 = \begin{cases} \dfrac{-2V}{k\pi \sinh \dfrac{k\pi w}{w}} & k \text{ odd} \\ 0 & k \text{ even} \end{cases}$$

$$A_m^1 = -A_m^2 = \begin{cases} \dfrac{-2V}{m\pi \sinh \dfrac{m\pi w}{\ell}} & m \text{ odd} \\ 0 & \end{cases}$$

(4)

for one direction of $\vec{H}$ and

$$A_k^1 = A_k^2 = \begin{cases} \dfrac{-2V}{k\pi \sinh \dfrac{k\pi \ell}{w}} & k \text{ odd} \\ 0 & k \text{ even} \end{cases}$$

$$A_m^1 = A_m^2 = \begin{cases} \dfrac{2V}{m\pi \sinh \dfrac{m\pi w}{\ell}} & m \text{ odd} \\ 0 & m \text{ even} \end{cases}$$

(5)

for the other direction. The resultant potential distributions (shown in Figure 2) predict a high field anisotropy comparable to that of the previous example. At lower fields, however, the field dependent distribution of current flow leads to an apparent negative magnetoresistance for one magnetic field orientation. This effect is shown in the graph of Figure 2 for an indium sample with adjoining edges bounded by superconducting electrodes.

To summarize therefore, spurious magnetoresistance effects can be avoided only if the distribution of current flow in the sample does not depend on the strength of the magnetic field. Highly conducting current and voltage electrodes can induce particularly severe current distortion because such electrodes inhibit the natural rotation of equipotential lines in the specimen as the strength of

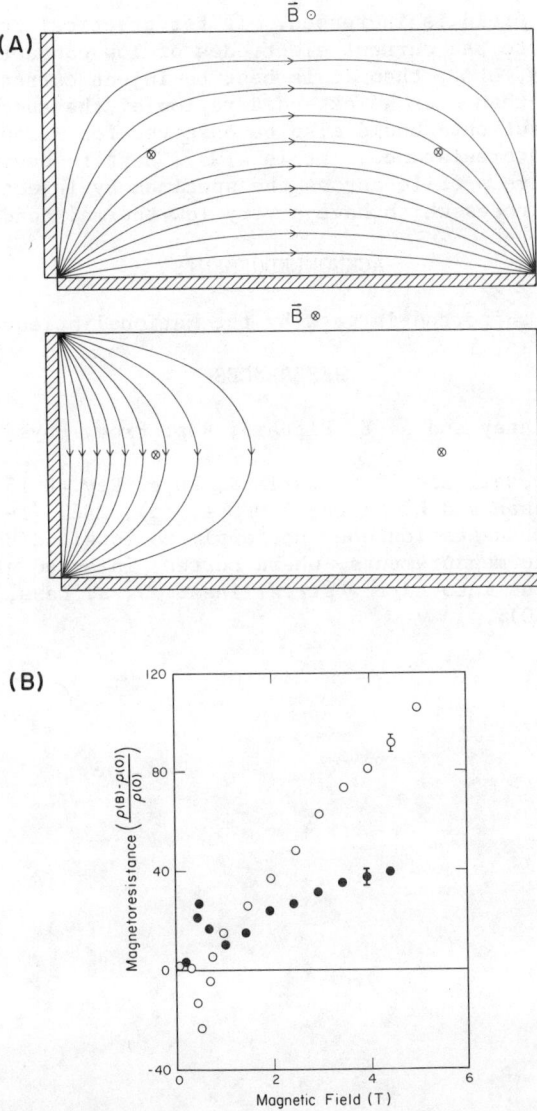

Fig. 2 (a) Diagram of equipotential lines in a thin plate sample for directions of $\vec{H}$ in the limit $\beta \to \infty$. High conductivity current electrodes are shown as hashed areas. The scaling between equipotentials is $V/10$. (b) Magnetoresistance data for two directions of $\vec{H}$ in a thin indium sample for $\beta \approx 100$. Superconducting current electrodes were attached to approximate the geometry of a.

the magnetic field is increased. If for practical reasons it is not possible to use current electrodes of low conductivity, e.g., brass, carbon, etc., then it is best to inject current at a single point rather than over an extended region of the specimen boundary.[4] Similar precautions should also be observed for measurement of the thermal magnetoresistance; it is always best to avoid distorting the temperature profile across the specimen by injecting heat through electrodes which have a very low thermal conductivity.

ACKNOWLEDGEMENTS

Research supported in part by the National Science Foundation.

REFERENCES

1. J. A. Delaney and A. B. Pippard, Rep. Prog. Phys. $\underline{35}$, 677 (1972).
2. J. B. Sampsell and J. C. Garland, Phys. Rev. B $\underline{13}$ 583 (1976).
3. H. H. Jensen and H. Smith, J. Phys. ($\underline{5}$), 2867 (1972).
4. This last suggestion does not apply to longitudinal magnetoresistance measurements, where current injected at a point would focus into narrow streamlines: J. S. Lass, J. Phys. ($\underline{3}$), 1926 (1970).

EFFECT OF INHOMOGENEITIES ON THE GALVANOMAGNETIC PROPERTIES OF PURE ALUMINUM

J. B. Sampsell and J. C. Garland
The Ohio State Univ., Columbus, Ohio 43210

ABSTRACT

The electrical and thermal magnetoresistance of polycrystalline and single crystalline aluminum have been measured in strong magnetic fields at liquid helium temperatures. The measurements are interpreted on the basis of current distortion effects induced by inhomogeneities in the samples. In single crystals, the inhomogeneities are most likely a result of sample imperfection such as voids, impurity clusters, or lattice defects. In polycrystalline specimens, however, there is an intrinsic inhomogeneity resulting from magnetic breakdown of the aluminum Fermi surface in some crystallites.

APPLICATION OF THE "EFFECTIVE MEDIUM" THEORY TO NORMAL-SUPERCONDUCTING MIXTURES

L. Morelli, R. Janik, and W. D. Gregory
Physics Dept., Georgetown Univ., Washington D. C.

ABSTRACT

The shape of the resistive transition of superconducting thin films as a function of temperature has long been a puzzle to explain. We find that, if we assume that a superconducting film breaks into normal and superconducting regions near T_c, we can use the "effective medium"[1] theory to calculate the film's resistance vs. temperature. We will show results of theoretical calculations and experimental data for Al and Sn films.

* Research supported in part by U. S. ERDA under Contract AT-(40-1)-3665.

[1] B. E. Springett, Phys. Rev. Lett. **31**, 1463 (1973).

HOPPING CONDUCTION AND SUPERCONDUCTIVITY IN GRANULAR ALUMINUM*

W. L. McLean, P. Lindenfeld, and T. Worthington
Rutgers University, New Brunswick, N. J. 08903

ABSTRACT

We have measured the electrical resistance as a function of temperature and magnetic field in superconductive granular aluminum specimens with room temperature resistivities up to 0.04 Ω-cm. In the normal state conduction is dominated by thermally activated hopping although the variation of log R with T cannot be described by a simple power law.

Superconducting fluctuations persist in fields up to 55 kG even above $2T_c$. At higher fields there is a negative quadratic magnetoresistance which we interpret in terms of the Zeeman contribution to the characteristic energy in the expression for the hopping conductivity.

INTRODUCTION

As part of a continuing investigation of granular aluminum we have measured the electrical resistance of some highly resistive specimens in transverse fields up to 9 T. In this report we concentrate on the properties of a specimen whose resistivity is 0.02 Ω-cm at room temperature, and an order of magnitude higher in the normal state at 1.3 K, with a superconducting transition at 1.76 K. All the features which we describe are confirmed by the data on a specimen whose resistivity is about twice as large.

The specimens are prepared by evaporation of 99.999% pure aluminum from an electron-gun source in about 10^{-4} torr oxygen, at a rate of 20Å/sec, and have a thickness of several microns. The substrate is glass in contact with a water-cooled copper block. Earlier work has established that the characteristics of our specimens are similar to those described by Deutscher et al.[1] We therefore expect them to consist of grains of aluminum about 30Å in diameter, in a matrix of Al_2O_3.

---

*Supported by the National Science Foundation.
Helium gas supplied by the Office of Naval Research.

RESULTS

Fig. 1 shows the resistance as a function of field at 3 K. We identify the initial rise in resistance with the suppression of superconducting fluctuations. The subsequent decrease follows the relation $R = R_N - AH^2$, which allows us to find R_N, the normal-state resistance in zero field.

The precision of the resistance measurement is of the order of 0.1 Ω. The uncertainty in R_N also depends on the extrapolation to zero field. Since the maximum in the resistance occurs at higher fields when the temperature is lowered, the range in field over which the resistance varies parabolically is then smaller, and the uncertainty in R_N greater. We nevertheless expect the values of R_N to have an uncertainty smaller than one per cent at all temperatures.

Fig. 2 shows R_N as well as the measured zero-field resistance, R_s, as a function of temperature. On the same graph we also show fits to three analytical expressions. Curves 1 and 2 follow the relations
$\sigma = \sigma_o e^{-(T_o/T)^{1/4}}$ and $\sigma = \sigma_o e^{-(T_o/T)^{1/2}}$ in accordance with the hopping models of Mott[2] and of Sheng et al.[3] respectively. The experimental curve is seen to deviate progressively from both of these curves as the temperature is lowered.

Curve 3, which fits the data, follows the expression
$\sigma = \sigma_m + \sigma_o e^{-(T_o/T)^{1/4}}$. It represents an attempt to allow for the presence of metallic threads or islands, comprising grains in metallic contact, imbedded in a medium for which the conductivity is described by Mott's formula. A similar formula with $(T_o/T)^{1/2}$ instead of $(T_o/T)^{1/4}$ also fits our low-temperature data, but, in contrast to the expression which we use, it does not fit the data at the temperature of liquid nitrogen and at room temperature.

Fig. 3 shows a graph of $(R_n-R_s)/R_s$ against $(T-T_c)/T_c$. On this graph lines with different slopes correspond to different dimensionalities, in accordance with the theory of Aslamazov and Larkin.[4] We interpret our results as showing that at sufficiently high temperatures the sample is zero-dimensional, i.e. that the grains are effectively decoupled. As the temperature is lowered the specimen approaches three-dimensional character, until coherent superconductivity is established at the transition temperature.

STRUCTURE

The expression which we use to fit the data for R_N as a function of temperature (curve 3 of Fig. 2) implies a metallic conductivity in parallel with a hopping conductivity. The limited temperature range of the measurements which we have made until now prevents us from knowing whether this is indeed the proper interpretation. It will be necessary to make measurements at lower temperature to see whether the resistivity approaches a constant value with the approach to zero temperature, characteristic of metallic conduction, or whether the resistivity rises indefinitely, indicative of localized electron states.

The experiment therefore does not decide whether bulk superconductivity is possible in specimens of this kind in the absence of continuous metallic paths. Nevertheless there are several indications that metallic contact from grain to grain throughout the sample plays at best only a small role.

First there is the magnitude of the resistivity, which is at least five orders of magnitude greater than that of the aluminum of which the grains are composed. A large resistivity would be possible if only a few metallic threads were responsible for the superconductivity, but measurements of the susceptibility of our specimens[5] indicate a Meissner effect characteristic of bulk superconductivity.

Secondly, the variation of the resistivity with temperature indicates that thermally activated hopping is the dominant conduction mechanism at low temperatures even though the contribution of metallic paths may not be negligible.

Thirdly, recent measurements[6] show that in high-resistance specimens like those discussed here there is no discontinuity of the heat capacity at T_c. This result once more rules out bulk metallic contact.

MAGNETORESISTANCE

Although there is as yet no rigorous description of the negative magnetoresistance, it seems likely to be due to the effect of the Zeeman splitting on the hopping transitions between grains. Qualitatively the fanning out of the degenerate energy levels by the magnetic field reduces the energy differences between some sublevels on adjacent sites and therefore increases the hopping probability.

ACKNOWLEDGMENTS

This work was inspired by our continuing collaboration with G. Deutscher. We would also like to acknowledge the help of Y.-H. Hsu and C.-H. Tang.

REFERENCES

1. G. Deutscher, H. Fenichel, M. Gershenson, E. Grünbaum, and Z. Ovadyahu, J. Low Temp. Phys. $\underline{10}$, 231 (1973).
2. N. F. Mott, Phil. Mag. $\underline{19}$, 835 (1969).
3. P. Sheng, B. Abeles, Y. Arie, Phys. Rev. Lett. $\underline{31}$, 44 (1973); B. Abeles, Ping Sheng, M. D. Coutts and Y. Arie, Adv. Phys. $\underline{24}$, 407 (1975).
4. L. G. Aslamazov and A. I. Larkin, Fiz. Tverd. Tela $\underline{10}$, 1104 (1968) [Sov. Phys.--Solid State $\underline{10}$, 875 1968)].
5. M. Gershenson and W. L. McLean, to be published.
6. T. Worthington and P. Lindenfeld, Bull. Am. Phys. Soc. $\underline{22}$, 52 (1977) and to be published.

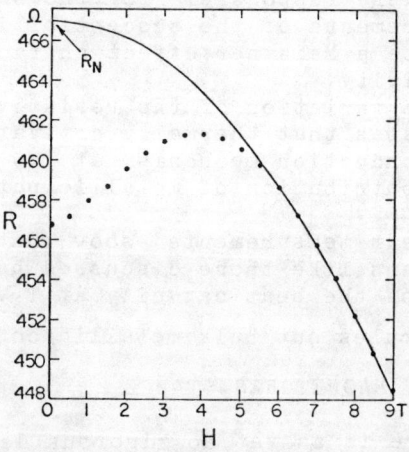

Fig. 1 Resistance as a function of field at 3 K. The solid line follows the relation $R = R_N - AH^2$.

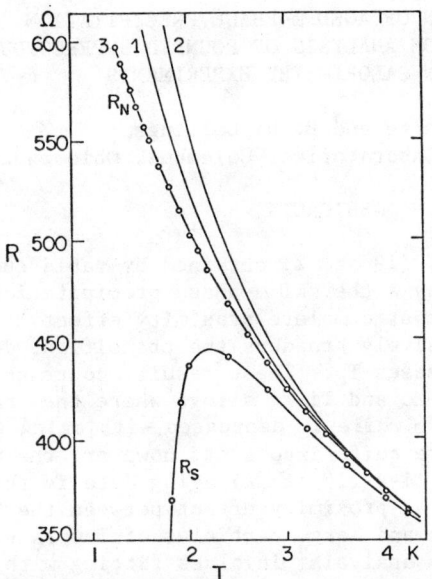

Fig. 2 R_N and R_S as a function of temperature. Curve 1 follows the relation $\sigma = \sigma_o e^{-(T_o/T)^{1/4}}$ and curve 2 the relation $\sigma = \sigma_o e^{-(T_o/T)^{1/2}}$. Curve 3 follows $\sigma = \sigma_m + \sigma_o e^{-(T_o/T)^{1/4}}$.

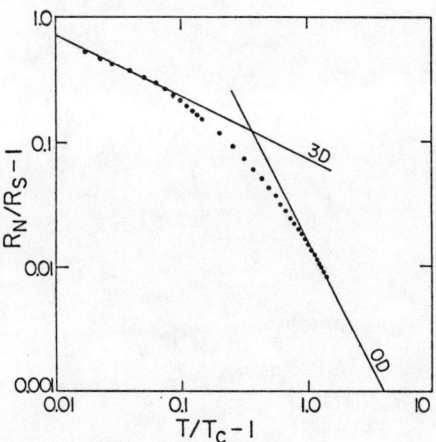

Fig. 3 $(R_N/R_S)-1$ as a function of $(T/T_c)-1$.

CHARACTERIZATION OF AGED ω-PHASE PRECIPITATION IN Ti-V (19 at .%) FROM ANALYSIS OF ROUNDED SUPERCONDUCTING TRANSITION CALORIMETRY EXPERIMENTS

J. J. White and E. W. Collings
Battelle, Columbus Laboratories, Columbus, Ohio 43201

ABSTRACT

The T_c of bcc Ti-V (19 at .%) obtained by rapid quenching is suppressed by some 2K by a thermal ω-phase precipitation through the operation of an almost complete proximity effect.[1] Subsequent aging at 300°C progressively broadens the transition, decreases its completeness, and increases T_c. These results contrast with equivalent experiments on Ti-Mo and Ti-Fe alloys where the transition remains sharp and complete while T_c decreases with aging time. Details of the ω-phase are quite important, however; the unique phenomenon in the case of Ti-V (19 at .%) aging data is the occurrence of an incomplete internal proximity effect between the high-T_c vanadium - rich bcc matrix and large particles of low T_c vanadium - poor ω-phase. The data analysis[2] involves fitting with a BCS model broadened by an asymmetrical Gaussian distribution of T_c's.

[1] E. W. Collings, P. E. Upton, and J. C. Ho, J. Less-Common Metals 42, 285 (1975).
[2] J. J. White and E. W. Collings, A.I.P. Conf. Proc. 34, 75 (1976).

PREPARATION AND CHARACTERIZATION OF LOW OPTICAL DENSITY, SUPERPARAMAGNETIC, SMALL-PARTICLE IRON OXIDE COMPOSITES

R. F. Ziolo, W. H. H. Günther and M. P. O'Horo
Xerox Corp., Webster Research Center
Xerox Square - 114, Rochester, N.Y. 14644

ABSTRACT

Two novel methods leading to the preparation of small particle γ-Fe_2O_3 composites have been developed. One method incorporates the thermal decomposition of $Fe(CO)_5$ in the presence of a porous SiO_2 matrix; the other method involves the wet chemical and heat treatment of an Fe^{3+} (or Fe^{2+}) loaded DVB crosslinked sulfonated polystyrene resin. Both methods lead to stable composites which consist of a uniform dispersion and size distribution of γ-Fe_2O_3, with particle size less than 200Å. The composites display superparamagnetic behavior and possess magnetic moments of about 6 to 8 emu/g consistent with the mass content of γ-Fe_2O_3. Additionally, composites 100μm thick are essentially transparent in the 600-800nm region.

ANOMALOUS ELECTRICAL RESISTIVITY AND MAGNETIC SUSCEPTIBILITY
TEMPERATURE DEPENDENCES IN Ti-V ALLOYS EXHIBITING REVERSIBLE
SOFT-PHONON-INDUCED STRUCTURAL INHOMOGENEITIES

E. W. Collings*
Battelle Memorial Institute, Columbus, OH 43201

ABSTRACT

Electrical resistivity studies of a series of Ti-V alloys have revealed anomalously large isothermal resistivities within the composition range 10-70 at.%V, and a negative resistivity temperature dependence within 20-33 at.%V. Alloys such as Ti-V exhibit electronically induced lattice instability which increases in severity as the electron/atom ratio decreases, and which manifests itself as a 2/3<111> longitudinal lattice displacement wave, the source of the ω-phase precipitation. Whether the negative resistivity temperature dependence were a result of scattering by phonons or by phonon-induced reversible athermal precipitation was not obvious. Analysis of the results of magnetic susceptibility temperature dependence measurements on the same system, however, yielded a reversible component whose appearance and disappearance paralleled the resistivity result. It follows that the anomalous resistivity is, at least in part, of macroscopic origin.

INTRODUCTION

According to the neutron diffraction results of Smith [1], the pure Group V bcc elements V, Nb, and Ta, support soft phonon modes. Furthermore, the electron diffraction results of Sass [2] indicate that this incipient bcc instability increases upon the addition of the Group IV elements Ti and Zr until (with increasing Group IV element concentration) martensitic transformation to hcp (α') takes place. The lattice instability takes the form of a 2/3<111> longitudinal displacement wave which, when of sufficient amplitude, effects a transformation of the bcc lattice to "ω-phase" which has a hexagonal unit cell. ω-phase thus forms as a precipitate during the quenching of Ti- and Zr-rich transition-metal-binary alloys. Its occurrence has been explored by Hickman [3] and reviewed by Sass [2], Williams [4], and Hickman [5]. The bcc-ω transformation and the crystallography of ω-phase have been considered in detail by Sass and his students [2], and lattice dynamical models of the transformation have been studied by deFontaine and co-workers [6]. The influence of ω-phase on the magnetic [7] calorimetric [8] and superconductive [9] properties of alloys such as Ti-V, Mn, Nb, Mo, have been considered by the present author. In this paper attention is focused on its effect on electrical resistivity.

As a consequence of lattice instability, ω-phase precipitation in a virtual lattice (identical atoms) should be reversible. During

---
*Supported by the Air Force Office of Scientific Research (AFSC) under Grant No. 71-2084.

the quenching of a real alloy, however, since the transformation may commence at moderately elevated temperatures, atomic diffusion which inevitably takes place, tends to 'fix' the precipitate. On continuing to cool the alloy below room temperature, subsequent <u>phonon-induced</u> transformation is no longer accompanied by diffusion, and is therefore reversible.

In the electrical resistivity of alloys, such as Ti-V, both the composition dependence and temperature dependence are anomalous. Over a small range of compositions $d\rho/dT$ is in fact negative below room temperature. This effect correlates with the ω-phase phenomenon; and the present paper addresses the question as to whether the anomalous temperature-reversible scattering is a result of an electron-phonon process, or scattering by evanescent (under thermal cycling) crystalline precipitate particles.

ELECTRICAL RESISTIVITY TEMPERATURE DEPENDENCE

The results of electrical resistivity (ρ) measurements on 19 Ti-V alloys at each of four accurately measured temperatures, were corrected where necessary to values at 300.0 K, 273.2 K, 200.0 K, and 77.3K; following which the slopes $d\rho/dT\big|_{273}^{300}$ and $d\rho/dT\big|_{77}^{200}$ were deduced and a set of resistivity-composition isothermals were plotted, Figure 1. Two features of the figure are (a) a pronounced peak in $\rho(c)$ at $c \sim 18$ at.%V, and (b) a negative $d\rho/dT$ for $20 < c < 30$ at.%V.

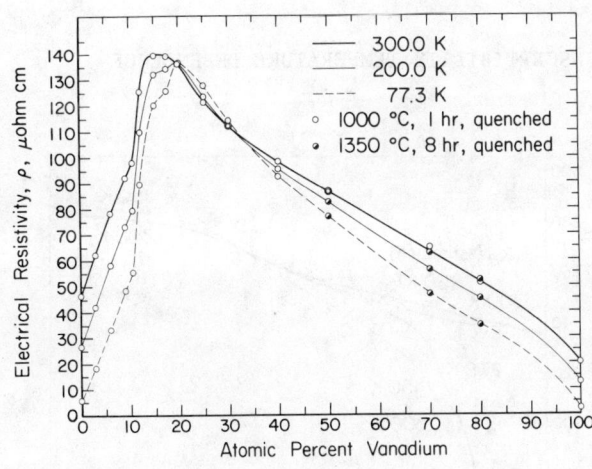

Fig. 1. Composition dependence of electrical resistivity in Ti-V alloys

Using a simple graphical analysis, associated with reasonable assumptions about the composition dependence of ρ_{normal} (i.e. $\rho_{ideal} + \rho_{resid.}$) the anomalous resistivity components corresponding to 300 K and 77 K were extracted from $\rho_{expt.}$ and plotted as in Figure 2.

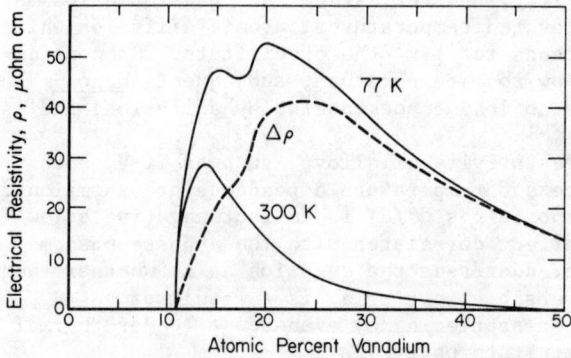

Fig. 2. Anomalous resistivity components at 300K and 77K and Δρ, the <u>increase</u> in anomalous resistivity 300K→ 77K

The anomalous peak in the isothermal $\rho_{300}(c)$ is taken to be a result of electron scattering from ω-phase precipitates in the as-quenched alloy, and corresponds well with the composition range over which precipitation has been observed in this system [10]. On cooling to 77 K, however, additional anomalous scattering develops which, when it exceeds numerically the <u>decrease</u> in ideal resistivity over the same temperature range, gives rise to the observed negative $d\rho/dT$. In an earlier paper [11] it was suggested that this increase in anomalous scattering with decreasing temperature was due to electron scattering by the 2/3<111> phonons associated with the bcc lattice instability. The results of a subsequent magnetic susceptibility temperature dependence study, to be discussed below, require that we re-assess this interpretation.

MAGNETIC SUSCEPTIBILITY TEMPERATURE DEPENDENCE

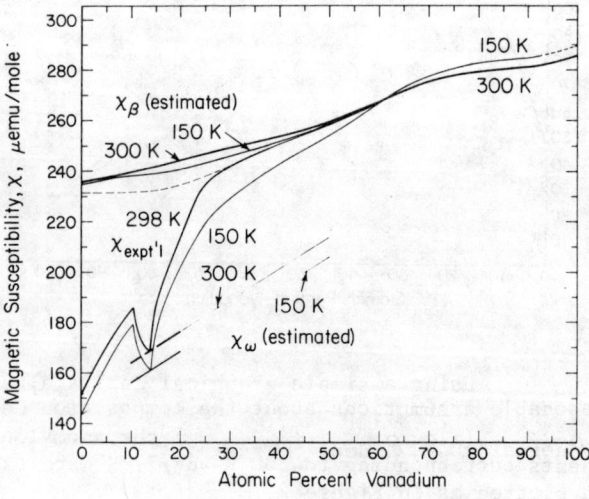

Fig. 3. Results of a series of magnetic susceptibility temperature dependence studies on Ti-V, replotted in χ versus %V format.

The magnetic susceptibilities (χ) of 17 Ti-V alloys were measured within the temperature range 77-1200 K. In analyzing the susceptibility temperature dependence data, special attention was given to the composition range 15-50 at.%V which included the two-phase $\omega+\beta$ region, within which $\partial\chi/\partial T$ was particularly strong. The total susceptibility of the two-phase mixture is of course given by $\chi = f_\omega \chi_\omega + f_\beta \chi_\beta$ where f represents mole-fraction. It was found that, after making reasonable estimates of $\partial\chi_\omega/\partial T$ and $\partial\chi_\beta/\partial T$ (Figure 3), the sum of $f_\omega(\partial\chi_\omega/\partial T) + f_\beta(\partial\chi_\beta/\partial T)$ was insufficient to account for the observed temperature dependence; and that as well as intrinsic temperature dependence it was necessary to invoke additional precipitation of ω-phase (i.e., reversible $\beta \rightleftarrows \omega$ transformation) to explain the observations. In other words the full equation

$$\frac{\partial \chi}{\partial T} = \underbrace{f_\omega \frac{\partial \chi_\omega}{\partial T} + f_\beta \frac{\partial \chi_\beta}{\partial T}}_{\frac{\partial}{\partial T}\langle\chi\rangle} + \underbrace{(\chi_\omega - \chi_\beta) \frac{\partial f_\omega}{\partial T}}_{\beta \rightleftarrows \omega \text{ precipitation}}$$

was required. f_ω is thus a reversible function of temperature and was estimated at 300 K and 150 K, respectively, using $f_\omega = (\chi_\beta - \chi)/(\chi_\beta - \chi_\omega)$, where χ is the experimental value, and χ_β, χ_ω could be read from Figure 3. The result of this evaluation is given in Figure 4.

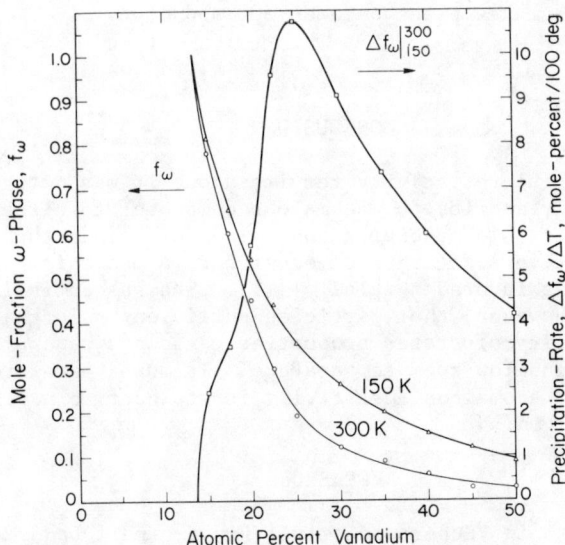

Fig. 4. Magnetically determined ω-phase abundance in quenched Ti-V alloys at 300 and 150K; and Δf_ω, representing the precipitation that occurs on cooling between these two temperatures.

The shape of the $\Delta f_\omega = f_{300} - f_{150}$ curve is seen to be remarkably similar to that of $\Delta\rho$ (Figure 2). In fact they superpose fairly well when appropriately scaled, Figure 5.

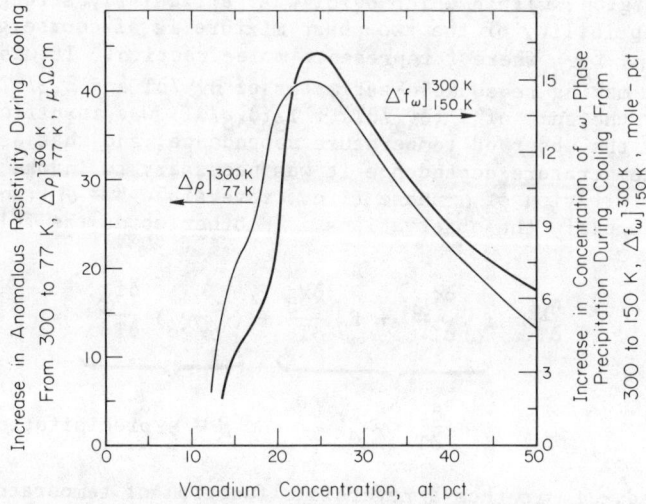

Fig. 5. Increase in anomalous resistivity, $\Delta\rho$, incurred on lowering the temperature between 300 and 77K, compared with the increase in ω-phase abundance which takes place between 300 and 150K.

CONCLUSION

Based on the resistivity results alone it was not possible to unambiguously ascribe the anomalous component to either soft phonons or reversible precipitation. In fact, since the former are such effective scatterers of electrons, a mechanism involving soft phonons was favored initially [11]. The susceptibility results show, however, that particles sufficiently large to have measurable density-of-states properties may occur reversibly with thermal cycling below room temperature. It must therefore be concluded that the anomalous resistivity is at least in part of macroscopic origin.

REFERENCES

[1] H. G. Smith in "Superconductivity in d- and f-band Metals", ed. by D. H. Douglass (AIP, Vol. 4, New York, 1972).
[2] S. L. Sass, J. Less-Common Metals 28, 157 (1972).
[3] B. S. Hickman, J. Inst. Met. 96, 330 (1968) and Trans Met. Soc. AIME 245, 1329 (1969).

[4] J. C. Williams, in "Titanium Science and Technology, ed. by R. I. Jaffee and H. M. Burte, Plenum Press, 1973, pp. 1433-1494.
[5] B. S. Hickman, J. Mater. Sci. $\underline{4}$, 554 (1969).
[6] D. deFontaine, Acta Met. $\underline{18}$, 275 (1970); d. deFontaine, N. E. Paton, and J. C. Williams, Acta Met. $\underline{19}$, 1153 (1971).
[7] E. W. Collings, J. Less-Common Metals $\underline{39}$, 63 (1975).
[8] E. W. Collings, P. E. Upton, and J. C. Ho, J. Less-Common Metals $\underline{42}$, 285 (1975).
[9] E. W. Collings and J. C. Ho, Solid State Comm. $\underline{18}$, 1493 (1976).
[10] K. K. McCabe and S. L. Sass, Phil. Mag. $\underline{23}$, 957 (1971).
[11] E. W. Collings, Phys. Rev. $\underline{9}$, 3989 (1974).

DETECTION OF RESIDUAL STRESS-INDUCED DOMAINS FROM THE TRANSITION ROUNDING OF ANTIFERROMAGNETS

J. J. White
Battelle, Columbus Laboratories, Columbus, Ohio 43201
S. N. Bhatia
Dept. of Physics and Astronomy, U. of Georgia, Athens, GA. 30601

ABSTRACT

Transport properties in magnetic materials are strongly affected by the onset of magnetic order.[1] However, there remains an inadequate description of the influence of residual stress-induced domains and other factors that contribute to transition rounding near T_c. It is thought that models that account for the rounding of high resolution specific heat measurements near T_c will contribute significantly to transport property interpretation. Analysis of specific heat data of nominal single crystals of $CoBr_2 \cdot 6H_2O$, $MnBr_2 \cdot 4H_2O$, and related systems[2] using the continuous heating method yield a jagged peak that is separable from the background noise. It is argued that the graininess of the transition temperatures of a stress-induced ensemble of domains has been detected. Related investigations of transition rounding will be discussed.

[1] See, e.g., G. S. Dixon, J. E. Rives, and D. Walton, in <u>Dynamical Aspects of Critical Phenomena</u> (Gordon and Breach, N. Y., 1972), p. 333.

[2] J. J. White, J. Phys. C <u>7</u>, L317 (1974) and cited refs.

AIP Conference Proceedings

		L.C. Number	ISBN
No.1	Feedback and Dynamic Control of Plasmas (Princeton) 1970	70-141596	0-88318-100-2
No.2	Particles and Fields - 1971 (Rochester)	71-184662	0-88318-101-0
No.3	Thermal Expansion - 1971 (Corning)	72-76970	0-88318-102-9
No.4	Superconductivity in d- and f-Band Metals (Rochester, 1971)	74-18879	0-88318-103-7
No.5	Magnetism and Magnetic Materials - 1971 (2 parts) (Chicago)	59-2468	0-88318-104-5
No.6	Particle Physics (Irvine, 1971)	72-81239	0-88318-105-3
No.7	Exploring the History of Nuclear Physics (Brookline, 1967, 1969)	72-81883	0-88318-106-1
No.8	Experimental Meson Spectroscopy - 1972 (Philadelphia)	72-88226	0-88318-107-X
No.9	Cyclotrons - 1972 (Vancouver)	72-92798	0-88318-108-8
No.10	Magnetism and Magnetic Materials - 1972 (2 parts) (Denver)	72-623469	0-88318-109-6
No.11	Transport Phenomena - 1973 (Brown University Conference)	73-80682	0-88318-110-X
No.12	Experiments on High Energy Particle Collisions - 1973 (Vanderbilt Conference)	73-81705	0-88318-111-8
No.13	$\pi-\pi$ Scattering - 1973 (Tallahassee Conference)	73-81704	0-88318-112-6
No.14	Particles and Fields - 1973 (APS/DPF Berkeley)	73-91923	0-88318-113-4
No.15	High Energy Collisions - 1973 (Stony Brook)	73-92324	0-88318-114-2
No.16	Causality and Physical Theories (Wayne State University, 1973)	73-93420	0-88318-115-0
No.17	Thermal Expansion - 1973 (Lake of the Ozarks)	73-94415	0-88318-116-9
No.18	Magnetism and Magnetic Materials - 1973 (2 parts) (Boston)	59-2468	0-88318-117-7
No.19	Physics and the Energy Problem - 1974 (APS Chicago)	73-94416	0-88318-118-5
No.20	Tetrahedrally Bonded Amorphous Semiconductors (Yorktown Heights, 1974)	74-80145	0-88318-119-3
No.21	Experimental Meson Spectroscopy - 1974 (Boston)	74-82628	0-88318-120-7
No.22	Neutrinos - 1974 (Philadelphia)	74-82413	0-88318-121-5
No.23	Particles and Fields - 1974 (APS/DPF Williamsburg)	74-27575	0-88318-122-3
No.24	Magnetism and Magnetic Materials - 1974 (20th Annual Conference, San Francisco)	75-2647	0-88318-123-1

No. 25	Efficient Use of Energy (The APS Studies on the Technical Aspects of the More Efficient Use of Energy)	75-18227	0-88318-124-X
No. 26	High-Energy Physics and Nuclear Structure - 1975 (Santa Fe and Los Alamos)	75-26411	0-88318-125-8
No. 27	Topics in Statistical Mechanics and Biophysics: A Memorial to Julius L. Jackson (Wayne State University, 1975)	75-36309	0-88318-126-6
No. 28	Physics and Our World: A Symposium in Honor of Victor F. Weisskopf (M.I.T., 1974)	76-7207	0-88318-127-4
No. 29	Magnetism and Magnetic Materials - 1975 (21st Annual Conference, Philadelphia)	76-10931	0-88318-128-2
No. 30	Particle Searches and Discoveries - 1976 (Vanderbilt Conference)	76-19949	0-88318-129-0
No. 31	Structure and Excitations of Amorphous Solids (Williamsburg, Va., 1976)	76-22279	0-88318-130-4
No. 32	Materials Technology - 1975 (APS New York Meeting)	76-27967	0-88318-131-2
No. 33	Meson-Nuclear Physics - 1976 (Carnegie-Mellon Conference)	76-26811	0-88318-132-0
No. 34	Magnetism and Magnetic Materials - 1976 (Joint MMM-Intermag Conference, Pittsburgh)	76-47106	0-88318-133-9
No. 35	High Energy Physics with Polarized Beams and Targets (Argonne, 1976)	76-50181	0-88318-134-7
No. 36	Momentum Wave Functions - 1976 (Indiana University)	77-82145	0-88318-135-5
No. 37	Weak Interaction Physics - 1977 (Indiana University)	77-83344	0-88318-136-3
No. 38	Workshop on New Directions in Mössbauer Spectroscopy (Argonne, 1977)	77-90635	0-88318-137-1
No. 39	Physics Careers, Employment and Education (Penn State, 1977)	77-94053	0-88318-138-X
No. 40	Electrical Transport and Optical Properties of Inhomogeneous Media (Ohio State University, 1977)	78-54319	0-88318-139-8
No. 41	Nucleon-Nucleon Interactions - 1977 (Vancouver)	78-54249	0-88318-140-1

QC
173.3
C6
1977

MAY 23 1979